Sensory and Consumer Research in Food Product Design and Development

Press

The *IFT Press* series reflects the mission of the Institute of Food Technologists—to advance the science of food contributing to healthier people everywhere. Developed in partnership with Wiley-Blackwell, *IFT Press* books serve as leading-edge handbooks for industrial application and reference and as essential texts for academic programs. Crafted through rigorous peer review and meticulous research, *IFT Press* publications represent the latest, most significant resources available to food scientists and related agriculture professionals worldwide. Founded in 1939, the Institute of Food Technologists is a nonprofit scientific society with 22,000 individual members working in food science, food technology, and related professions in industry, academia, and government. IFT serves as a conduit for multidisciplinary science thought leadership, championing the use of sound science across the food value chain through knowledge sharing, education, and advocacy.

A John Wiley & Sons, Ltd., Publication

Sensory and Consumer Research in Food Product Design and Development
Second Edition

Howard R. Moskowitz, PhD
President and Founder, Moskowitz Jacobs, Inc., White Plains, New York

Jacqueline H. Beckley, MBA
President and Founder, The Understanding & Insight Group LLC, Denville, New Jersey

Anna V.A. Resurreccion, PhD
Distinguished Research Professor, Department of Food Science and Technology
The University of Georgia, Griffin, Georgia

 | Press

A John Wiley & Sons, Ltd., Publication

This edition first published 2012 © 2012, 2006 by Blackwell Publishing Ltd. and the Institute of Food Technologists

Blackwell Publishing was acquired by John Wiley & Sons in February 2007. Blackwell's publishing program has been merged with Wiley's global Scientific, Technical and Medical business to form Wiley-Blackwell.

Editorial offices: 2121 State Avenue, Ames, Iowa 50014-8300, USA
           The Atrium, Southern Gate, Chichester, West Sussex, PO19 8SQ, UK
           9600 Garsington Road, Oxford, OX4 2DQ, UK

For details of our global editorial offices, for customer services and for information about how to apply for permission to reuse the copyright material in this book please see our website at www.wiley.com/wiley-blackwell.

Library of Congress Cataloging-in-Publication Data
Moskowitz, Howard R.
   Sensory and consumer research in food product design and development / Howard R. Moskowitz, Jacqueline H. Beckley, and Anna V.A. Resurreccion. – 2nd ed.
      p. cm. – (Institute of food technologists series)
   Includes bibliographical references and index.
   ISBN 978-0-8138-1366-0 (hard cover : alk. paper)   1. Food–Sensory evaluation.   2. Commercial products–Testing.   I. Beckley, Jacqueline H.   II. Resurreccion, Anna V. A.   III. Title.
   TX546.M68 2012
   664′.07–dc23
                                                        2011035804

A catalogue record for this book is available from the British Library.

Wiley also publishes its books in a variety of electronic formats. Some content that appears in print may not be available in electronic books.

Set in 10/12 pt Times by Aptara® Inc., New Delhi, India
Printed and bound in Malaysia by Vivar Printing Sdn Bhd

1   2012

Titles in the *IFT Press* series

- *Accelerating New Food Product Design and Development* (Jacqueline H. Beckley, Elizabeth J. Topp, M. Michele Foley, J.C. Huang, and Witoon Prinyawiwatkul)
- *Advances in Dairy Ingredients* (Geoffrey W. Smithers and Mary Ann Augustin)
- *Bioactive Proteins and Peptides as Functional Foods and Nutraceuticals* (Yoshinori Mine, Eunice Li-Chan, and Bo Jiang)
- *Biofilms in the Food Environment* (Hans P. Blaschek, Hua H. Wang, and Meredith E. Agle)
- *Calorimetry in Food Processing: Analysis and Design of Food Systems* (Gönül Kaletunç)
- *Coffee: Emerging Health Effects and Disease Prevention* (YiFang Chu)
- *Food Carbohydrate Chemistry* (Ronald E. Wrolstad)
- *Food Irradiation Research and Technology* (Christopher H. Sommers and Xuetong Fan)
- *High Pressure Processing of Foods* (Christopher J. Doona and Florence E. Feeherry)
- *Hydrocolloids in Food Processing* (Thomas R. Laaman)
- *Improving Import Food Safety* (Wayne C. Ellefson, Lorna Zach, and Darryl Sullivan)
- *Innovative Food Processing Technologies: Advances in Multiphysics Simulation* (Kai Knoerzer, Pablo Juliano, Peter Roupas, and Cornelis Versteeg)
- *Microbial Safety of Fresh Produce* (Xuetong Fan, Brendan A. Niemira, Christopher J. Doona, Florence E. Feeherry, and Robert B. Gravani)
- *Microbiology and Technology of Fermented Foods* (Robert W. Hutkins)
- *Multivariate and Probabilistic Analyses of Sensory Science Problems* (Jean-François Meullenet, Rui Xiong, and Christopher J. Findlay)
- *Natural Food Flavors and Colorants* (Mathew Attokaran)
- *Nondestructive Testing of Food Quality* (Joseph Irudayaraj and Christoph Reh)
- *Nondigestible Carbohydrates and Digestive Health* (Teresa M. Paeschke and William R. Aimutis)
- *Nonthermal Processing Technologies for Food* (Howard Q. Zhang, Gustavo V. Barbosa-Cánovas, V.M. Balasubramaniam, C. Patrick Dunne, Daniel F. Farkas, and James T.C. Yuan)
- *Nutraceuticals, Glycemic Health and Type 2 Diabetes* (Vijai K. Pasupuleti and James W. Anderson)
- *Organic Meat Production and Processing* (Steven C. Ricke, Michael G. Johnson, and Corliss A. O'Bryan)
- *Packaging for Nonthermal Processing of Food* (Jung H. Han)
- *Preharvest and Postharvest Food Safety: Contemporary Issues and Future Directions* (Ross C. Beier, Suresh D. Pillai, and Timothy D. Phillips, Editors; Richard L. Ziprin, Associate Editor)
- *Regulation of Functional Foods and Nutraceuticals: A Global Perspective* (Clare M. Hasler)
- *Sensory and Consumer Research in Food Product Design and Development, second edition* (Howard R. Moskowitz, Jacqueline H. Beckley, and Anna V.A. Resurreccion)
- *Sustainability in the Food Industry* (Cheryl J. Baldwin)
- *Thermal Processing of Foods: Control and Automation* (K.P. Sandeep)
- *Water Activity in Foods: Fundamentals and Applications* (Gustavo V. Barbosa-Cánovas, Anthony J. Fontana Jr., Shelly J. Schmidt, and Theodore P. Labuza)
- *Whey Processing, Functionality and Health Benefits* (Charles I. Onwulata and Peter J. Huth)

WILEY-BLACKWELL

A John Wiley & Sons, Ltd., Publication

Contents

Preface

How do we begin? All beginnings are hard. They may be sweet, but they are hard. Perhaps a personal memory will help introduce our effort. People questioned Howard and me (Jackie) with polite, and sometimes not so polite, skepticism when we talked about measuring a person's emotions that might be associated with food. This was a decade ago, in 2001, before this book, before all three of us grew a bit wiser by 10 years. The skepticism was not just about food, but about vision to take approaches from the comfortable world of food and drink and move it into new-to-us fields as diverse as charitable giving and retailing. Well, today, as we put the final touches on this second edition, the skepticism has gone away. The ideas of a decade ago, the ideas in the first edition of this book, are showing up in many areas. Whether the field is called neuromarketing or behavioral economics, the ideas in the first edition published in 2006 have become a foundation, anchoring the building efforts of others. Our dear friend and longtime colleague, Dr. Harry Lawless, emeritus professor at Cornell University, humorously called the first edition of this book "an articulation of a type zero error—you need to know what you know and don't know to effectively understand type 1 and 2 errors."

But, of course, there's always more. When we began this work, prodded by Mark Barrett of Wiley, we anticipated that there would be more data in the world or a business than *anyone* could understand. And we were determined to bring new science to our field. Anna joined us, and so we had our solid foundation. From statistics to experimental design, from applications to products and onto ideas, we proceeded to elaborate our vision. We hope that the tools in this book, the visions of what could be, the core fundamentals, and the new applications will power us moving forward. We look forward, to yet another edition, the third one, where the core fundamentals of knowledge mapping, value diagramming, and robust, yet fast, Internet testing will be accepted and used widely, bringing our field of sensory and consumer research to the forefront of the food industry.

Author biographies

Howard R. Moskowitz, PhD, is the president of Moskowitz Jacobs, Inc., founded in 1981. Dr. Moskowitz is an experimental psychologist in the field of psychophysics (the study of perception and its relation to physical stimuli) and inventor of world-class market research technology.

Moskowitz graduated from Harvard University in 1969 with a PhD in experimental psychology. He graduated from Queens College (City University of New York), Phi Beta Kappa, with degrees in mathematics and psychology. He has written/edited 26 books, published over 400 refereed articles and conference proceedings, lectures widely, serves on the editorial boards of major journals, and mentors numerous students worldwide. He was named the 2010 winner of Sigma Xi's prestigious Walter Chubb Award for innovation in research across scientific disciplines, a unique honor showing the value of Mind Genomics™, the new field of knowledge he founded. Simply put, Mind Genomics™ is the "Inductive Science of Everyday Life" with the goal of advancing science and business, knowledge and application.

In 2009, Dr. Moskowitz cofounded iNovum to bring the science of Mind Genomics™ and Addressable Minds™ to world industries. iNovum's goal is to commercialize the award-winning science, to reignite the American Dream, to export that dream around the world, and at the same time to improve the education and life prospects of young people of the next generation.

In his 42-year-long career since receiving his PhD from Harvard University, Dr. Moskowitz's science/commercial/visionary work has won numerous awards. These include the 2001, 2003, 2004, and 2006 ESOMAR awards for his innovation in web-enabled, self-authored conjoint measurement, and for weak signals research in new trends analysis and concept development. Self-authored concept technology brings concept/package design development and innovation into the realm of research, substantially reducing cost, time, and effort for new product and service development. In 2005, he received the Parlin Award from the American Marketing Association, its highest award. In 2006, Dr. Moskowitz was awarded the first Research Innovation Award by the Advertising Research Foundation and was also inducted into the Hall of Fame of New York's Market Research Council.

Among his contributions to market research is his 1975 introduction of psychophysical scaling and product optimization for consumer product development. In the 1980s, his contributions were extended to health and beauty aids. In the 1990s, the concept development approach was introduced to pharmaceutical research. His research/technology developments have led to concept and package optimization (IdeaMap®.Net, MessageMap® for pharma), integrated and accelerated development (DesignLab®), and the globalization and democratization of concept development for small and large companies through an affordable, transaction-oriented approach (IdeaMap®.Net).

Dr. Moskowitz's latest efforts focus on four key areas:

(i) *Mind Genomics™ and Addressable Minds™*: Using experimental design of ideas to understand how people respond to everyday situations and products, to what particular, granular-level mind-set segment a person belongs for each situation, and then determining the array of life-relevant mind-set segments to which a specific person belongs (sequencing the genome of the person's mind).

(ii) *Rekindling the American Dream through the Institute for Competitive Excellence (ICE) at Queens College, City University of New York*: Promoting experimental design of ideas and products to understand customer requirements for products and services, innovating using such information, and increasing American competitiveness through such systematized knowledge.

(iii) *Experience optimization*: Using experimental design to understand and optimize customer experience.

(iv) *The law and psychophysics*: Using experimental design of messaging for juror selection (with law firms) and for package/shelf/web design (dynamic landing page optimization).

Jacqueline H. Beckley, MBA, is founder and president of The Understanding & Insight Group, Denville, NJ. As a business innovator, Ms. Beckley has developed exacting methods for measuring consumer response better, faster, and more clearly. She has directed the creation of some of the earliest database systems for retrieval and storage of consumer information. These systems continue to be refined to advance new product development processes more quickly and accurately. Pinpoint accuracy in knowing consumer wants and how to apply that knowledge to specific products and services is what sets Ms. Beckley heads above others in the field. The approaches that were discussed in the first edition of this book have proven to lead business people and researchers alike to clarity in the front end of the product development process. It works every time!

She combines extensive scientific training with broad exposure to social sciences and fine arts to craft a strategic approach to product and business development. Its results are pragmatic, flexible, creative, and authentic. Along with traditional methods and web-based resources, this integrated approach has fostered leading products that have defined their business categories.

Clients praise her skillful understanding of complex concepts and her astute ability to translate them into practical applications that can be implemented quickly.

Previously, Beckley held positions within industry and consulting, including director of consumer perception at Nabisco, Inc., vice president for Peryam & Kroll Research, group manager of sensory research and R&D for the Quaker Oats Company, and research scientist for Amoco Chemical Company.

Anna V.A. Resurreccion, PhD, is a distinguished research professor in the Department of Food Science and Technology at The University of Georgia. Her current research interests focus on methods in consumer and sensory evaluation, development of value-added products and functional foods, consumer-based optimization of food formulations and processes, and shelf-life determination; these are the topics on which her 597 scientific and technical publications are based. She has authored 157 peer-reviewed journal articles, seven chapters, and two books on consumer sensory evaluation in product design and development; presented 232 papers published as abstracts or proceedings; and another 208 articles in the field of food science and technology. She was former chair of the product development division of the Institute of Food Technologists (IFT), served as Associate Scientific Editor of the Journal of Food Science, and is currently on the editorial board for the Journal of Sensory Studies, among a number of scholarly journals. Dr. Resurreccion is, clearly, one of the world's most recognized researchers in the field of consumer and sensory science and food quality evaluation for her exemplary achievements and would certainly rank among the world's top outstanding scientists in sensory and quality evaluation in academia.

Dr. Resurreccion's achievements in international peanut product research in Asia, Europe, Africa, and the United States integrate consumer and sensory science with chemistry and engineering, and has far-reaching effects on the design of food products that prevent blindness and vitamin A deficiency in children and on minimizing health problems due to one of the most potent carcinogens, aflatoxin. Her innovative research on consumer-based optimization of innovative food processes to maximize bioactive components in food has led to major breakthroughs in analytical chemistry of phenolic compounds and antioxidants, and for the peanut industry, a patent ready for licensing by a food company, to deliver substantial health benefits to consumers, in products with high consumer acceptance. Dr. Resurreccion has garnered over $12.4 million as principal or coprincipal investigator on several prestigious competitive research awards from the US Agency for International Development and the National Institute for Food and Agriculture.

Dr. Resurreccion is the recipient of numerous international, national, and regional awards, including the Professional Scientist Award (Southern Region of the Institute of Food Technologists), the Gamma Sigma Delta Senior Faculty Award of Merit (University of Georgia Chapter), the International Award from the University of the Philippines College of Home Economics Alumni Association, and the Outstanding Professional Scientist Award for Food Science and Nutrition from the University of the Philippines Alumni Association. The Institute of Food Technologists (IFT) awarded her the distinction of Fellow of the IFT in 2000, distinguished lecturer for IFT for 3 years (2002–2004), and the prestigious Bor S. Luh International award for her outstanding achievements in the field of food science and technology. Furthermore, she is one of no more than three University of Georgia faculty members to receive two D.W. Brooks Faculty Awards, for Excellence in Research and then for Excellence in International Agriculture. Last year, The University of Georgia awarded her the title of Distinguished Research Professor, the highest recognition awarded by the university to its faculty. Dr. Resurreccion is the consummate research professor and, without question, throughout her entire career has truly exemplified sustained excellence in food science research and education.

Acknowledgments

Howard:
There are many people to thank for this second edition of our book. Most of all, however, I'd like to thank my coauthors, Jacqueline Beckley and Anna Resurreccion. It was the three of us who made this book possible, took the journey, and now we bring you our vision of the future, hopefully your future as well.

But there is, of course, more. There is my beloved wife, Arlene, the driving force who continues to inspire. There are the "boys," now men with families of their own, Daniel and David, who prod me to continue developing, never giving me a chance to rest on my laurels, never permitting me to grow stale.

Linda Ettinger Lieberman, my editorial assistant, keeps it all straight, all the time, even when things seem dauntingly complex.

And, of course, the others, most precious of whom is my colleague and friend of 30 years, Bert Krieger, executive vice president of Moskowitz Jacobs, Inc. It is you, Bert, who made many of these ideas come alive with clients. You've prodded, pushed, developed, and improved on my thoughts, in the office and of course at our now-cherished weekly breakfasts.

Thank you each of you, those named and unnamed, who have shared the struggle and who made the pain bearable, the joy worthwhile.

Jackie:
This second edition would not have been possible without the assistance of clients and friends in the field of consumer understanding. Additionally, I want to thank John Thomas for his graphics, Leslie Herzog, Jennifer Vahalik, Nancy Nemac, and Nathalie Tadena for their assistance in updating information, and Linda Ettinger Lieberman at Moskowitz Jacobs, Inc. who has made this one sound even better than the first edition!

Anna:
First, I wish to thank my coauthors, Howard Moskowitz and Jackie Beckley, for their vision that made this book possible, their unsurpassed creativity and innovation that have brought forth a new frontier in product design and development, and their dedication to make this new, bigger, and better second edition through to publication.

I acknowledge a number of people who worked behind the scenes, Lotis Francisco, Jocelyn Sales, and Paula Scott; they provided me with invaluable editorial and technical assistance. To Linda Ettinger Lieberman, whose magic got us to the finish line.

I dedicate my role in this book to Rey, the love of my life. Forty-four years ago, he taught me how to write.

1 Emerging corporate knowledge needs: how and where does sensory fit?

INTRODUCTION

We begin this second edition of our book with history, as we did before. Why? It's simple. The history of the field tells us a lot about how people think, what problems they faced, what methods they developed, what institutions they created, and what they considered to be worthy of studying and doing. History is not, in the words of Henry Ford, "one damned thing after another." Rather, history embeds within it keys to what we do and why we do it. History is of paramount importance in the world of sensory science because knowing how the field developed tells us a lot about why we do what we do.

During the past 30 years, companies have recognized the consumer as the key driver for product success. This recognition has, in turn, generated its own drivers—sensory analysis and marketing research, leading first to a culture promoting the expert and evolving into the systematic acquisition of consumer-relevant information. Styles of management change as well. At one time, it was fashionable to laud the "maverick executive" as a superior being, perhaps the management equivalent of the expert. Over time, we have seen this type of cowboy machismo declining into disrepute. Replacing this maverick decision-making has been an almost slavish adoption of fact-based decisions, and the flight from knowledge-based insight into the "soulless" reportage of facts.

How does corporate decision-making affect a discipline such as sensory analysis, which has only begun to come into its own during the past four decades? If one were to return to business as it was conducted in the 1950s and 1960s, one might discern a glimmer of fact-based decisions among the one or two dozen practitioners of what we now call sensory analysis. These individuals—scattered in corporations, working quietly in universities, executing food acceptance tests for the US military, and a handful of others scattered about in other countries around the world—were founding the field that now provides this type of fact-based guidance for product development and quality assurance. In the early years, many of the practitioners did not even know that they were creating a science that would emerge as critical, exciting, and eminently practical. These pioneers simply did the tests the best they could, attempted to understand how people perceived products, and in the main kept to themselves, hardly aware of how they were to affect the food industry in the years to come. Many of these pioneers were bench chemists and product developers. They just wanted to know what their work products tasted like, smelled like, especially when the work product was a new food.

Sensory and Consumer Research in Food Product Design and Development, Second Edition.
Howard R. Moskowitz, Jacqueline H. Beckley, and Anna V.A. Resurreccion.
© 2012 Blackwell Publishing Ltd. Published 2012 by Blackwell Publishing Ltd.

As the competition among companies to secure market share in consumer goods relentlessly increased, and as the consumer continued to be bombarded with new products, it became increasingly obvious to many that consumer acceptance would be increasingly paramount. Whereas before one might hear such excusing platitudes as "people always have to eat" as an excuse for complacent mediocrity, one would now hear catch phrases such as "consumer tested" or "significantly preferred." Companies were catching on to the fact that the consumer had to actually like the product. The privations of World War II and before were fading in memory. The supply economy was giving way to the demand economy. The consumer, surfeited with the offerings of countless food manufacturers, could pick and choose among new products that often differed only in flavor or in size from those currently available. In the face of such competition by fellow manufacturers, it became necessary for the marketer and product developer to better understand what consumers would actually buy, and in so doing perhaps understand what consumers really wanted.

The end of the twentieth century saw the professionalization of product testing. What had started out 50 years before as a small endeavor in corporations to "taste test foods" as one step in the quality process became a vast undertaking (e.g., Hinreiner, 1956; Pangborn, 1964). Company after company installed large market research departments reporting to marketing and sensory analysis departments reporting to R&D. Whether this was the optimal structure was unclear. Often, the two departments did similar studies. The express purpose of these often-competing departments was to ascertain what consumers wanted, and feed back this information in a digested, usable form to those who either had to create the product at R&D or those who had to sell the product. The era of fact-based decision-making was in full swing. Decisions would no longer be made on the basis of the response from the president's "significant other" (whether husband, wife, child), but rather would be made on the basis of well-established facts, such as the positive reaction by consumers who would test the product under conditions that management would trumpet as being "controlled and scientific." Such fact-based decision-making would be introduced into all areas dealing with consumers, first as a curiosity, then as a luxury, and finally as a desperate necessity for survival. For the food and beverage industries, the emergence of fact-based decision-making would bring new methods in its wake.

THE ERA OF THE EXPERT, AND THE EMERGENCE OF SENSORY ANALYSIS OUT OF THAT ERA

The real business-relevant beginnings of sensory analysis occurred in the 1950s and 1960s, and can be traced to the quantum leap in business thinking provided by Arthur D. Little Inc. (ADL), in Cambridge, Massachusetts. ADL was a well-known consulting company, with one division specializing in agribusiness. In the 1940s, a group of enterprising consultants at ADL developed the Flavor Profile, a then-revolutionary idea to quantify the flavor characteristics of foods (Cairncross & Sjostrom, 1950; Little, 1958). The Flavor Profile was precedent shattering on at least two fronts:

(i) *Systems thinking*: No one was thinking about flavor in this organized, quantifiable fashion. It was certainly unusual to even think of a formalized representation of flavor. Researchers had thought about flavors for years, but the formalization of a descriptive method was certainly new.

(ii) *Anyone could become an expert—albeit after training*: The expert reigned supreme, in brewing, in perfumery, etc., but to have the experts created out of ordinary consumers by a formalized training program was new thinking.

Sensory analysis as an empirical discipline emerged from the application of expert judgments in formalized evaluation. Before the Flavor Profile (Caul, 1957), the expert judgment would certainly be called upon and relied upon as the last word. The notion of consumer acceptance, or consumer input, was not particularly important, although the successful product would be touted as filling a consumer need. The Flavor Profile formalized the role of the expert in the situation of disciplined evaluation. The expert was given a new task—evaluate the product under scientific conditions. ADL won numerous contracts on the basis of their proclamation that the Flavor Profile could assure so-called *flavor leadership* for a product.

At about the same time as ADL was selling its Flavor Profile, the US Government was winning World War II. The popular aphorism attributed to Napoleon Bonaparte that "an army travels on its stomach" guided the development of new methods. The US Quartermaster Corps recognized the importance of food to soldiers' health and morale. The slowly emerging scientific interest in measuring responses to food, appearing here and there in industry, took strong root in the military. Measuring soldiers' food preferences became important because the commanders could often see firsthand the effects of food rejection. Unlike the executives sitting at the heads of food companies, the commanders walked among their troops. Failure to feed the troops meant a weakened army and the real prospect of a lost battle or even war. Food acceptance became a vital issue, and its measurement a key military task (Meiselman & Schutz, 2003).

The confluence of sensory analysis in the food industry and the military recognition of the importance of consumer-acceptable food produced in its wake the sensory analysis industry. The industry did not emerge overnight. It emerged slowly, haltingly, like all such new creatures do, with false starts hampered by wrong decisions, but in its own way matured. Expert panel approaches begun by ADL matured to more quantitative, statistics-friendly methods such as quality data analysis (QDA) (Stone *et al.*, 1974). Military interest in food acceptance led to advances in sensory testing, and the 9-point hedonic scale (Peryam & Pilgrim, 1957) to actually measure the level of acceptance. The US Government funded research into food acceptance (Meiselman, 1978) and eventually got into the funding of taste and smell psychophysics, especially at the US Army Natick Laboratories where Harry Jacobs built up a cadre of young scientists interested in the sensory evaluation of foods (Meiselman & Schutz, 2003). Other organizations such as the Swedish Institute for Food Preservation Research in Gothenburg (now Swedish Institute for Food Research) pioneered research methods and applications as well as recording the literature from the burgeoning field (Drake & Johannsen, 1969).

Industrial organizations adopted methods for product testing, and the field grew and prospered. The field heralded its maturity through journals and conferences. The first major international symposium involving sensory analysis took almost 50 years from start of the field in the 1940s. This Pangborn Symposium held in Jarvenpaa, Finland, just outside of Helsinki, attracted more than 200 participants. The organizing committee headed by Dr. Hely Tuorila had expected this conference to represent a one-off event, but the palpable excitement shared by the participants soon changed the committee's mind. Eleven years later, the same conference, in its fifth convening, held in Boston, attracted more than 700 participants.

Popularity increased so that from being held every third year the conference is now held every second year. Allied conferences, such as Sensometrics, also developed, to the point where the Sensometrics Conference is held on the years that the Pangborn Symposium is not. The field was well on its way. Scientific decision-making in the food industry had given rise to a new discipline.

The ensuing years would be good to the field of sensory. The Pangborn conferences would be the first specific conferences. They would give impetus to more US-based conferences such as the SSP (Society of Sensory Professionals). Of course, once these meetings began to occur, the floodgates opened. There would be meetings in Europe, Latin America, and Central America. It is always a good thing when meetings proliferate. At some point, they rationalize and the better ones survive, but the first meetings of the various sensory organizations pump the necessary emotional and intellectual nutrients into the field.

The success of the Pangborn Symposia, along with their continuing increase in attendance in the face of decreasing attendance at other conferences, deserves a short digression that can also shed light on the growing field of sensory analysis and the pent-up needs of the members. When the era of the expert was in its heyday, there were no conferences to speak of, and the professionals in sensory analysis were few, scattered, and scarcely aware of each other, all laboring away in, as John Kapsalis had often said, "splendid isolation." The Pangborn Symposium brought these individuals together in a concentrated, 4-day format, somewhat longer than that provided by the more conventional professional organization such as IFT (Institute of Food Technologists). At least six things occur at such extended meetings:

(i)   *Masses of people with very similar interests interact in a confined location.* The participants meet with individuals who are, by and large, sympathetic to them. Rather than participating in specialized symposia where the sensory specialists come together, albeit as a minority, in the Pangborn Symposium they come together with many of the same purposes. This mass of people is an intellectual hothouse.

(ii) *Easy meetings occur so that like-minded people can reach out to each other.* The interpersonal nature of the meeting cannot be overemphasized. Many people have known each other for years, so the close and long meeting allows these people to renew acquaintanceship.

(iii) *Density plus time plus fatigue reduce interpersonal barriers.* The surrounding density of people at the meeting and the continued stimulation over time from seeing people with common interests leads to fatigue, real reduction of barriers, and increased professional intimacy.

(iv) *Long meetings create shared memories.* The 4-day period suffices to imprint many positive memories of interactions on the participants. The scientist lives in the future, propped up by memories and propelled by hopes.

(v)   *Information intake and exchange allows people to take each other's measure.* The plethora of posters, talks, and meals shared together allows people to come and go at their convenience, spend time looking at other people's work in an unhurried situation and, in general, get comfortable with each other. They size up each other, challenge, share, form opinions of character, of promise, and of expectations for each other's future. In a sense, people learn about each other in a way no journal article could ever hope to imitate.

(vi) *The laying on of hands, from the older to the young, occurs more readily in this environment.* The young researcher can get to meet the older, more accomplished researcher on a variety of occasions, some professional and some social. This opportunity to meet each other in the field produces in its wake a cadre of inspired young professionals who can receive the necessary reinforcement from their older role models in this artificially created, short-lived "hothouse of kindred souls." One should never underestimate the value of interpersonal contacts in science, and the effect on the morale, motivation, and joy of a younger scientist who is recognized and encouraged by an older role model. The Pangborn Symposium was set up, perhaps inadvertently, but nonetheless successfully, to produce that motivation and "laying on of hands" over its extended, 4-day time.

THE MANIFOLD CONTRIBUTION OF PSYCHOPHYSICS

Psychophysics is the study of the relation between physical stimuli and subjective experience (Stevens, 1975). The oldest subdiscipline of experimental psychology, psychophysics makes a perfectly natural, almost uncannily appropriate, companion to sensory analysis. The study of how we perceive appearances, aroma, tastes, and textures of food might easily be a lifelong topic of psychophysical research. Indeed, many of today's leading sensory analysts have been grounded either in formal education in psychophysics or at least have enjoyed a long-term interest in the details of psychophysics. Psychophysics did not start out as the conjoined twin of sensory analysis, although to many novices in the field the intertwining of the two areas seems unusually tight and quite meaningful.

Psychophysicists are natural complements to sensory analysts, but with a slight change in focus. Sensory analysts study the product, using the person as a bioassay device. Knowledge of how we perceive stimuli does not help sensory analysts do their job better in terms of the specifics, but does give the analyst a broader perspective in which to operate. Psychophysics uses stimuli as probes to understand how the sensory system processes external information. Historically, and for a great many years, psychophysics confined itself to the study of "model systems," such as sugar and water or simple chemical odorants. In their desire to be pure, these psychophysicists valued systematic control over real-world behavioral meaning. Psychophysics of taste and smell followed psychophysics of hearing and vision, wherein the stimulus variability could be controlled by the researcher and then channeled into systematic stimulus variation.

Psychophysics expanded its scope, however, in the early 1970s as a group of young researchers moved out from academia to the applied world. During the 1960s, psychophysics underwent a renaissance, initially promoted by S.S. Stevens at Harvard University but later taken up by others worldwide in a variety of fields. These young researchers found that they could use Stevens' method of magnitude estimation to measure the perceived intensity of stimuli. Stevens had provided the tool, and young researchers, such as Linda Bartoshuk, William Cain, Donald McBurney, Herbert Meiselman, Howard Moskowitz, and others, would use the magnitude estimation method for direct estimation of sensory magnitudes, applying it to model systems first, and then to more behaviorally meaningful stimuli such as foods, beverages, the environment, etc. (e.g., McBurney, 1965). Bartoshuk, Meiselman, and Moskowitz all began their careers with some involvement at the US Army Natick Laboratories, in Massachusetts, working with Harry Jacobs. Natick would stimulate each to look at the application of psychophysics to food problems, a stimulation that would have lifelong consequences for these researchers and for their contributions to the field.

THE EMERGENCE OF STATISTICAL THINKING IN SENSORY ANALYSIS

Quantitative thinking has long been a *leitmotif* of sensory analysis, from its early days, a half-century ago, through today, in academia and in industry, both in the United States and abroad. Indeed, with the founding of the Sensometrics Society (www.sensometrics.org) and the burgeoning number of quantitatively oriented papers in sensory analysis presented at the different symposia, one might almost conclude that sensory analysis could not exist as it does had it not been based on statistical methods. The question is: Why this reliance on quantitative methods? (Why are numbers so important?)

To answer this question, we have to consider the hedonics or likes/dislikes, the intellectual history of sensory analysis, and the nurturing influences of both science and business. Sensory analysis deals with the response of people. People are, by definition, variable. They lack the pleasing uniformity that delights a scientist. Subjective data are messy. When it gets down to likes and dislikes, the pervasive variation across people becomes almost unbearable to some, who want to flee back to a world of ordered simplicity.

When we imagine what it was like a half-century ago or longer, we notice first that many of the sensory professionals were chemists or other individuals in corporations who did not fathom that they were inventing a new field. Chemists are not accustomed to variability. They are familiar with regularity in nature, with variability constituting an unwanted secondary influence to be dispensed with, either by controlling it or ignoring it. When dealing with the issues involving food and the subjective reaction to these foods, the natural inclination of a chemist is to ask simple questions, such as magnitude of intensity and magnitude of acceptance. Not having any other intellectual history, such as sociology, the early practitioners relied on simple quantitative methods by which to make conclusions. Therefore, it should come as no surprise that the statistics used by those chemist/sensory practitioners would be simple inferential statistics. It was not the nature of the problem that influenced those practitioners, but the nature of the world view. The intellectual history and quantitative predilections of such practitioners would be those of chemists thrust into a world far beyond that which had formed their intellectual character years before.

Fifty years later, cadres of chemists were no longer the main practitioners of sensory analysis. Instead, the practitioners were people with newer, more informed, sophisticated world views, coming from statistics, experimental psychology, and other fields. The predilections of these professionals for measurement and modeling would be more profound, because they were nurtured on world views that could handle variability, rather than perceiving it as an intractable nuisance. Not content to find differences between samples, these new practitioners had been schooled to search for relations between variables and for representing these relations either in terms of equations or in terms of maps (e.g., Heymann, 1994). They were looking for laws, or at least generalities, not coping with the often more profound and equally disquieting issue of "how do I measure this private sensory experience?"

What does all this have to do with sensory analysis? Quite simply, quantitative thinking has emerged as a major facet of sensory analysis, and not just the ability to do analyses of variance. Most meetings with sensory analysts have some portion of the meeting devoted to quantitative methods. Indeed, quantification using "modern methods" has become so very popular that researchers in sensory analysis have formed a group, the aforementioned Sensometrics Society, to promote the approach. Sensometrics is growing and thriving, embracing more and more adherents and acolytes each year, and of course, in the process, providing what has turned out to be virtually a treasure chest of analytical tools.

A sense of the growing power of quantitative approaches in the field can be readily seen from the nature of conference presentations. Whereas four decades ago interest focused on new methods for removing variability in analysis of variance, today interest focuses on methods for representing data and gleaning insights. Four decades ago, the researcher involved in quantitative methods was happy to show that some effect occurred, as revealed by significant treatment effects in analysis of variance. The focus for new methods lay in the ability to provide added types of analysis, cautiously remaining, however, within the framework of inferential statistics, descriptive statistics, and kindred approaches. The notion of insights in the data as empowered by statistics would have to wait three decades for the birth of available, easy, cheap, and powerful computing. The PC revolution also revolutionized statistics, as the more adventurous and inquiring statisticians began to explore other methods with this available computing power, such as mapping.

Increased *quantitation*, especially beyond the more conventional tests of differences among products, generated at least three outcomes:

(i) *Infused intellectual vitality*: The sensory analyst, armed with these new techniques, felt empowered to advance beyond a simple service role and do more scientific work. Whereas before, the sensory analyst was many times relegated to "tray pusher" despite the protestations of being a professional, all too often that is exactly what happened. The ability to collect data, then create maps, equations, reveal novel relations among product stimuli, apply this information to many types of stimuli produced a sense of pride in one's capability.

(ii) *Increased ambition in the corporate world*: The ability to understand aspects of products through high-level statistics led to the realization that this information was valuable to the business. Knowing the strengths and weaknesses of the in-market competitors gave the sensory analyst some degree of power to influence business decisions. This power led to increased ambition, or at least to a desire for greater roles in corporate decision-making.

(iii) *New currency for interchange with fellow scientists at meetings*: Whereas in the 1960s and 1970s, the birthing years of sensory analysis, there was little really to talk about except one's hope for the future, now in 2010, with high-level statistical analyses, there is always something to talk about at meetings. Having a thriving, robust corpus of statistical methods allows the researcher to analyze data in many different ways and to present the data and the analysis at conferences. Different types of analyses are always more interesting to scientists than, say, the consumer acceptance of yet another flavor of dessert pudding. This statement is not meant to denigrate the old data but rather to emphasize that, as the sensory scientist became familiar with statistical techniques, that familiarity led to new ways of analyzing data, which would become the basis for presenting papers and posters at meetings. Simple research, of the disciplined, well-executed type promoted by Rose Marie Pangborn, doyenne of sensory analysis in the 1960s through 1980s, and those of her associates, could never have produced this "currency" for scientific meetings. It would take a new generation of skilled, quantitatively oriented professionals to leap the barriers that circumscribed and limited sensory analysis for so many years.

ROSE MARIE PANGBORN—FROM FOCUS ON EXPERTS TO FOCUS ON CONSUMERS

The early history of sensory analysis is a history of studies with small numbers of subjects and a focus on their ability to detect differences and describe perceptions (Amerine, Pangborn, & Roessler, 1965). To some degree, this focus came from the intellectual heritage shared by the chemists and product developers who found themselves in sensory analysis jobs, even before the field was recognized. They turned to the literature and found the work of perfumers, flavorists, winemakers, brew masters, and the emerging science promoted by consultants at ADL, as described previously. It did not take the researchers long to conform to the standard that sensory analysts were developing. The field was to focus on the description of sensory characteristics (descriptive analysis) and perhaps on the discrimination of small differences. The descriptive efforts were part of the Linnaean tradition, which was prevalent first in biology, then in psychology, and then in sensory research. Linnaeus confronted the unknown world by describing it. Description was a natural task. It seemed reasonable that one could learn about the product properties by first elucidating them. Experimental psychologists just a half-century ago had done the same by describing the characteristics of sensory experience in the psychological school of "structuralism." Edward Bradford Titchener had laid the groundwork at Cornell by the methods of introspection. Sensory analysts took these methods and ran with them (Boring, 1929).

Decades later, and with the influence of business objectives as motivation, sensory researchers evolved away from pure descriptive analysis to understanding consumer behavior. Descriptive analysis was fine but not particularly cogent in a highly competitive business world. One could, of course, link descriptive analysis to ongoing product quality, as many researchers did and did effectively. However, the bigger picture demanded from the sensory analyst that he or she concentrate on the consumer. It was acceptable to "keep one's foot in the profiling world" as stated by more than one researcher, as long as the sensory researcher dealt with consumers. The focus on consumers would grow in the 1980s but emerge very strongly in the 1990s to constitute the prime direction. One reason was the call of business—those employed by corporations had to stay relevant or lose their jobs and their raisons d'être. Another, and a more subtle reason, was the premature death of Rose Marie Pangborn, a founder in the field and a purist. Pangborn trained many of the students at the University of California, Davis, and in some ways single-handedly crested the academic field. Pangborn was part scientist, part teacher, 100% rigorous, but with an inspiration to introduce her students to the scientific method. She encouraged purism on the part of her students, many of who went into descriptive analysis. While she lived, many of her students maintained an unspoken level of scientific purity through descriptive analysis, even though Pangborn was more sympathetic to psychophysics than to descriptive analysis. From descriptive accounts of her classes, Pangborn was clearly a mother figure, but one who spared no criticism when her student departed from the path of rigid, pure, and puritanical science. After her death, however, the rigid purity that she so strongly espoused and the elevation of methodological correctness and orthodoxy became less evident. The unique force of her professional personality waned as it must wane after one's death. Those fortunate students who had gained her respect through tightly controlled descriptive analyses were somewhat freer to pursue consumer research, and many did so. Thus, through the fortuitous combination of business influence with the focus on fact, and driving sales, and the passing of Pangborn's influence, the sensory analyst was liberated to focus more on consumers.

DESTROYING OLD MYTHS IN THE CRUCIBLE OF THE MARKETPLACE

Having been influenced by science, sensory analysis would also be influenced by marketing. This nascent discipline was caught in another emerging current, the whirlpool of business, filled as it was with currents, counter-currents, cabals, capriciousness, and yet at the same moment unbelievable opportunities. Business required different ways of thinking than science did, and the direction in which sensory analysis grew in the fertile ground of business was quite different. Sensory scientists often began their careers with dreams of understanding the way products work, at least at the subjective level. Business issues soon disabused industry-based sensory scientists of many such idealistic visions. The business world demands obedience, demands delivery, demands success. Sensory scientists could practice their field and craft, but under the strict auspices of a research director, held accountable for splashy product introductions, unerring product quality, and profitable market success.

Therefore, it should come as no surprise that in the crucible of the marketplace the sensory scientist should change course. What had been in the 1950s and 1960s a slow dance between scientists studying sensory perception and business-oriented researchers studying products changed to a set of silos that would inevitably discourage cross-fertilization. ADL Flavor Profile, so carefully constructed by Cairncross, Sjostrom, Caul, and others during the 1940s and 1950s, had matured into a big business, supporting infrastructures in ADL and in laboratories of their corporate clients. The introduction of their descendent methods such as the QDA method (Stone *et al.*, 1974) in the 1970s and the Spectrum method in the 1980s (Munoz & Civille, 1992) found fertile, protected ground. However, it would be some years before scientists would publicly scrutinize the method (Zook & Pearce, 1988). In the meanwhile, sensory analysts quickly flocked to profiling methods, leaving psychophysicists and their research methods far behind. The story would not end there as we will see later. However, it is worth noting that the 1970s and the 1980s witnessed the diverging paths apart of sensory analysis and psychophysics. What had originally been a conjoined, developing, and occasionally intimate relation in the 1960s with psychophysics invited to food science meetings turned somewhat colder a decade or two later on. A great deal of the polarization came from the need of sensory analysts to do routine, ongoing profiling work. The success of sensory analysis in industry came at the price of increased demands on the sensory analyst to do maintenance work. That success turned sensory analysis away from its psychophysical roots, as the practitioners in the field enjoyed their acceptance, but paid the price in corporate demands on their time.

THE INEVITABLE SLIDE INTO TURF WARS

Turf wars for control of primary research among consumers characterized much of the relation between the growing field of sensory analysis and the incumbent field of marketing research. Both disciplines had responsibility to understand the consumer, but came at their tasks from radically different directions. As discussed previously, sensory analysis came from the tradition of physical and chemical science, and indeed many of the early practitioners of sensory analysis during its *terra incognita* stage were bench scientists involved in product development. They knew the products well, but the subjective perceptions less well. We can contrast this group of explorers with their somewhat counterparts sitting in marketing, the so-called marketing or consumer researchers. These individuals were rarely, if ever, trained in science, tended to be professionals who studied social science (and now business), and were, in general, not particularly comfortable in high-level mathematics. They did understand inferential statistics and generally could trace their intellectual heritage to sociology, or at least acted as if they had come equipped with a sociological background. They were interested in market performance of the product and had no sympathy for the product itself except as the topic of research. They focused on how the consumer bought the product or accepted the product, but for the most part the products could be substituted for each other, willy-nilly, without making any particular impact to the way these market researchers analyzed their data.

From the perspective of top management, sensory analysis and market research deal with many of the same issues. Indeed, in 1974, then Professor Erik von Sydow, head of the Swedish Institute for Food Preservation Research (SIK) in Gothenburg, said that the eventual roles of the sensory analyst and the market researcher would merge to become one product-focused role. It would take more than 30 years for von Sydow's insight to take hold, but in the mean time the similarity of function and the desire to provide valuable corporate feedback about products had an unexpected outcome. That outcome was an ongoing turf war lasting more than three decades, which in its wake created barriers and silos that only today are being torn down.

Ironically, the turf wars came about because both groups wanted to do a good job in product research and now in what is colloquially called "consumer insights." The sensory analyst, poorly prepared at first to battle in the corporation, retreated to scientific methods, to esoteric charts from newly developing methods, and to presentation of himself or herself as the low-cost supplier. The sensory analyst fighting these turf wars was poorly equipped to make his or her case as a strategic partner in marketing, primarily because the personality of the sensory analyst in those early days of the turf wars (1980s) was focused on science and validation of oneself, not on success in a corporation. In contrast, the marketing researcher did not carry around the burning desire to found a science, and to be judged acceptable and worthy by professionals in other sciences. There were no self-avowed physical or biological scientists working in marketing, as there were in the biology and psychophysics of taste and smell. Hence, the marketing researcher was unconstrained by many agendas. Some marketing researchers had academic aspirations and would teach on the side as adjuncts in the university, but for the most part the marketing researcher focused on doing a good job. Smart enough to acquire a discretionary budget to hire outside suppliers, the market researcher became a purchasing agent for talent and information, and was able to use some of the better brains in the industry to work on projects and provide necessary insights. Sensory analysts, however, unaccustomed to a budget to "outsource" their efforts, did not ask for, nor did they receive, this outsourcing budget. Rather, they grew organically in size, overhead, and responsibility in the organization. They were content to fight the turf wars by showing that they could do everything internally, or at least claimed to be able to do so at a lower cost. It was now a classic fight between the outsourcing model and the internal capabilities model. In business, this is the ever-present tension between "buy" versus "build." Does one buy a capability in the way the market researcher buys, or does one build a capability as the sensory analyst builds? When these two approaches vie for the same corporate task—insights about the product—turf wars break out.

WHERE ARE WE HEADING TODAY—AND WHY ARE WE HEADING THERE?

Where is sensory analysis going? When we look at the number of practitioners in the industry or the number of papers published by academics, we might feel justifiably proud that here is a field that is burgeoning. The life force is almost palpable at meetings, with young researchers actively seeking to show their work to their older counterparts. All the signs of life are about us. Yet, there is some trouble brewing. Many of the young researchers are heavily involved in measuring rather than in thinking. The plethora of new technical methods, the ease and availability of computation, and the willingness of companies and funding institutions to sponsor research all combine to nurture a thriving business in "stimulus assessment" (namely, applied product testing).

On the other side of the coin is the recognition that the younger researchers do not have a chance to think. Their very success depends upon using some of the latest research techniques to grind through data. The young researchers are caught in a race with methods. Each group wants to be the first to use new computer analysis techniques. Each young researcher wants to be the first to win approval by showing prowess at these new techniques and often sacrifices the slow, methodical, often not apparently productive thinking for the frenetic pace of analysis.

We might look at the field of sensory analysis in the way that the poets write about their world—a world of nature becoming increasingly sophisticated, losing its way, losing contact with its origins. We can see some problems emerging in our world. These problems, often disguised as opportunities, are rapidity of data collection, the plethora of tools, and the abundance of conferences. These influences pull us in two directions. One direction is more professionalization, better science, far more rapid advance in knowledge. The other direction is narrow specialization and the creation of sensory professionals instead of true scientists. Perhaps that polarization and dichotomy are inevitable and come to all fields, such as sensory analysis, that have the fortune to survive their own childhoods.

MIND-SETS AND HOW THE SENSORY PROFESSIONAL MIGHT COPE WITH DATA

How do different sensory researchers cope with data since they have been confronted with data and data analytic methods for a half-century or longer? An interesting organizing principle for people was propounded in the Crave It!® Study, but might have application here. Beckley and Moskowitz (2002) have suggested from a set of large-scale conjoint analysis studies that consumers fall into three mind-sets when it comes to how they respond to concepts about foods and beverages (see Chapter 6 for a discussion of the method). One group, called *Elaborates,* responds strongly to descriptions of the sensory characteristics of food and responds strongly when these are romanced. A second group, called the *Imaginers*, likes the characteristics of food, but also wants other things such as ambiance, emotion, and brand. Imaginers respond to nonsensory cues as well, although they are strongly affected by the sensory ones. The third group, *Classics*, likes foods in the traditional way.

According to Beckley, perhaps the same typing occurs for sensory researchers. Watching more than 600 researchers at the Dijon Pangborn Symposium (2001) and more than

700 researchers at the Boston Pangborn Symposium (2003) led Beckley to note that the same typology emerged for research papers and posters. Some researchers went profoundly into the data and could be called *Data Elaborates*. Others incorporated a variety of nondata sources not strictly in the study but using current trends and could be called *Data Imaginers*. Still others remained on the straight and narrow path and could be called *Data Classics* because they maintained the conventional analytic techniques, with constraints, applying those techniques simply to a new data set.

WHERE ARE WE TODAY? MIND-SETS ABOUT ONE'S ROLE IN THE SENSORY ANALYSIS WORLD

Mind-sets represent another way to approach the history of sensory analysis and the relevance of its mission in business and science. Mind-set refers to the predisposition of the individual, to the way the individual responds to external stimuli and to the nature of actions that the individual engages in. We saw different mind-sets in the previous section, regarding one's treatment of data. How about mind-sets for one's own job in the sensory world?

The importance of mind-set cannot be overstated. By understanding a person's mind-set, it becomes possible to make sense of how the person makes choices in the world. For the world of evaluating ideas for products and products themselves, mind-sets provide an organizing principle to cope with the ever-present variability one observes in data. Mind-sets provide a way to deal with, and perhaps even harness, that variability.

One might consider all sensory analysts to be similar, and perhaps divide them by their scientific background and ways that they solve research issues. Another approach comes from the way that the sensory analyst thinks about his job, his responsibilities to his employer, and to his field. This way of dividing the professionals emerges from a study of the mind-set of employees, reported by Ashman and Beckley (2002) as the "professionalism study." The professionalism study was conducted twice. The goal of the study was to better understand what it meant to be a sensory professional. Ashman and Beckley discovered, probably not surprisingly, that the sensory analyst does not constitute one simple persona. We might have expected this. Sensory professionals appeared to fall into one of three different segments (see Table 1.1), on the basis of their pattern of responses to a variety of concepts that portrayed the sensory professional:

Segment 1—the Academic: This segment, comprising 26%, are not necessarily academics as in university professors, Rather, this segment exists and flourishes as well in industrial settings. For the most part, sensory analysts in Segment 1 want to keep up with the literature, want to keep abreast of the newest and best methods. They often come from academia, which is not surprising. They show little real interest in the applications of the method to practical, business problems.

Segment 2—the Helpful Staff: This segment, comprising 44%, better reflects what people have thought the sensory analyst to be. The Helpful Staff segment takes little risk. Segment 2 seems to want clean and neat studies. One might liken the Helpful Staff segment to the middle manager. The Helpful Staff can be found in many companies. They are the backbone of the field.

Segment 3—The Business Builder: Characterized by an understanding of how sensory analysis can help build a business. From a total of 137 respondents, the Business Builders

Table 1.1   Utilities values for 24 concept elements describing the sensory professional.

Tentative title	Total	Seg 1 Academic oriented 26%	Seg 2 Helpful staff 44%	Seg 3 Business grower 30%
Number of respondents	**137**	**36**	**60**	**41**
Elements driving the Academic segment				
Recognized as an expert in his or her field	5	15	−3	7
Maintains thorough knowledge of technical literature	0	12	−12	8
Maintains close liaison with other practitioners in the field	3	11	−2	3
A team player	4	9	8	−5
Actively promotes new and innovative approaches through the organization	5	9	5	1
Elements driving the Helpful Staff segment				
Adept at applying knowledge and follow-through on the tasks required to complete the job	4	−1	9	2
Shows others how to integrate product, consumer, and market knowledge in the project	5	8	7	1
Elements driving the Business Builder segment				
Applies creativity and critical thinking to move the business forward	−1	−8	−4	9
Provides an opinion and guidance in critical situations	3	−1	3	9
Remains committed, with a drive to succeed	1	−7	1	8
Irrelevant elements or elements that detract from sensory professionalism				
Actively provides point of view in professional discussions	5	7	3	7
Often accepts a leadership role	−3	−15	−1	6
Remains authentic to his or her personal values while considering the values of others or the values of the organizational culture (politics)	−1	−5	−3	5
Provides vision and resourcefulness	2	3	0	5
Provides a role model for individuals new to the field	2	1	2	4
Takes action when discovering that something was done wrong or inappropriately by a functional group	0	−8	4	1
Makes difficult decisions under pressure	−2	−11	1	1
Passionate about listening to the needs and ideas of others	0	0	1	−1
Oriented toward new possibilities, open to change and new learning	0	4	−1	−3
Uses coaching and negotiation to motivate coordinated action to achieve goals	−5	−6	−6	−3
Continues to seek out new internal and external ways to do business	0	−3	5	−5
Personally tries out new and innovative approaches	−2	5	−1	−8
Publishes articles in various journals and books	−19	8	−41	−10
Shows humility in presenting his or her ideas while accepting constructive criticism and contrary opinions without being defensive	−7	−3	2	−24

Note: The study was run using the method of conjoint analysis (see Chapter 6). The utility is the conditional probability that a respondent will agree that the statement describes a sensory professional.

comprised 30%. The existence of this Business Builder segment was not expected, because for the most part sensory analysts who participated in the study to understand mind-sets did not come from a marketing or business background. Rather, they came from scientific backgrounds. The Business Builder is an integrator who always keeps an eye on the business implications.

Whether a person falls into a single segment and stays there all his or her career, or whether the person changes from one segment to another as a function of changes in job and responsibility remains an interesting topic for further research. Certainly, however, the segmentation of the sensory professional by mind-set gives one food for thought, especially as it mirrors the nature of the different types of behaviors in the field. One might expect a dynamic tension in the field as the *Business Builders* go about pulling the company into the future, the *Helpful Staff* dutifully and loyally contributing, all the while as the *Academics* stand back, take matters a little more slowly, spend more time and "worry more" about the appropriateness of the tools used.

As noted previously, the sensory analyst continually deals with data in one form or another. The sensory analyst, generally challenged to provide newer, better, more actionable answers to problems, all too often feels overwhelmed by the never-ending, two-pronged assault of business problems and new techniques. Such assaults promote growth by the sheer demands they make. The classification of sensory analysts on the basis of the way they approach problems, the work product they generate, and the data they collect provides a novel, provocative and possibly fruitful area to study this emerging profession, one that hints of deep cross-currents. The dynamic tension between mind-sets, the ever-changing demands of business, and the maturation of sensory scientists from idealistic novices to battle-hardened professionals promise to make sensory analysis a field worth watching, and a potentially good home base in which to spend part or even all of a career.

REFERENCES

Amerine, M.A., R.M. Pangborn, & E.T. Roessler. 1965. *Principles of Sensory Evaluation of Food*. New York: Academic Press.

Ashman, H. & J. Beckley. 2002. *The Mind of the Sensory Professional*. Unpublished manuscript.

Beckley, J. & H.R. Moskowitz. 2002. Databasing the consumer mind: The Crave It!®, Drink It!®, Buy It!® & Healthy You!® Databases. Presented at the Institute Of Food Technologists Convention, Anaheim, CA.

Boring, E.G. 1929. *Sensation and Perception in a History of Experimental Psychology*. New York: Appleton Century Crofts.

Cairncross, S.E. & L.B. Sjostrom. 1950. Flavor profiles—a new approach to flavor problems. *Food Technology* 4:308–311.

Caul, J.F. 1957. The profile method of flavor analysis. *Advances in Food Research* 7:1–40.

Drake, B. & B. Johansson. 1969. Sensory Evaluation of Food. Annotated bibliography, supplement 1968–1973. *Vol. 1, Physiology and Psychology; Vol. 2, Methods, Applications, Index*. SIK-Rapport Nr. 350. Svenska Instituet for Konserverings Forskning, Göteborg, Sweden.

Heymann, H. 1994. A comparison of free choice profiling and multidimensional scaling of vanilla samples. *Journal of Sensory Studies* 9:445–453.

Hinreiner, E.H. 1956. Organoleptic evaluation by industry panels—the cutting bee. *Food Technology* 31(11): 62–67.

Little, A.D. 1958. *Flavor Research and Food Acceptance*. New York: Reinhold Publishing Corporation.

McBurney, D.H.A. 1965. Psychophysical study of gustatory adaptation. *Dissertation Abstracts* 48:1145–1146.

Meiselman, H.L. 1978. Scales for measuring food preference. In: *Encyclopedia of Food Science*, eds. M.S. Petersen & A.H. Johnson, pp. 675–678. Westport, CT: AVI.

Meiselman, H.L. & H.G. Schutz. 2003. History of food acceptance research in the US Army. *Appetite* 40:199–216.

Munoz, A.M. & G.V. Civille. 1992. The spectrum descriptive analysis method. In: *Manual on Descriptive Analysis Testing for Sensory Evaluation*, ed. R.C. Hootman. West Conshohocken, PA: ASTM.

Pangborn, R.M. 1964. Sensory evaluation of foods: A look backward and forward. *Food Technology* 18:63–67.

Peryam, D.R. & F.J. Pilgrim. 1957. Hedonic scale method of measuring food preferences. *Food Technology* 11:9–14.

Stevens, S.S. 1975. *Psychophysics: An Introduction to Its Perceptual, Neural and Social Prospects*. New York: John Wiley & Sons.

Stone, H., J.L. Sidel, S. Oliver, A. Woolsey, & R. Singleton. 1974. Sensory evaluation by quantitative descriptive analysis. *Food Technology* 28:24–34.

Zook, K. & J.H. Pearce. 1988. Quantitative descriptive analysis. In: *Applied Sensory Analysis of Foods*, ed. H.R. Moskowitz, pp. 43–72. Boca Raton, FL: CRC Press.

2 Making use of existing knowledge and increasing its business value—the forgotten productivity tool

How you gather, manage, and use information will determine whether you win or lose.

Bill Gates, *Business @ The Speed of Thought*

INTRODUCTION

The data you own as a company—tracking, insights, sensory, trends, marketing, product development, market research, supply chain, manufacturing, quality, operations, sales, and retail successes and failures—represent the single biggest productivity tool for food development. Most companies today, large or small, have access to more data, lists, and information than they effectively make use, or even could make use of, on a regular basis. Turning this information into knowledge that will lead to actionable results is the challenge.

The key step in turning information into knowledge is to understand the context in which the data were originally calculated, reframe the data for the current situation, and then creatively manage the data quickly, effectively, and affordably for greater clarity of perspective. And, of course, do all of these things quickly and at lowest possible cost since time and money factor into all successes in today's supermarket, grocery store, or food service establishment.

This chapter discusses why barriers occur, suggests one approach for making existing and new data more actionable, and then demonstrates how to use existing information for focused problem-solving and decision-making.

LEARNING FROM THE PAST

The clever retexturing and reanalyzing of what is known within a company is the ultimate source of so-called white spaces, that area of undefined, yet tantalizing, present possibilities. We have often heard teachers say, "Students often dislike history because they do not see it as relevant to their lives or necessary to know" (History Matters, 2006). The classic Greek philosopher Heraclitus is said to have declared, "Lovers of wisdom must open their minds to very many things" (Von Oech, 2001).

Sensory and Consumer Research in Food Product Design and Development, Second Edition.
Howard R. Moskowitz, Jacqueline H. Beckley, and Anna V.A. Resurreccion.
© 2012 Blackwell Publishing Ltd. Published 2012 by Blackwell Publishing Ltd.

Why do the foregoing thoughts pertain to food development today? After 100 years of sustained food development, we have entered an era of more than enough supply of just about every food item. Whereas there are issues with food distribution in some of the least well-developed countries, by and large there are more than enough choices of any type of food group available in the grocery or superstores across the United States and in other developed countries. This overabundance of choice, in which a thousand items in a food store have now multiplied to more than forty thousand (Trout, 2000), provides the food industry with a much more complex scenario in which to develop new products for markets that have become demand driven.

Survey results suggest that the areas of innovation and product development can rank first and second on the list of factors that CEOs consider as sources of competitive advantage (PriceWaterhouseCoopers, 2002; Sopheon, 2002; Schultz, 2009). The question is no longer "innovation" but rather "innovation—how?"

NPDP, the new product development process (Lynn, 2000) teaches us the concept of systematic product development, wherein understanding previous successes and failures is important. But few food companies have practiced, or even installed, a robust process of conducting "postmortems." There are no examples of industry commitment to celebrating the best of the year, as with theater, music, and movies (the Tonys, Grammys, and Oscars, respectively). Those who want to understand what the industry thinks to be the best can study the selections, and by so doing continue to evolve.

At the beginning of 2000, we saw small groups of people within food companies whose job was to deal with the company's business strategy. There were even fewer individuals whose function focused on product development strategy. In the last 10 years we have seen a growth in the discipline of products research, which encompasses strategic consumer understanding directed from a product perspective. Prior to the last decade, Procter & Gamble alone stood as an example of this holistic approach to product and consumer understanding. Today, gathering more of a "seat at the table" are these groups, generally located in the research and development departments of companies who practice consumer research/consumer insights (examples are H.J. Heinz, Hershey, Gallo, Frito-Lay, Quaker/Gatorade/Tropicana, and General Mills). Yet, this is still a very new area.

Reasons abound to explain why thorough understanding of a product business situation has not been respected as an ongoing discipline in food companies. At least eight reasons come to mind:

(i) *Human nature*: There is the ever-present difficulty embracing that which one has not created. This is the so-called NIH (not invented here), or NIMBY (not in my backyard). An example that comes to mind is the project manager at a grain-based company who tried to interest a group of associates in her findings and could not get them to pay attention, since they had not been part of the exercise.

(ii) *The old standby of lack of time*: Time compression is a very common factor in business today. There is always a rush to get a project started and meet a timetable that someone else has set. As a result, doing things thoroughly from the beginning is thought to be a "luxury." For example, a breakfast snack went from concept to manufacture within a short period of time, avoiding any product design or evaluation. It failed in the test market and then was "tweaked" for two more years before the project was killed, allowing the product to die the abysmal death that was built in at the start because of poor planning.

(iii) *Failing to recognize what the company already has learned*: Few organizations appreciate how much knowledge they actually have and tend to feel that the history they have is too dated or does not apply to the particular initiative. The result is a lack of recognition of the value of past learning. One well-known company studied the same idea over the course of 4 years, four times, but with different consulting firms being asked to come up with the "true" answers. After four different PowerPoint presentations, the answer was still the same but with slightly different words and segments.

(iv) *Recognizing what is important*: Big shifts in thinking or behavior are fairly easy to spot. When these changes come about from subtler interrelated forces, they are more difficult to observe. These changes, called "weak signals," are easy to dismiss as trivial or nonquantitative. The popular story about the development of Post-it® Notes is a classic example of a weak signal that took time to watch and understand and then capitalize upon.

(v) *Go do*: Americans have been a *go-do,* go-get-it-done society. Except for a few academics, the study of the past seems to interfere with this quintessential American human nature of pioneer spirit and the inward nagging desire to get out and do something. As a result, a knowledge-collecting phase of a process appears slow, overly deliberate, and fairly archaic. The nature of most product development people is "to go make a recipe or a formulation." That is, simply do *something, anything*, to reduce the anxiety of a goal unfulfilled. One consequence is that the disciplined organization, analysis, and thus assessment of data and learning from the past just does not look like or feel like work. Examples are the conversations from many product development teams dealing with the joys (and pain) of bringing up a line to manufacture a product. The act of learning is not accorded the same degree of interest, nor does it merit much emotion. Somehow piecing together knowledge of what is known is not viewed with the same level of excitement and joy.

(vi) *Information technology (IT) for wiser product development is still emerging*: Other than technology solutions that speak to the strength of knowledge management, there are few examples of how to apply a structured approach to knowledge mining in order to *directly and measurably* benefit the product development effort. It is known that around 80% of the real knowledge in a food company is nonquantitative (Hawver, 2004). Thus most IT knowledge management systems are missing a lot of the critical information on any given subject. This situation is right at an inflection point. The purchase in 2009 of SPSS, a popular data analysis tool, by IBM suggests that this area of opportunity had been "discovered" (IBM, 2010).

(vii) *Universities do not teach integration across disciplines*: There are few people in the food development area who have been taught to appreciate the value of knowledge of the past, how to understand it, how to make use of it, and how to think about a problem or business situation once the information is presented in one place. There are few courses in the science curriculum that educate individuals with respect to critical thinking and connecting science with business. So where then would the skill set come from—a skill set that could foster a deep-seated commitment to understanding, thinking, and creatively influencing others with respective "historical" knowledge about a business or product class? Certainly, that skill would not emerge out of today's curricula.

(viii) *Power from knowledge is often hoarded, not shared*: Keeping track of how decisions were made in the past and what knowledge a company has is very empowering. When senior managers want to control situations and "spin" information to suit their purposes or to hide mistakes, another person's access to and understanding of the knowledge hoard within a company can reduce the perceived power base of that senior manager.

VALUING AN AVAILABLE ASSET

Within most food companies, the most stable group of individuals (i.e., people who have been in their roles for a significant period of time) are those individuals who are part of product research or food technology. Reasons today suggest that the path for rapid growth that is embraced by the marketing/business side of a company is just not a part of the career path for the other, slower-moving, more dutiful research groups. Whereas academic achievement of leaders of research groups has grown in the past 30 years, not one leader of a research group in a food company has emerged as the leader of a food organization. So when the head of the organization "tops out" at vice president or chief technology officer and is not felt to have the necessary skills to move higher within the organization, it makes sense that those individuals choosing to follow the research path will experience a slower rise up the corporate ladder. And whereas downsizing and rightsizing have been the norm for most of modern food product development, greater stability is found in the food research teams than in any other part of the organization.

It is logical that the major drivers for knowledge management, strategic product development, and user/product understanding should be in the product development groups. But this is not where the proponents of knowledge integration have been found. The brand champion, marketing, has generally played the role of knowledge proponent. Over the last 10 years, we have started to see product development with the products research groups; sensory or consumer insight groups begin to assume the role of knowledge ownership. The ownership role and the responsibilities that the role evokes have not come easily. Specific training with proprietary, organized programs (individualized with names like technical brand manager training or consumer product design) have started to facilitate this change. *The transformation of the sensory scientist to a products researcher or a knowledge agent in the company is an unexploited opportunity. Thus, the roles of product development and its supporters, like sensory, have been framed in the past as necessary—the means to the end—but never critical for strategic decision-making regarding the business . . . until now.*

We see examples of the problems faced by consumer and sensory researchers in Figures 2.1 and 2.2, which present timelines and available tools from the past. The consumer and sensory research timeline acknowledges and applauds techniques, tools, and analytical assessments. Most of these elements are tactical. These are the elements needed to assure that the products were good enough to survive in a "picky" world. So if a person can be defined by what he does, then both sensory and consumer research have been tactical associates, together assisting the even more tactical product development effort. However, see what began to emerge (Figure 2.3) with the evolution of the Internet and many off-the-shelf collaborative tools. Change is occurring, albeit quite slowly.

If tactical behavior is what has been rewarded and accepted over the years in food development and sensory and consumer research, then it begins to become clear why there are so many more product failures (some claim up to 90% failure) than there are product successes. Have the pragmatic needs of producing foodstuffs actually created a vacuum of critical thinking? Has the amazing success of the US food industry created the situation where there is a lack of honest understanding about why products actually succeed or fail?

Robert Cooper, the developer of Stage-Gate™, conducted research in the area of improving the odds of winning at new product research. A study conducted by Cooper in 2002 suggests that the highest success rate occurs during the development, testing, and commercialization phases, whereas the lowest success rate occurs during exploration (6–13% success at this phase). Effective use of knowledge within the company definitely could improve the three poor-performing areas of exploration: preliminary market assessment, detailed market study, and prototyping. Failure in these categories across industries occurs an astounding 48–74% of the time (Cooper, 2002), an enormous proportion when the reader stops to think about the implications. Early stage understanding simply is given short shrift. To that end, Cooper has begun to suggest an improvement in Stage-Gate™ called Nex-Gen systems (Cooper, 2008).

It is important to understand why marketing discipline (brand management) has had more appreciation for the study of brands than its counterpart, food development, has had for the study of products. Brands have been studied and appreciated because they have generally been designed to last a long time. Many of today's well-known food brands have been around for over a hundred years. In contrast, the products themselves have been viewed as much more transitory and therefore less valuable. The consequence of this perspective is that there has been less appreciation and consideration for specific food products, either inside or outside of a company.

However, things are changing. Longinotti-Buitoni suggests that this paradigm is evolving when we understand the message of selling a "dream." The author writes:

> Senses are the vital ability to bring the environment inside us so we can relate to it. They offer the chance to live and feel the palpitation of time. They keep life from fading away, linking us to the past in ways far more dynamic than any rational memory. A sight, a sound, a smell, a texture, or a taste can bring to life, in the most resonant ways, moments of pleasure or pain that are long gone.
>
> Longinotti-Buitoni, 1999

This is what food development and sensory and consumer research can bring to the innovation dialog. If they do not, others have and will continue to do so (Lindstrom, 2005).

The need today for the product development organization, all the way from product management and design down to sensory and consumer research groups, is to take their integral role in helping the rest of the organization to understand. The understanding is fundamental: *why* the success or failure has occurred, and the *strategies* that will cut paths around the issues going forward. Linking well-studied brand attributes to the thoroughly studied product attributes and understanding the connecting threads between the two is an absolute necessity for successfully moving through the development process.

A very clear and developed path uses patterned thinking, systems thinking, personal mastery of these disciplines, and construction of physical, visual, mental models. This path may well constitute the cornerstone of evolved sustained development and innovation (Barabba & Zaltman, 1991; Davis, 1993; Buzan & Buzan, 1994; Pickover & Tewksbury, 1994; Senge, 1994; Webster, 1994; Reinersten, 1997; Hirschberg, 1999; Mahajan & Wind, 1999; Rathneshwar *et al.*, 1999; Struse, 1999/2000; Hamel, 2000; Cooper, 2001; Macher & Rosenthal, 2003; Vriens, 2003).

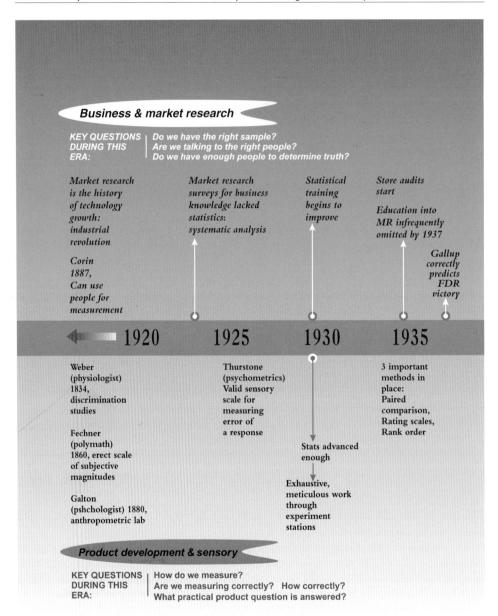

Figure 2.1 Timeline from before 1920 until 1960.

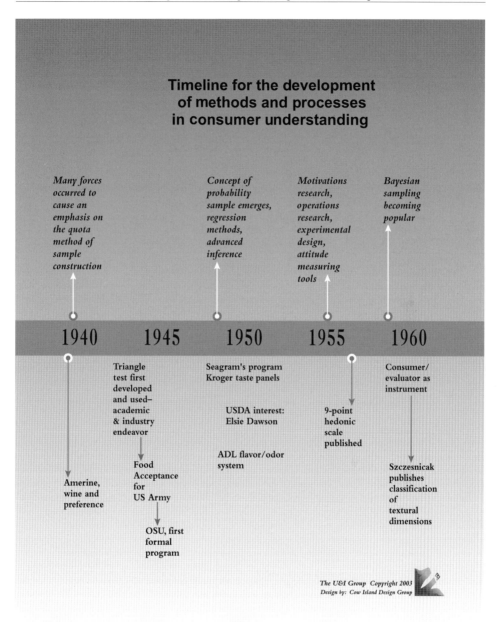

Figure 2.1   (Continued)

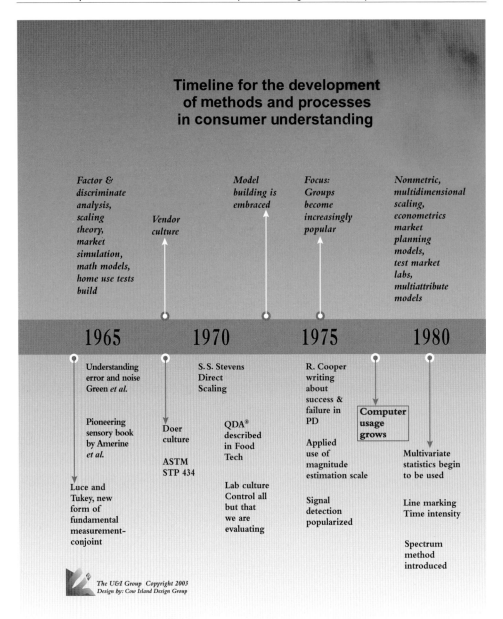

Figure 2.2   Timeline from 1965 to 2003.

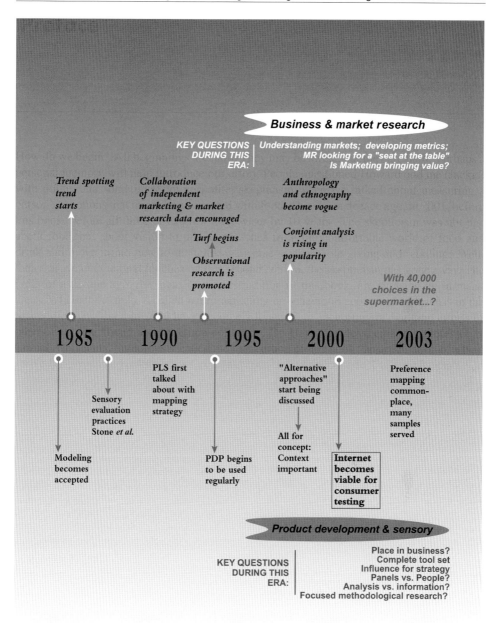

Figure 2.2   (Continued)

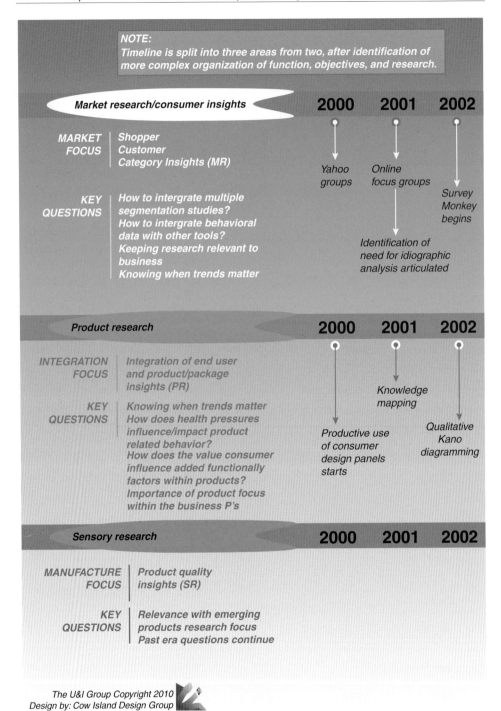

Figure 2.3 Timeline from 2000 to 2009.

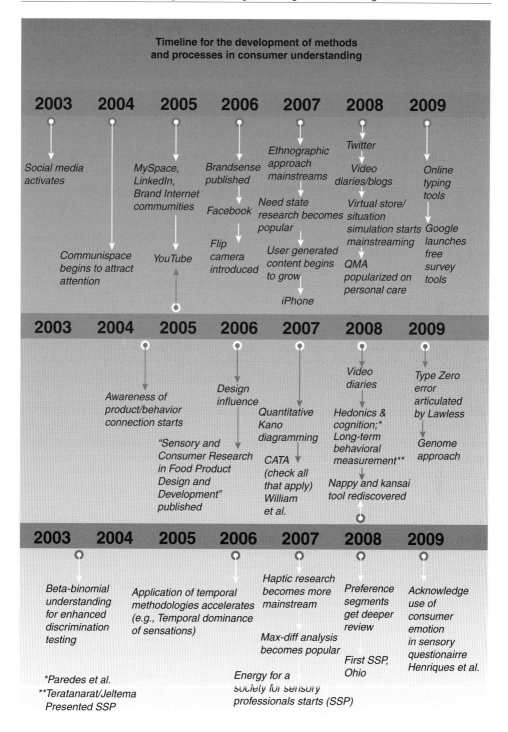

Figure 2.3 (Continued)

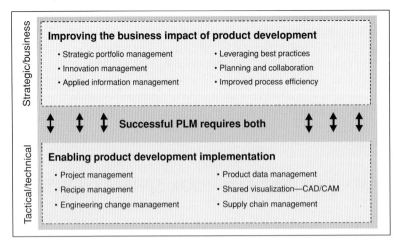

Figure 2.4   Illustrating the roles of product development in the PLM process (Sopheon, 2002). Used with permission.

EMBRACING KNOWLEDGE

Product development management and sensory and consumer research must embrace knowledge, not just suffer it soullessly in a *pro forma* fashion. Classically, brand management has been given the job of being the keeper of brand knowledge, carrying that brand promise to the consumer. Whereas this is a traditional role, there are groups of businesspeople who suggest that product development is really a series of strategic decisions (Sopheon, 2002) and therefore should work with issues formerly considered to be only in the marketing arena.

Figure 2.4 illustrates the intertwining of product development within a product lifecycle management (PLM) process and how that adds impact to the business.

The more *knowledge-centric* a company is, the more productive can be its product development process. Figure 2.5 shows an example that illustrates four groups of processes

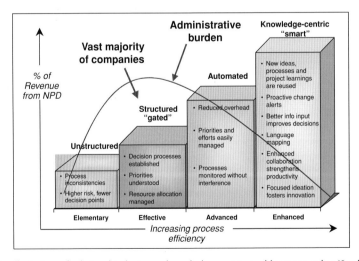

Figure 2.5   Illustration of relationship between knowledge-centric and business value (Sopheon, 2002). Used with permission.

a company can put into place: (1) unstructured, (2) structured "gated," (3) automated, and (4) knowledge-centric "smart." Most companies operate at level 2, in which they have a decision process established, priorities understood, and resources allocation managed. The benefit of moving further along the process efficiency path to level 4, knowledge-centric, is increased revenue from new product development efforts. The increased revenue comes about because all parts of product development have heightened knowledge of what needs to occur proactively for the company. This knowledge empowers them to "sense and respond." The response appears to be virtually intuitive, although it is not.

INTEGRATING MORE THAN YOUR SPHERE OF INTEREST

To be successful, companies might find it useful to increase the pace of change within their organizations, open up the decision-making process, and relax conventional notions of control. Relaxed control, faster change, and open processes foster the nourishing environment for the product development team; everyone does better. Product development scientists and sensory and consumer professionals perform far better and contribute more and help the company take advantage of business opportunities. The much-heralded economist Joseph Schumpeter called this process "Creative Destruction" (McMillan, 2004). The secret of being able to creatively destroy is *to know what part to eliminate* during the destruction and what parts are absolutely critical to maintain.

The product development teams with their sensory and consumer specialists are the individuals within the company who might best direct these changes. Their stability, background with products, and day-to-day experience solving development quandaries are all "pluses." But they need to understand the entire business construct, else the wrong parts can get recombined. Tuorila and Monteleone (2009) mentioned that multidisciplinary collaboration could help sensory food scientists further their tools for consumer research. Frito-Lay understood exactly what they needed to do with their product elements to drive consumers who had become disenchanted with Snackwell's cookies to begin to purchase Baked Lay©'s, a salty snack. Frito-Lay's launch of Baked Lay©'s was the beginning of the decline for the Snackwell©'s brand. Snackwell©'s never really recovered from that assault (Riskey, 1996).

The use of cross-functional teams within the new product development (NPD) process is well established. The front-end team for NPD needs to own the knowledge assimilation process. The largest issue is maintaining a given subject knowledge base for the company as individual teams move their respective projects through the process. Peter Senge in *The Fifth Discipline* clarifies the issue well:

> The first challenge concerns the need for sustained effort. It is relatively easy to get people to be interested in new ideas like systems thinking and mental models. Sustaining effort at "practicing the disciplines" is another matter—especially since the practice goes on forever. In building learning organizations, there is no "there," no ultimate destination, only a lifelong journey.
>
> Senge, 2006

Alongside the struggles for sustained effort and corporate memory, a third issue arises with cross-functional teams in contemporary food development. This issue is the level of disengagement of the regular team member. The food industry was one of the first places for large-scale consolidation. Whereas the thought that "people always need to eat" motivated a lot of people to enter the multifaceted food "field," the food industry has gone through the same tribulations as the rest of American industry. One overarching outcome is that whereas well-qualified, trained professionals still populate the food firms, they constitute an increasingly battered group of individuals who confront the worries of job loss and corporate loyalty with their struggle to balance work stress and other life issues (Ashman, 2003).

The data from research studies suggest that when teams lack real trust, owing to these different struggles, many behavioral events happen that can lead to quality problems in one's business work product (Lencioni, 2002). Keeping the momentum of any given project high for as short of a duration as possible helps battle-weary teams to complete their quest to collect and use complex knowledge needed for successful product development and marketing. It has been suggested that fast-cycle capability actually leads to flexibility, which then becomes a powerful tool in rapidly changing situations (Smith & Reinertsen, 1998). Figure 2.6 presents an adapted model of business behavior required to sustain these goals.

Solutions for carrying knowledge forward once it has been collected rely upon databases and software. At this writing (2010), there continue to be some limitations in this area, but those will probably diminish over time with increasing focus on the problem, along with improving technology and process such as cloud computing and downloadable applications on smart electronics. Zahay, Griffin, and Fredericks (2003) pointed out that there is currently no standardization in this category of software.

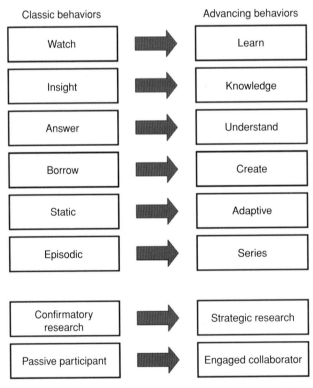

Figure 2.6 Changing behaviors in a changing research world (created by The U&I Group LLC and Stacey Cox).

Recent literature has presented attempts to develop standardized, user-friendly software for food product development, such as Mahajan *et al*.'s design for fresh produce (Mahajan *et al*., 2007). The limitations that exist relate to what is being automated (process management, information control) as well as what is not being captured (all of the qualitative, merging of the qualitative, and how to integrate data other than words, and using data across time). The biggest issue for companies is that they all have created slightly different organization structures, preventing standardization and the efficiencies that come with a common structure. A consequence of such variability is that, rather than standardization, there are few enterprise-wide examples of tools for PLM. An enterprise-wide tool would work in a flexible way with the dizzying array of databases and archives and functional groups possessed by the various companies (e.g., marketing and product development).

Despite the lack of standards, companies recognize the need to archive knowledge. Most companies are developing systems that archive on the basis of "silos." Users in different silos have access to only that information that the company believes they should have, rather than having access to all of the information. We know of systems that have been developed on a custom basis (Breitbart, 2004; Gates, 1999). There are new tools, some of which emphasize more fluid programming to deal with the flexible and less structured information found in most front-end processes (see www.Sopheon.com, www.ide.com). Given the need, new solutions will be developed, but those solutions will take time. Another approach is to design into the knowledge-gathering system strategies that move qualitative information into a more quantifiable form (Friedhoff & Benzon, 1989). That in itself is a major undertaking and deserves attention as a profitable, future direction for companies.

ENABLING BROADER UNDERSTANDING OF INFORMATION TO CREATE KNOWLEDGE

With all these situations, people, training, and technology barriers to utilizing knowledge a company has to cope with, why bother? The simple answer is that, in truth, there are no other options. Corporate survival today, and the profitable use of talent including that of sensory scientists, depends upon making the most of the human ability involving knowledge. But it goes beyond just the knowledge. Success today involves moving forward with this knowledge into an increasingly competitive world.

One key method to use a company's knowledge base and the wisdom of its people is known as *knowledge mapping*. We will discuss this in detail because it is both currently very useful in many situations and holds the promise of evolving into far more powerful techniques in the future.

VISUALIZATION OF KNOWLEDGE (KNOWLEDGE MAPPING) TO UNDERSTAND COMPLEXITY

We present here a stepwise approach to take to develop a culture within food development in which knowledge of the past is incorporated into each new project. The approach is easy to implement, low cost, and provides for record archiving for future reference, all of which are key factors for use by increasingly beleaguered corporations.

Table 2.1 Sources of data used in knowledge mapping.

	Internal	External
Tacit	Knowledgeable people within the organization. Fact based and opinions/"intuition." Suppliers to the organization. Stories about events within the organization.	Outside sources of knowledge on the subject from business, academic, and other individuals. "Facts" and factoids that are commonly available.
Explicit	Data, documents, reports, research results, written information, completed studies, commercials, ads, video, images, packaging from all parts of the organization (product development, consumer insights, sensory insights, analytical, financial, marketing, strategy, etc.).	Academic and business-oriented publications. Media (written, audio, video). Sources that are both nonfiction and fiction.

Knowledge mapping is constructed using the tenets of a learning organization (Senge, 2006). There are five key components. They are systems thinking, personal mastery, mental models, building a shared vision, and team learning. Knowledge mapping is, perhaps, the most robust way to rapidly capture both implicit (tacit) and explicit information (see Figure 2.6 for examples). Knowledge mapping combines the practices of thorough data selection and provides easy data summarization, idea organization, and concept linkage.

It is clear from Table 2.1 that knowledge mapping selects a broad array of information to use. The information is both deep and wide. As a way to cast as wide a net as possible, information contained in a knowledge map may comprise more *objective* quantitative sources along with what is considered more *subjective* sources. Examples of subjective sources are tacit knowledge from experts, thoughts from current and past employees, and current articles gleaned from the media. Other sources of information might be the content of interviews with executives who have points of view.

Knowledge mapping uses a number of tools designed to help acquire knowledge and create new insights. Some, but certainly not all, of the tools are:

(i) Room with empty walls.
(ii) Rolls of white or brown butcher paper.
(iii) Post-it® Notes. Each participant (so-called knowledge mapper) has his or her own color note.
(iv) The knowledge (in the preceding paragraph as well as information from Table 2.1).
(v) Various writing instruments that allow individuals to pick the writing instrument that works best for them.
(vi) Tables on which to write.
(vii) Participants—at least two; 10–12 people are better.
(viii) Three to four hours.
(ix) Clear definition of what the question is that must be answered and mapped.
(x) A facilitator who has experience with achieving collaborative knowledge sharing.

Tools without content, without structure, and without steps are simply things. The process comes alive when the facilitator follows defined steps, with those steps designed to make the knowledge emerge from the mind of the individual participant into the public view of the group of participants. In so doing, the knowledge has been mapped, made explicit, and morphed from a set of inchoate thoughts in one person's mind to a formalized idea that can be discussed by others. The ideas become clearer, the insights sharper, and the opportunities emerge.

We see the six key steps for implementing a knowledge mapping exercise in Table 2.2. An ultimate goal of knowledge mapping is the creation of the concept of "common knowledge." This is a concept that has been proposed by Nobel laureate Robert Aumann: "two individuals cannot forever agree to disagree." As people's assorted beliefs, formed in rational response to different bits of private information, gradually become common knowledge, these assorted beliefs change and eventually coincide. In turn, in today's era of individuals who have

Table 2.2 The knowledge mapping process.

Step	Issue	Reason
Step 1	**What is the need for knowledge?** What is the specific business question(s) that need to be answered?	Ground mapping in specific business-based issue. Clarify what needs to be known.
Step 2	**Review the research data.** Go through all information that has been brought into session. See Figure 2.6. The data is summarized as "sound bites," small phrases or sections of information that encapsulate the general thoughts or feelings from specific pieces of data (*not summaries*, primary data). All team members participate in the organization of the data. Each member gets a different color of Post-it® Notes so the inputs can be distinguished by contributor.	Allows group to know what is known and what is unknown. Visualize the information so that they can "experience" information, not just think about it in words. See how the different inputs from team members integrate with each other.
Step 3	**Lay out what is known.** The data that has been summarized in Step 2 on Post-it® Notes is organized by topic area on a very large sheet of paper. The team members who organize the "sound bites" work together (in twos) to find an orderly pattern to the clusters and develop some arrangement of hierarchy for the information.	Allows review of the information by two team members, usually one representing the science part of the business and one connected with the business part of the organization.
Step 4	**Understand what is not known.** The team reviews the map as laid out by the team representative. As the team moves from subject to subject, they become aware of aspects of the information that are missing either because it has been forgotten or it is not known.	This phase looks a lot like informed brainstorming since the team knows all of the data (at the same time) and can speculate, as a group, what implications may ensue.
Step 5	**Make connections.** The team begins to build a conceptual network of where things link so that they begin to understand the complexity, or lack thereof, of the question they are trying to address. They know what the company knows about the subject and have a fairly good idea of what is unknown. Anomalies become clearly apparent.	High-level thinking and processing is occurring due to the graphic visualization of information. Team members are working with information and are focused on a common goal, which is also time limited.
Step 6	**Identify the gaps.** The team, as a unit, knows why certain questions and areas of inquiry are gaps. They clearly can articulate what they do not know about a given subject, what have been guesses that have succeeded or failed, and what absolutely needs to be understood in order to make progress answering the question posed in Step 1. They have also clarified whether Step 1 is the correct question(s) or not.	The team knows as a unit what it needs to find out and why and can be much more focused on outcomes that are meaningful. Issues that are clustered around anomalies can be addressed specifically rather than appearing to be the focus of one team member or another.

continuous partial attention (Ball, 2010), it is important that we move to business relationships in which "you know what I know, you know that I know it and I know that you know it." In game theory this is known as "complete information." At that point, the individual knowledge a person has can begin to be merged into the corpus of general information for the benefit of common knowledge.

THE IMPORTANCE OF MAKING INFORMATION REAL

In today's work environment, commonly used business tools such as Microsoft Word™, PowerPoint™, and Google™, along with extensive archiving systems, can readily fool one into thinking that it is very straightforward, almost trivial, to have plans, to know what one is doing without really thinking about what the data, words, or information means.

Yet, it takes a short exercise to show what's known and what's not known. When food development teams take a few hours to map the knowledge they possess for a given issue, what becomes immediately clear are as follows:

  (i)   The team knows a lot.
  (ii) The team does not know what they have and know.
 (iii) There is more information and knowledge in common than not.
 (iv) The subject or question for which they are trying to develop the initiative may not be as clear or as approachable as the team leader had imagined at the outset.

WHY KNOWLEDGE MAPPING WORKS, AND WHAT IT REALLY ACCOMPLISHES

Knowledge mapping works because it forces people to "look" at what they are thinking about. Friedhoff and Benzon explained in *Visualization* that "visual thinking is real" (Friedhoff & Benzon, 1989). Whereas they suggested that many years ago, visual thinking had been treated with skepticism by experimental psychologists, it has had a rebirth with the advent of neurophysiology and computerization. They point out that a persistent finding is that visual thinking differs from individual to individual, distributed normally in the same way as, say, intelligence. Friedhoff and Benzon suggest that some think more visually than others, a supposition that makes sense in light of what we know about the distribution of abilities and styles.

It is important to keep in mind the large amount of space that the human brain already devotes to vision, since it raises the question about why the brain would be designed this way. Perception research tells us that wherever we look, in the retina, in the mechanism of color perception, in stereopsis, in the "wiring" of the visual cortex, we find the visual system to be organizing its inputs to aid behavior. The visual brain is not a passive machine but a creator that sorts through all that is ephemeral in search of enduring signs. It is just plain difficult to "fake" knowledge mapping when we "default" to vision as the arbiter of truth (Changizi, 2009).

Physiology is one thing, practice another. Knowledge mapping must have a process to be successful. Ehrenberg (2001) suggests five rules of data reduction that appear to work wonders in turning data into information. These rules let us eyeball and then check our data by enabling our fragile short-term memory to cope when we take in and relate the different numbers or other pieces of information. The five rules are: (1) order by some aspect of size, (2) round drastically, (3) average for visual focus, (4) layout and labeling to guide the eye, and (5) brief verbal summary. With the intersection of neuroscience and human neural processing, we are beginning to understand better why processes need to be put in place. This science is evolving rapidly, yet it is providing us with explanations such as Dehaene has articulated in his popular book on reading:

> The act of reading is so easily taken for granted that we forget what an astounding feat it is. The mystery thickens when we consider that we read using a primate brain that evolved to serve an entirely different purpose.
>
> Dehaene, 2009

The process of "sound biting" the knowledge forces each participating knowledge mapper to make editing decisions regarding what aspects are really salient about the information being reviewed. Perhaps and even more important, the sound bite forces one to think about what the group must clearly understand in order to get the information clearly into their minds. This approach has a more commonplace practice with Twitter™ or IM feeds. Placing the Post-it® Notes up on the brown paper attached to the walls is somewhat electric because it adds "action" to static data. The team process of reclustering, recombining information into meaningful "buckets" of information is a form of choice and decision-making. The process transforms the individual judgment from something owned by one person to something acknowledged and understood by the entire team. Taking the time to review the entire map with the team summarizes the data. It soon becomes very clear to the participants what is known by the company, what is unknown, where gaps in knowledge and data exist, and where there is a systematic behavior that can be correlated with the data. At the end, the process turns into a very public form of consensus building that builds this consensus around the data, the "knowledge," while at the same time allowing and even celebrating the individual's interpretation of the now-public data (really now-public knowledge).

Once constructed, a specific knowledge map allows others to "view" data in a way they never would have done, not, in fact, in a way to which they never would have access any other way. New participants as well as mapping participants now "experience" the data. The outcome is rapid understanding of problems or issues. Individuals who have not been part of the original session can recluster the "data" easily in order to suit their own needs. As others view the data, it becomes clear that the way we interpret information varies across people. The differences among people in the way information is looked at and processed help the collaboration. When managed through a collaborative team environment, the specific output and those person-to-person differences generate much richer problem-solving and idea generation. Senge has suggested that the most productive learning occurs when managers combine the skills of advocacy and inquiry (Senge, 1994). Knowledge maps foster both behaviors.

Knowledge by itself and the ensuing insights can be transitory. How does the company keep what it has learned and discovered through mapping? It should not be surprising that a major focus recently is on the creation of methods to preserve and vet data to enhance the knowledge and insights. One simple method retains the native map, as created in the session, along with other materials for a project. The file cabinet approach can become cumbersome since the paper sources fall apart after a time and may get mixed up with other, nonrelevant materials.

Technology comes to the rescue here, and other approaches utilize technology solutions. Transferring the map into a graphic software system, such as those provided by Inspiration^{TM} (www.inspiration.com) or Visio^{TM} (www.msn.com), allows the large physical maps to become graphic, electronic images of the knowledge. These graphics records can be retrieved at any time, reviewed electronically, or printed out for more in-depth review of the information. As new knowledge is collected, it can be added to the existing map so that the living map shows knowledge is growing and evolving. The maps can be made into physical entities, printed out as "map" and viewed by the team for the duration of a project. They can then be stored much as map books are archived. Information retrieval here becomes simpler when the system works with "key words" that can be set up at the time the archive is created. Additionally, digital photography has allowed these maps to be photographed at high-resolution levels, knitted together in a panorama mode, and recreated again either as a full-scale map or as a reduced image. This graphic is then electronically available and can be viewed from a computer or projected onto a wall. With good photographic resolution, the map becomes visually real. Figure 2.7 is a photographic example of a map; the real physical map was 4 feet high × 15 feet long.

Figure 2.8 is further illustration of the knowledge map—now however presented as a graphic representation of a segment of the knowledge map in Figure 2.7—in a form that can be retained electronically. Figure 2.9 shows how the representation of knowledge mapping can be shown as a two-dimensional electronic array of pieces of knowledge—so-called knowledge bubbles. Figure 2.9 represents the entire content of the information found in the specific "knowledge bubble" identified as "Anxiety"—a small portion of the knowledge mapped in Figure 2.8.

We create the world we see. The visualization system of knowledge mapping provides a strategic product development team with the resources to develop a theory, with which they can then proceed to the learning phase (Senge, 1994). There is broad-based integration of data from a team perspective that allows different problem-solving abilities to all merge. The knowledge mapping process we have described is simple to use, reuses data a corporation already has, and builds upon the concepts of a learning organization. It allows for the visualization of close or possibly linked data and provides for electronic databasing. IDEO, a design firm that is recognized as one of the long-term innovators of the 1990s and has

Figure 2.7 Example of a knowledge map. The actual map covered 60 square feet—4 feet high × 15 feet long.

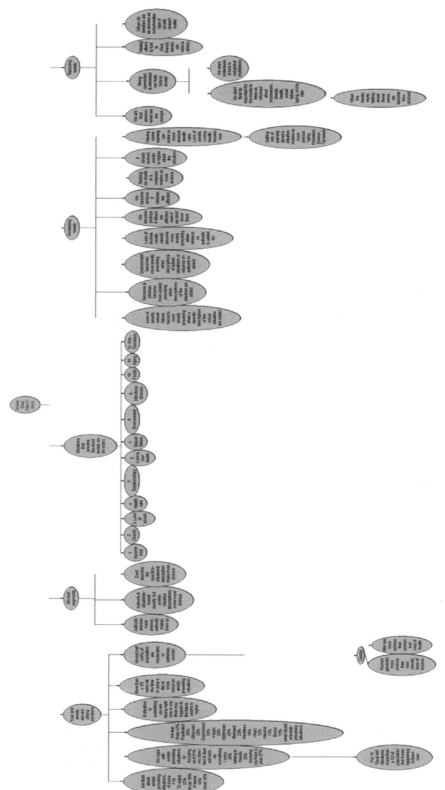

Figure 2.8 Representation of a knowledge map dealing with general issues facing consumers in the early part of the twenty-first century. Expansion of knowledge found in upper middle of Figure 2.7.

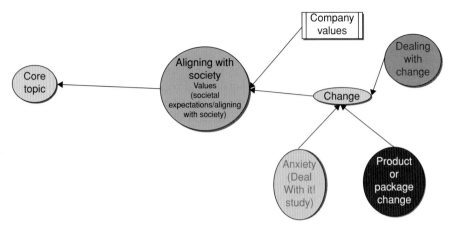

Figure 2.9   Information that is the detail for the bubble in Figure 2.8 called "Anxiety" (Deal With It!© study).

continued into the twenty-first century, extols the benefits of the power of such spatial memory (Kelly, 2001).

AN OVERVIEW—HAZARDS AND REMEDIES

With today's IT, and in light of the massive competitive threat faced by companies, why are new methods such as knowledge mapping so difficult to introduce to companies? So what is the barrier to implementing this powerful and holistic approach? Very simply, these new methods constitute changes in the way things are done. Although this approach (and other similar approaches) links the real world we live in with a visual digital system that stores knowledge, it is new and therefore foreign. Yet, in a world with knowledge sources changing (Klinkenborg, 2010; Isaacson, 2009), being sure of what one knows, and if they really know, becomes essential.

Understanding what is known and following that pattern is a very old approach to the world. Some call it science. But graphically picturing knowledge is new. And the profound impact it can have on food development and the pace that food scientists, sensory scientists, and consumer researchers solve problems is remarkably unexpected. It's fitting to close with a quote from Peter Drucker:

> Because its purpose is to create a customer, the business has two basic functions: marketing and innovation. Marketing and innovation produce results, all the rest are costs.
>
> Drucker, 1973

REFERENCES

Ashman, H. 2003. Understanding the impact current events have on consumer behavior and how to use your knowledge of "today" to make better business decisions "tomorrow". Future Trends Conference (IIRUSA), November, Miami, FL.

Ball, A.L. 2010. Are 5001 friends one too many? *New York Times* May 30, Sunday Style Section, p. 1.

Barabba, V. & G. Zaltman. 1991. *Hearing the Voice of the Market: Competitive Advantage through Creative Use of Market Information*. Boston: Harvard Business School Press.

Breitbart, D. 2004. Personal communication to the author.

Buzan, T. & B. Buzan. 1994. *The Mind Map Book—How to Use Radiant Thinking to Maximize Your Brain's Untapped Potential*. New York: Dutton Books.

Changizi, M. 2009. *The Vision Revolution*. New York: Ben Beenbella Books.

Cooper, R. 2001. *Winning at New Products: Accelerating the Process from Idea to Launch*. Cambridge: Perseus Books.

Cooper, R. 2002. *Product Leadership—Creating and Launching Superior New Products*. Cambridge: Perseus Books.

Cooper, R. 2008. Perspective: the Stage-Gate idea-to-launch process—update, what's New and NexGen systems. Annotated paper from 'The Stage-Gate idea-to-launch process—update, what's new and NexGen systems.' *Journal of Product Innovation Management* 25(3):212–232.

Davis, R. 1993. From experience: the role of market research in the development of new consumer products. *Journal of Product Innovation Management* 10:309–317.

Drucker, P. 1973. *Management: Tasks, Responsibilities, Practices*, p. 61. New York: Harper & Row.

Dehaene, S. 2009. *Reading the Brain*. New York: Viking.

Ehrenberg, A. 2001. Data, but no information. *Marketing Research* 13:36–39.

Friedhoff, R.M. & W. Benzon. 1989. *Visualization: the Second Computer Revolution*. New York: Harry N. Abrams.

Gates, W., III. 1999. *Business @ the Speed of Thought—Using a Digital Nervous System*. New York: Warner Books, Inc.

Hamel, G. 2000. *Leading the Revolution*. Boston: Harvard Business School Press.

Hawver, C. 2004. Personal communication to the author.

Henriques, A.S., King, S.C., & Meiselman, H.L. Consumer segmentation based on food neophobia and its application to product development. *Food Quality and Preference* 20(2):83–91.

Hirschberg, J. 1999. *The Creative Priority*. New York: HarperBusiness.

History Matters. 2004. History Matters home page. http://historymatters.gmu.edu.

IBM. 2010. IBM to Acquire SPSS Inc. to Provide Clients Productive Analytics Capabilities. www.ibm.com/press/us/en/pressrelease/27936.wss (May 30, 2010).

Isaacson, W. 2009. How to save your newspaper. www.time.com February 5.

Kelly, T. 2001. *The Art of Innovation*. New York: Doubleday.

Klinkenborg, V. 2010. *Further thoughts of a novice e-reader*. New York Times May 30, p. 7.

Lencioni, P. 2002. *The Five Dysfunctions of a Team*. San Francisco: Jossey-Bass.

Lindstrom, M. 2005. *Brandsense—Build Power Brands through Touch, Taste, Smell, Sight and Sound*. New York: Free Press.

Longinotti-Buitoni, G.L. 1999. *Selling Dreams: How to Make Any Product Irresistible*. New York: Simon & Schuster.

Lynn, G. 2000. *New Product Certification Workshop*. New Jersey: Product Development Management Association Certification Committee.

Macher, J. & S. Rosenthal. 2003. "Managing" learning by doing: an empirical study in semiconductor manufacturing. *Journal of Product Innovation Management* 20:391–410.

Mahajan, P.V., F.A.R. Oliveria, J.C. Montanez, & J. Frias. 2007. Development of user-friendly software for design of modified atmosphere packaging for fresh and fresh-cut produce. *Innovative Food Science & Emerging Technologies* 8(1):84–92.

Mahajan, V. & J. Wind. 1999. Rx for marketing research. *Marketing Research* 11:7–13.

McMillan, J. 2004. Quantifying creative destruction: entrepreneurship and productivity in New Zealand. Productivity: Performance, Prospects and Policies by the Treasury, December, Wellington, New Zealand.

Pickover, C. & S. Tewksbury. 1994. *Frontiers of Scientific Visualization*. New York: John Wiley & Sons.

PriceWaterhouseCoopers. 2002. Innovation's the leading competitive advantage of fast growth companies. White paper, June 24, New York.

Rathneshwar, S., A.D. Shocker, J. Cotte, & R.K. Srivastava. 1999. Product, person and purpose: putting the consumer back into theories of dynamic market behavior. *Journal of Strategic Marketing* 7: 191–208.

Reinersten, D. 1997. *Managing the Design Factory*. New York: Free Press.

Riskey, D. 1996. Keynote speech. American Marketing Association Attitude and Behavioral Research Conference, January, Phoenix, AZ.

Senge, P. 1994. *The Fifth Discipline: the Art & Practice of the Learning Organization*. New York: Currency Doubleday.

Senge, P. 2006. *The Fifth Discipline*. New York: Doubleday.

Schultz, H. 2009. Starbucks Corporation fiscal 2009 annual report, pp. 1–3.

Smith, P. & D. Reinertsen. 1998. *Developing Products in Half the Time*. New York: John Wiley & Sons.

Sopheon, PLC. 2002. Improving the business impact of product development. White paper.

Struse, D. 1999/2000. Marketing Research's top 25 influences. *Marketing Research* Winter/Spring:5–9.

Trout, J. 2000. *Differentiate or Die: Survival in Our Era of Killer Competition*. New York: John Wiley & Sons.

Tuorila, H. & E. Monteleone. 2009. Sensory food science in the changing society: opportunities, needs and challenges. *Trends in Food Science & Technology* 20(2):54–62.

Von Oech, R. 2001. *Expect the Unexpected (or You Won't Find It)*. New York: Free Press.

Vriens, M. 2003. Strategic research design: using research to gain and sustain a competitive advantage. *Marketing Research* 15:20–25.

Webster, F.E., Jr. 1994. *Market-Driven Management*. New York: John Wiley & Sons.

Zahay, D., A. Griffin, & E. Fredericks. 2003. Sources, uses and forms of data in the new product development process. Paper presented at PDMA Conference, October 4–5, Boston, MA.

3 Understanding consumers' and customers' needs—the growth engine

This chapter presents the case against thinking too much, at least thinking without profoundly understanding one's topic. In an article written for the *Dallas Times Herald-Washington Post,* Malcolm Gladwell, best-selling author of *The Tipping Point,* suggested that a lot of how the consumer responds is fairly reactive, emotional, and not extensively thought out, an idea that most scientists who study people would like not to believe (Gladwell, 1991).

In the food industry we have just begun to understand today's consumer in a profound way to guide product development and marketing. There has been no lack of perceptual maps that, through mathematical legerdemain, place products and product attributes together in a geometrical space, hoping by this process of geometry to create insights in the viewer's mind. There has been no dearth of focus groups conducted through research or marketing to "listen" to the consumer, to probe, and somehow to get nuances and more insights that will guide development and marketing. And there are many well-designed studies conducted by a research guidance group, or market research group, or today's fad, the so-called consumer insights. In other words, there's a "lot of knowledge" out there, somewhere. The problem is that the knowledge isn't necessarily what provides the insight, or at least a lot of the profound insight.

Really trying to *understand the consumer and their interaction with food* actually constitutes a relatively new concept, different from the standard stock in trade methods of the past 30–50 years. Honest hearing of consumers, which involves hearing them somehow articulate their precise series of trade-offs, is not simple.

The world today is a very complex place. The complexity creates a demand. It requires that we look very clearly at what it really takes to meet a person's needs or wants. The struggle for food businesspeople *today* is to *wrap their heads* around a world that can be characterizing as giving a prospective buyer more than forty thousand offers in a supermarket. The struggle is to understand that the "slam dunks" that got the food company powerhouses to their glory in the 1980s (almost 25 years ago) are gone.

Companies are not blind, nor are professionals cavalier about the difficulties that they face. The consumer or customer is acknowledged everywhere as being the key to all packaged goods and service-oriented companies. But then what? Effectively assessing individual experiences and making sense of them is a skill. It's now a matter of connecting with people to learn about them. These approaches need to be flexible. Furthermore, these approaches need to be context based. They must be relevant and understandable to today's marketer and

Sensory and Consumer Research in Food Product Design and Development, Second Edition.
Howard R. Moskowitz, Jacqueline H. Beckley, and Anna V.A. Resurreccion.
© 2012 Blackwell Publishing Ltd. Published 2012 by Blackwell Publishing Ltd.

food developer, as well as actionable, to use a hackneyed but appropriate phrase. People buy and eat food in situations, in contexts that may be as important as the food itself.

DISCOVERING A CONSUMER

Reviewing the textbooks for market and consumer research and sensory evaluation from different times during the last 40 years suggests something surprising. For the better part of those 40 years, the customers and consumers of the input from food development have not really been people. They have been *test subjects*. So is it any surprise that we are now just beginning to embrace the idea of the consumer?

A sense of how researchers treat consumers and sensory judges can be gained by looking at the respective perspectives of consumer researchers and sensory researchers. The lack of a sense of "personhood" becomes increasingly evident when we read what is specified by the two disciplines (Table 3.1). We cannot help but be struck by the detached, clinical point of view. As Pruden and Vavra so clearly stated in 2000:

> Traditionally, marketing researchers have approached consumers with only one purpose in mind: to collect information. They expect those consumers they sample to respond because they have been asked to do so.... Trained in the world of opinion polls and advertising tracking studies, marketing researchers revere cooperative respondents but only as a source of information.
>
> Pruden and Vavra, 2000

Table 3.1 Goals and definitions in consumer research and sensory research.

Consumer/market research perspective	Sensory research/sensory evaluation perspective
Stated goal: Marketing research is the systematic gathering, recording, and analyzing of data about problems relating to the marketing of goods and services (def. AMA, 1961).	*Stated goal*: Sensory analysis of food was developed to reduce the rejection of food by the potential consumer. Sensory analysis rests on a thorough knowledge of sensory physiology and an understanding of the psychology of perception. Careful statistical design and analysis is essential. Correlation with physical and chemical data must be understood (Amerine, Pangborn, & Roessler, 1965).
Definition: Marketing research is the function that links the consumer, customer, and public to the marketer through information—information used to identify and define marketing opportunities and problems; generate, refine, and evaluate marketing actions; monitor marketing performance; and improve understanding of marketing as a process. Marketing research specifies the information required to address these issues, designs the method for collecting information, manages and implements the data collection process, analyzes the results, and communicates the findings and their implications (AMA, 1995).	*Definition*: Sensory evaluation is the scientific discipline used to evoke, measure, analyze, and interpret reactions to the characteristics of food and materials as they are perceived by the senses of sight, smell, taste, touch, and hearing (Sensory Division Web site, February 2004). *Alternatively*: Sensory evaluation is a science of measurement. Like other analytical test procedures, sensory evaluation is concerned with precision, accuracy, sensitivity, and avoiding false positive results (Meiselman, 1993).

Note: These definitions were approved by the American Marketing Association Board of Directors and are included in the *Dictionary of Marketing Terms*, 2nd edn, ed. Peter D. Bennett, published by the American Marketing Association, ©1995. The definition of marketing first appeared in *Marketing News*, March 1, 1985.

When the end point of traditional market and sensory research is unbiased, that is, clinically pure results, engaging in the understanding of humans can become messy. The unplanned, unexpected occurs with surprising regularity. Research with people is full of anomalies and apparent disconnections in results (Einstein, 1991; Lawless, 1991; Moskowitz, 2002; Pruden & Vavra, 2000; Rainey, 1986).

More than two decades ago, Professor Rose Marie Pangborn, early practitioner and doyenne of the field, pointed out that convincing academics and administrators of the importance of behavioral research (nee "consumers") and its scientific value was not easy (Pangborn, 1989). The subsequent and easier default strategy was to treat the entire process as a clinical exercise, eliminate that intractable humanness that makes the results a bit sloppy, and report clean, statistically supportable results.

Despite the tendency to *clinicize* the research process, it's becoming less possible. There's a simple reason: the change in the way business regards the consumer. Today, as never before, consumers and customers are trumpeted as being at the heart of product development and innovation. Many business leaders agree with Kathman, who insisted that leadership brands must have empathy, a term typically reserved for people rather than for ideas. Kathman suggests that brands must "connect" with their consumers, not simply by meeting their rational needs but by addressing the emotional context of the need as well (Kathman, 2003). This drumbeat has continued with many popular books suggesting much of the same philosophy. What surprises here is the shift from people recognizing the value of consumers and customers to even inanimate ideas doing so.

When we begin to understand *where* this consumer-centric idea comes from, and how deeply it's embedding itself in today's vocabulary, we may begin to become a fair bit wiser. This knowledge of today's regard for customers informs a great deal. It begins to be clearer to us why product development and sensory and consumer research have struggled with the concept of consumer-centrism in the last 10 years or so, and why many problems have emerged as a consequence. Let's explore this notion a little more, and see what emerges for us.

The literature according to two experts, Robert Cooper (2002) and Abbie Griffin (1992), suggests that the popularization of the "consumer" concept emerged from two areas: Quality Functional Deployment (QFD) and popularization of the product development process (PDP, or NPDP for new product development process).

In the 1970s, QFD began in Japan. By the end of the 1980s, QFD had become part of business practices for many manufactured goods. Industries such as automotive and copiers began QFD, but soon the process moved into the components, such as the underlying metals and other aspects. The early 1990s saw a number of academic publications on the subject (Griffin & Hauser, 1993). The whole goal of QFD is to help "translate customer needs and wants into a technique, concept, or design." QFD was not an unalloyed success; many companies had trouble implementing QFD. Yet it was valuable, and remains so.

Moving to the PDP, we see a different path, and one that has more success, perhaps because it was softer, less technical, and involved thinking rather than doing. The concept here is the so-called voice of the customer. The term has stuck. Procter & Gamble (P&G) has been credited with having the first written plan for the process in the mid-1960s (Davis, 1993). The P&G process was called "Twelve Steps to Test Market."

Robert Cooper began a lifelong understanding of the PDP that began with publication of papers in the middle 1970s. In 1986, Cooper published the first edition of *Winning at New Products* by proposing a systematic approach to product development, which he named Stage-Gate™ a few years later (Cooper, 1988, 2001, 2002). The process, as it is laid out today, comprises five stages and five gates. There is an additional early stage, called "discovery" (Figure 3.1), which emphasizes ideas, ideation, and early engagement with the consumer.

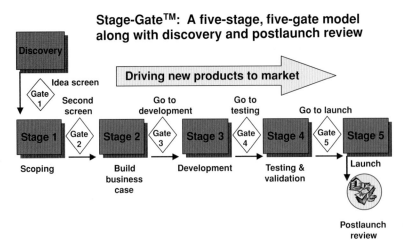

Figure 3.1 Example of R. Cooper's Stage-Gate™ process. Stage-Gate is a registered trademark of the Product Development Institute.

Looking at what probably happened to advance consumer-centric concepts, what becomes clear is that the discipline of the consumer is coming from fields that are neither what we know to be sensory research nor what we know to be consumer research. We are dealing with a different world, a world of less rationality and more emotion, a world that violates the laws of rationality on which a lot of choice behavior is assumed to operate.

The notion of a different world, a less rational one, is being recognized by more people as time goes on. For example, Dijksterhuis recognizes and even writes about the fact that dealing with the anomalies of consumers where one presumes rationality but must allow for unconscious emotional and affective motivations can lead to professional controversy (Dijksterhuis, 2004). Professional controversy is not pleasant, even though it eventually leads to increased knowledge and wisdom among all practitioners.

The outcome of conjoining of "neat and tidy research styles" with the messiness of human behavior results in professional discomfort. The discomfort manifests in avoidance, denial, direct conflict among professionals, or the problem of not being able to summarize data simplistically for the literature.

For the field of consumer understanding in product development, embracing of the consumer in a more holistic way is taking a longer time than expected to work itself thoroughly into the product development system, since it is not part of the academic curriculum for either marketing (where market research/consumer research originated) or food science or psychology (where both market researchers and sensory analysts can be trained). For example, in 2003 fewer than 15% of the sensory professionals reported using nontraditional methods to acquire consumer knowledge (Ashman & Beckley, 2003). Since then, further studies suggest that number has increased over 24% (Beckley, 2007).

THE SO-CALLED FUZZY FRONT END

Popular literature has termed the discovery phase of product development "fuzzy." It may be that this fuzziness is simply a matter of an orphan field—consumer understanding. When we

Table 3.2   Aspects of consumer and sensory research that may hinder consumer connections.

What standard market research practices might actually decrease the likelihood of finding emerging customer/consumer needs?	What standard sensory research practices might actually decrease the likelihood of finding emerging customer/consumer needs?
Creation of silos that reduce opportunity to take an integrative approach with company data.	Value of scores above everything else.
Artificially distinguishing among types of marketing research approaches such as qualitative and quantitative.	Overdependence on objective measures.
Separation of market research interviewing from modeling of data.	Overdependence on average scores.
Discontinuity of marketing research and decision support systems.	Impact of "scoring" on interpretation of data.
Lack of adaptive experimentation approaches (driven by rigidity of supplier lists and benchmarking).	"Throw" a few practices at every question, thereby ending up with a tool that is not a good fit for the decision.
Focusing on reporting of "safe" results instead of expanding strategic impact.	Abandoning responsibility of the interpretation of the data in favor of blind statistics.
Tools and infrastructure built to support limited interactions with anonymous samples of consumers.	Rely on reporting of data rather than analysis and understanding of anomalies.
	Long-standing, ongoing internal focus.
	Emphasis on being the low-cost supplier ("tests are us").

Source: Mahajan and Wind, 1999; personal conversations with leading sensory authorities, 2004; Struse, 1999/2000.

now realize that neither consumer research nor sensory have really been owners of consumers, it is understandable why confusion exists. Furthermore, a thorough understanding of the frameworks that comprise the foundations of both consumer research and sensory science now makes it seem clear why there is such a high failure rate of idea conversion at the front end of the PDP (Cooper, 2001). Additionally, the recognition of the importance of closer understanding of the consumer allows us to see why techniques developed for market research or sensory science purposes do not, and perhaps cannot, today translate into the discovery phase very well at all (see Table 3.2). The structure for discovery doesn't work because market research and sensory science are governed by other mind-sets, which demand simplicity and order.

Nothing stands still. Consumer and sensory researchers do not, and in fact cannot, remain unaffected by the *zeitgeist* and by the undesired effects of corporate failures, despite the fact that researchers believe themselves to be purveyors of knowledge in the interest of business. These researchers are inevitably affected by business trends and vagaries.

The past decade with its turmoil has brought a renewed focus on knowledge of consumers. Qiu (2004) reported only moderate agreement (neither agree/disagree) among marketing professionals about the ability of companies to address consumer needs. Whereas a decade ago or more the research business would have been above this fray, at the time of this writing (2010) we see that the problem has become more severe. Professionals dealing with the consumer are ready to change. In fact, 21% of sensory professionals and 40% of marketing researchers feel that understanding consumers/customers or one-to-one marketing is a critical business need (updated statistics, Ashman & Beckley, 2003; Struse, 1999/2000). The need to understand is complemented by the recognition that speed is a factor, not just correctness and not just adherence to time-hallowed protocols. Dovetailing with that recognition is the call through popular literature indicating the need for producers to identify consumers' changing needs quickly in order to develop successful products (Sharma & Rawani, 2009).

Of course, many professionals who keep their eyes and ears attuned to trends recognize the need for better consumer knowledge, done faster and with more meaningful results. We're not talking here about the platitudes one hears in meetings, reads in magazines, sees broadcast in industry calls for change. Rather, we're talking here about a groundswell, about talking about the emergence of a real gap, the lack of capability to profoundly understand consumers. True training at the graduate level still does not appear to exist across the spectrum of schools that train insight professionals and may not exist for the near future. The training requires an integrated approach to true learning for this complex subject (Tuorila & Monteleone, 2009. That is just not a part of the university mind-set today. It is fascinating to compare the real reason for the gap in understanding the consumer in this chapter to issues discussed in Chapter 2 regarding reuse of company data. They are the same.

LOOKING AT THE INDIVIDUAL IN CONTEXT—HAZARDS AND REMEDIES

When all intellectualizing about purchase behavior, choice, predictable preference behavior, and the like is set aside, the fundamental goal of understanding consumers is simple. It's to offer them goods and services that they will want to purchase, use, and enjoy. *How* to motivate companies to begin to purchase from vendors who might provide approaches that give them deeper consumer understanding is not quite as simple. The goal is straightforward, the path strewn with the emotions, prejudices, and just plain stubbornness of a field that grew as "Tests R Us."

Commercial interests in the vendor/supplier market have always driven much of the push with respect to what approaches are the right choices to understand behaviors and motivations (Meiselman & MacFie, 1996). Casale expressed concern that this "problem" is actually a bigger one today than it was in the past; too many research firms suggest that they are selling solutions when all they are really selling is product bundles. Real solutions are not necessarily limited to a stated research objective, but rather they get to the heart of the underlying business need in a highly integrated fashion (Casale, 2003).

Understanding consumers and customers as individuals is not simple. The reality is that understanding consumers is not as "clean" a process as many other commercially available methodologies. Klemmer states it succinctly: "field testing with untrained judges under uncontrolled conditions will usually produce large variability in results and thus poor reliability" (Klemmer, 1968). And it is this point of view that has held sway among consumer/sensory researchers for a long time.

To get close to consumers, one must go to them. On the surface, having people come to a central location or take a survey over a computer or the Internet separates the consumer/sensory researcher from truly knowing his or her subjects. The consequence may be subtle or may be severe. One consequence is that without knowing the consumer in a close, profound way, one will end up faced with theories or hypotheses that might not ring true for the individual or actually generalize well.

There are reasons why people don't get closer to their consumers. One is fear of loss of control of one's theories on a given subject. That's enough to cause one to avoid the situation.

Whereas most people trained as scientists have learned the "scientific method," the prospect of practicing real-time hypothesis testing can be a frightening experience. Additionally, it can be hard work:

> Buyers often have a great deal of difficulty in translating abstract, higher-order goals (e.g., enhancing productivity in the office workplace) into specific, lower-order decision criteria such as the particular features to be sought from desirable alternatives.
>
> Ratneshwar *et al.*, 1999

Yet, there is a body of work that suggests the closer one gets to the consumer, the better off one will be. At the early phases of product development projects, it is a strategic imperative to take fifty-thousand-foot thinking (Cooper, 2001) and employ "camping out," "fly-on-the-wall," "day-in-the-life-of," field research (Zikmund, 2000). The objective is to focus on the customer, try to identify his or her problems, unmet needs, and even perhaps discover unarticulated needs.

But what methods should one use? If "camping out" is so good, then what tools work with this brave behavior? Most writers suggest that there is no set collection of methods to be recommended. Rather, the research involves working closely with the customers, listening to their problems, and understanding their business, operation, and/or workflow.

However, there are new methods, or more precisely "new" but "old" methods. For example, at the early "scoping" phase, when one needs to learn rather than to test, the approaches embodied by anthropology, especially ethnography, are increasingly valuable. The keen interest in this whole area is growing; as an example, around 12% of sensory professionals reported using observational techniques in 2005, but only 3 years later in 2007, double that number, 24%, reported using observational methods (Beckley, 2008).

Those who appear to be succeeding best in this area are the companies that take to heart the need to invest in the true education of individuals up and down the product development value chain (product developers, their product researchers, and the business team members and management). These are the people who need to link arms to succeed (Paredes, Beckley, & Moskowitz, 2008). Yet it's not the development value chain that is really benefitting. Rather, it's the agency with a sense of "what's going on" that benefits. In practice, we're likely to see anthropology and indeed a whole array of such approaches used by and promoted by the advertising agency or large global market research conglomerate. There's a business in unarticulated consumer needs; these outside professionals know it and are capitalizing on it.

UNDERSTANDING CONSUMERS' NEEDS IN RELATION TO PRODUCT DEVELOPMENT

Concepts

Concept development is an area that has traditionally been owned by the consumer market researcher and most definitely not by the sensory scientist or product development scientist or manager. Concept development and evaluation of concepts has been a classical part of marketers' training. Few food science curriculums spend much time on the subject (Institute of Food Technologists, 2004).

One bright spot in idea generation and concept development in the world of food science programs is the team-based activity that may or may not support the Product Development

Competition. This competition is held every year at the annual meeting and exposition of the Institute of Food Technologists (www.ift.org/annualmeeting). The competition is a focused experience filled with learning while doing. Teams of undergraduates and graduates in food science work with product ideas, pushing these ideas through the concept phase, and into protocepts or test products embodying the concept. The experience of going through all phases is valuable. We can expect the experience to imprint an opportunity for consumer research on the mind of the young participants. Traditionally, the participants learn how to develop concepts, screen the ideas among consumers, and identify which ideas are winners, which are losers. Later on, and in most cases, this process of screening ideas will be handed over to the company's market research department, which will dutifully and properly screen concepts, get scoreboard ratings of concepts on attributes, and report back to the bench scientists regarding which particular concepts won and which lost.

With the advent of the PDP, all aspects of the food and drink business are putting greater emphasis on ideation and concept development (Table 3.3). Product development must play an integral role in the development of concepts; compelling new ideas require clear understanding of what the product-based levers might be and how they need to be developed (Cooper, 2001). Pine and Gilmore refer to this as "valuable distinctions" (Pine & Gilmore, 1999).

Chapter 6 of this book provides insight into approaches for concept screening that provide a more efficient as well as a more detailed understanding about why concepts succeed or fail. For this chapter, it is important to keep in mind the role of the product developer, rather than the method. The scientist, product developer, sensory researcher, and marketing researcher should get practical experience in concept development and concept testing. The more effort expended by the company and professional to understand consumer needs and then to convert those into concepts, the more likely the path to product development will be correct, with better, more promising ideas.

Now, to some specifics. The most important specific is that it is vital to remain *consumer-centric* during the development of concepts. Professionals tend to get all tied up in their own words. They "know" what will work and what will not—or at least they think they know. However, the truth of the matter is that they do not. Professionals must realize that the concept is an idea, not a PhD thesis, with dense explanation of even denser ideas. The truth of the matter is that in a test and in the outside world, most consumers will simply read a very small part of that "concept," despite the efforts to craft a masterpiece. Most of the concept will need to be experienced through the product or brand promise, as well as the delivery through the product (and by that we mean what is in the package or the bottle, or the package or experience itself).

Table 3.3 Where average-performing and "best-practice" companies can be found in the ideas (concept) and feasibility stages.

	Type of company	
Stage	**Average (%)**	**Best practice (%)**
% of ideas that get to the stage of feasibility assessment	23	5
% of ideas that get to the stage of development	10	2.5
% of resources spent on idea and feasibility stages	12.5	20
% of sales from products introduced in the past 5 years	25	49

Source: Sopheon, 2002. Used with permission.

There's more to the specifics. One additional message is worth delivering. And that there's art and there's science, magic and rationality. It is important for professionals in product development to realize that part of concept development involves moving into areas they may consider "unscientific," those so-called touchy-feely areas that call out to emotions and to less-than-rational thinking. Whereas there is a wide range of science involved in the creation of concepts, the creation and evaluation of these concepts is unfamiliar to the food developer, a vast *terra incognita*, at once intriguing and frightening. The lack of familiarity makes people uncomfortable and engenders concern, distrust, and perhaps even fear, surprisingly (see Moskowitz, Porretta, & Silcher, 2005).

Understanding consumers' needs and wants

I don't have diddly squat. It's not having what you want; it's wanting what you got.
Singers Cheryl Crow and Jeff Trott (2002), providing
an understanding of what a want is in today's world

During the early periods of today's contemporary food world with its processing capabilities, the manufacturing (supply chain) and the distribution were identified as being as the most critical parts of the growth equation for products. Things have changed, however, as the abundance of products and services in most sectors has driven the market to a more demand-based economy. That economy, in turn, has experienced its peaks and valleys of economic stress (Kash, 2002). In this type of an economy, it is critical to offer products that people either need or want, and to make sure that the products are offered in the way that people want them (Byron, 2010). For example, desirable specific features are those that relate to people in a meaningful fashion, that help people to define themselves, and that then earn a role in the category of objects that people will to use (Gutman, 1982). *Plainly speaking, figure out what consumers want to buy and try to give it to them.* For example, in their review of new dairy products Bogue and Ritson accomplished this by utilizing sensory information and data about consumer preferences (Bogue & Ritson, 2005).

But how do we go about finding these features? They are not written on people's foreheads. And we can identify them from where people live, no matter how many segmentation studies are run. There are many organizing principles offered by psychologists that enable us to make sense of how people live and make choices. Here are six among many:

(i) *Study a person's values.* Values have been shown to be a powerful force in governing the behavior of individuals in all aspects of their lives (Rokeach, 1968). Since there are many fewer values than there are objects that a person can choose, the nature of the sorting mechanism that an individual goes through in the selection process is critical to understand.

(ii) *Study a person's reality/realities.* Personal constructs make a difference—we make our own reality. In the 1950s, there was a group of social psychologists who believed that people had constructs that allowed them to live their lives (*constructive alternativism*). A key outgrowth of this system was introduced by George Kelly in his theory of personal constructs (PCT). PCT posits that people act like scientists in the way they evaluate the world around them: formulating, testing, verifying, and updating hypotheses (Kelly, 1955; Anonymous, 2004a, 2004b). Thinking within the organizing principle of PCT leads to ways to interact with consumers, such as the repertory grid method wherein the consumer is presented with a set of ideas with instructions to identify how they differ and how they are the same.

(iii) *Delve into the person's unconscious.* The unconscious holds a very important key. Another body of psychologists, behaviorally oriented, approached human decision-making with a belief that people are more unaware of their behaviors. Whereas often people seem to be acting rationally and respond with rational reasons, such rationality is, in fact, not part of the regular thought process. For instance, people are emotional (Bucklew, 1969; Laros & Steenkamp, 2005).

(iv) *Uncover the hierarchy of needs (Maslow).* The concepts of human development offered by Abraham Maslow (1968–1999) have contributed to one of today's most influential theories. Maslow's thinking about the hierarchical nature of motivation informs much of the more holistic work and approaches to people (consumers) today. Maslow suggested that a person's inner nature was part unique and part species-wide. According to Maslow, science could illuminate and help us discover the inner nature. What is valuable in Maslow's world view and approach is that the research to understand the levels of a person's mind and motivation lays the groundwork for *linkage theory* (Gutman, 1982, 1991, 1997; Reynolds & Gutman, 1988). Linkage theory suggests that there is always an end goal for a person, whether acknowledged or not. Such linkages can be studied to understand how individuals experience products, situations, and services (Griffin & Hauser, 1993). The approach to understanding linkages is called "means–end chain," where "means" is equated to the product and "ends" is related to the end goal or personal value stage.

(v) *Move beyond the rational to the emotional.* Look at emotion as a form of cognition. Psychologists who study cognition (often sensory or market researchers) have recently begun to acknowledge that emotion is a cognitive process. Both Rook and Dijksterhuis have recently written about the "intellectual tension" that still exists between the constructivist and emotionalist schools. The tension and actual conflicts oftentimes lead astray the otherwise productive discussion of testing methods. The tension moves the discussion from an in intellectual discourse about foundations to conversations about who is right versus who is wrong (Dijksterhuis, 2004; Hunt, 1994; Rook, 2003).

(vi) *Focus on the consumer in terms of what the consumer "needs."* The concept of "need state" research is a marketing research construct. Need states came into vogue after 2005. There is little else to write about need states; the field needs much more data for general patterns to emerge.

Redefining consumers, customers, needs, and wants in light of organizing principles

Given the foregoing set of organizing principles, let us look at how consumer research is structured. This structure, consumer-centric rather than product-centric, will play an increasingly important role for marketers. The structure is also starting to play an important role for product developers working in a consumer-centric environment:

(i) A (customer) need is a description, in the customer's own words, of the benefit to be fulfilled by the product or service (Griffin & Hauser, 1993).

(ii) A (customer) want is the underlying desire for a product or service. This definition allows us to be more specific, more granular, when applying psychological theory to product development and to consumer/sensory research. Needs can be better defined and linked more closely to functional attributes. In contrast, wants link more with the emotionally based factors of products and services.

(iii) A consumer can be a statistic, a hypothetical figure symbolizing millions of people (Longinotti-Buitoni, 1999).

(iv) A customer, in contrast, is an identifiable person with a name, who is highly selective with distinct tastes and desires. Furthermore, each customer has unique situations that need to be understood specifically (Leonard & Rayport, 1997).

 (v) Discussions about needs and wants generally progress to a point where the conversation focuses on the unarticulated want or need. Leonard suggests the reason for the lack of articulation:

> Customers are so accustomed to current conditions that they don't think to ask for a new solution—even if they have real needs that could be addressed. Habit tends to inure us to inconvenience; as consumers, we create 'work-arounds' that become so familiar we may forget that we are being forced to behave in a less-than-optimal fashion—and thus are incapable of telling market researchers what we really want.
>
> Leonard and Rayport, 1997

"Not getting it"

Often in the world of practical, industrial research, one intuits that there is a need. Yet, somehow the researcher either cannot understand the need or cannot figure out what the want is. Just because companies are in the business of creating and selling products and services does not mean that they understand the consumer. The consumer is a moving target here.

When the unexpressed want or need *is not due* to context, research suggests that this may often be due to the errors of the interviewer and not problems with the interviewee. An interviewer brings in biases. It's natural. The frame of reference filters or biases the interaction itself and the interpretation of that interaction. Those biases exert a significant impact on the discovery phase of the PDP. In numerous studies by one of the authors (Beckley), applying a systematic approach to consumer hearing shows that consumers are extremely clear and specific about what they want. Oftentimes, it is just that the listener is not ready to recognize the message being sent (Jarvinen, 2000; Moskowitz, 2004).

Moving to the whole experience

Research continues to inform us that most of thinking, reactions, and behaviors are enveloped in our unconscious mind (Birdwhistell, 1970; Mehrabian, 1971; Zaltman & Schuck, 1998). The Marketing Science Institute lists "greater insight into the customer experience" as one of its top research needs. But how is this to be made systematic and believable to scientists and businesspeople when even Pine and Gilmore, in their very popular book *The Experience Economy* (Pine & Gilmore, 1999), suggest that providing an experience is currently an art form? And Rolf Jensen's soliloquy in *Dream Society* (1999) should cause chills to go up and down one's spine when he announces that there is a market for feelings:

> The consumer buys feelings, experiences, and stories. This is the post-materialistic consumer demanding a story to go with the product. Food that is good quality, tasty, and nutritious will no longer be sufficient. It must appeal to the emotions with a built-in story of status, belonging, adventure, and lifestyle.
>
> Jensen, 1999

So what really is an experience in food, and what could this experience look like if it were to be captured in text? Figure 3.2 (on the following page) provides an example of an "experience" report. This particular example attempts to lay the framework for the following:

(i) Why the product was developed?
(ii) What business factors relate to the products developed?
(iii) How the product performs relative from a consumer's (or rather a customer's) perspective?

The structured approach to experiencing presented in Figure 3.2 shows us a way to study experiences and to decide whether or not a product achieves a specific experience "goal." The approach is neither magical nor mysterious. It applies principles of marketing, consumer insight, competitive scanning, product attributes (sensory factors), and current social environment. It is integrated.

Contrast this holistic approach with other efforts to summarize experience. Looking at most product-based summaries in trade magazines today, the reader typically is served up a simple, not necessarily enlightening overview of what the product is, what its ingredients are, how much it sells for, and what it tastes like. And that's all:

> But why would the target consumer buy the product? What was the thinking of the company that produced and marketed the product? Why would this company commit business funds for this venture instead of some other venture?

Furthermore, in an era of more than enough food or calories in developed countries, when you are a product developer or sensory or consumer researcher and you are defaulted to saying, "the answers to those just-posed questions is marketing's job," then you have now answered why 70–90% of the new food products introduced continue to fail in the marketplace:

> In a demand economy, when a person is part of the product creation cycle, then it is that person's job to understand the product in the marketplace. It's the person's duty, really, to engage in an intellectual discussion with research and business management about why they think a product will or will not possibly succeed. It is everyone's job to think about how the customer will react to the offer. And it is definitely not the job just to "do the job."

In northern New York State, Robert McMath has made a living in recent years offering a museum of failed, often ridiculous products (at least by today's standards), and then asking the question, "What were they thinking?" (McMath, 1998). There is no longer reason to default to thinking that the *they* are anyone other than the product developer and his or her consumer insight and sensory research professionals.

From professionalism in doing the job to professional in understanding the consumer

In recent years, some companies have experienced a quiet revolution inside their insights group, those knowledge workers dealing with people. The revolution is about perspective, a chance from doing a job to truly understanding. What has started with some companies, and needs to continue, is for the researcher to look at a person in terms of how the person

EXPERIENCING: A & W Root Beer Float

Nostalgia. Soda Fountains. Large chilled glass mugs. Poodle skirts. Then add smooth vanilla ice cream, bubbly root beer, foam collecting around the rim of the glass and around your lips and mouth. Classic root beer was created in the mid-1800s by Philadelphia pharmacist Charles Hires, the original root beer was a (very) low-alcohol, naturally effervescent beverage made by fermenting a blend of sugar and yeast with various roots, herbs and barks. The story around root beer floats suggest that either Robert Green of Philadelphia or Fred Sanders of Detroit figured out how to get a beverage with a foamy head, a root beer and ice cream taste which actually takes the small wonder of air bubbles nucleating to large bubbles of carbon dioxide to make one of the original molecular gastronomy creations.

This innovative product produced a trend that led at some point to hundreds of brands of root beer being produced in the US, moved a previously traditional beverage and herbal medicine into mainstream and was a huge hit during prohibition. But now, a root beer float in a bottle?

Why go to something ole' timey and nostalgic? Since 1999 when Pine & Gilmore ("The Experience Economy") drove home the idea that people (customers/consumers) might be willing to pay more for an experience, and most business people were willing to acknowledge that the US economy had left the supply economy of the 20th century for the demand economy of the 21st century (see Kash, The New Law of Demand and Supply), the idea of experience creation has been the goal of most packaged goods firms.

Combine that knowledge with the third consecutive year of negative volume gains for the carbonated soft drink industry, and you begin to see why the Dr. Pepper Snapple Group, the owner of beverage rights to A&W bottled root beer, thought it might be a grand idea to create an experiential beverage called A&W Float. While Dr. Pepper Snapple has struggled to gain share against the megaliths of Coke and Pepsi, going through a "demerger" this year, they manufacture many core products like A&W, Dr. Pepper, and Snapple that have consumer bases that LOVE THEM. Using a front end process called Discovery Innovation Group, Dr. Pepper Snapple found that the idea of an A&W Root Beer Float in a bottle was one of the highest scoring ideas they had found. The challenge – create a product that lives up to the promise!

Understanding: (What is going on in the marketplace?)

The idea of carbonated beverage or juice mixed with dairy ingredients just doesn't sound good to Americans. This idea has seen a fair amount of use in other countries (Europe, Asia) but here – the idea gets a big YUCK. So it is very interesting to see how Dr. Pepper Snapple was able to engage the concept of a float (ummm . . . . root beer (carbonated beverage) plus ice cream (last time we check, a dairy food) and produce a winning concept – this is a product that was created for "a root beer float lover who has a busy life, yet desires a simple indulgence that has a good flavor." A product that has "bottled the flavor of an ice cream float without the hassle of preparing one." There is a promise of "rich, creamy, one-pour preparation with a little taste of heaven" (www.floats.com).

There is a lot of pressure on the bottlers of CSDs (carbonated soft drinks). Between issues with sweeteners (sugar vs. high fructose corn syrup), cost of ingredients (corn sweeteners, fuel, packaging, etc.), nutrition (full sugared products being one of the stated "evils" that has led us to our US obesity issues, and concern about sustainable manufacturing (plastic bottles, large use of plastic feedstock and everything that comes with it), it is really tough to make significant business gains – "significant headwinds" is the word that The Dr. Pepper Snapple Group, Inc., mentioned during their earnings conference call in June 2008.

The numbers show us that as of July 2007, CSDs declined 3.8% (IRI, Nielsen). As Michael Bellas, CEO of Beverage Marketing Corp. explained, "This category is losing to everybody – it's losing to water, to energy, to RTD teas and enhanced waters. So until these other categories start to slow down and quit feeding off of CSDs, it's going to be a difficult time for them to show positive growth" (Bev World, 4/2008). Total size of the CSD US market for 2007 was estimated to be $14.3 B. (Nielsen, 2007).

Figure 3.2 Example of a report on a food product experience.

So what about Dr. Pepper Snapple? It is a company that is changing. Last year, vocal shareholders suggested that there would be more benefit to Cadbury Schweppes, the former owner of the beverage group now called Dr. Pepper Snapple Group. Due to tight equity markets, it was necessary to handle the spin-off as a demerger as compared to a sale or straight spin-off. This leaves DPS with about 15.2% of the market compared to Coca-Cola at 43% and PepsiCo at 31.6% (Beverage Marketing Corp., 2008). A&W is the second largest root beer behind Mug (Pepsi) and Barq's (Coke). Knowing that some change was coming in the company, launching a highly innovative liquid "treat" would make sense. DPS has been utilizing a sped up idea/product creation process. While this may not have been the pathway for the A&W Floats, innovative products like this combined with a drive for more rapid product and market development with innovations sometimes coming within 16 weeks between concept and commercialization is certainly the message Wall Street likes to hear.

Key: The consumer might be able to engage in an idea for a beverage treat that is easy to make and doesn't require either the ingredients that might or might not be in the refrigerator, pantry, or freezer. Having something that is special and has a lot of stories and excitement behind it might just create a magic moment with the kids or the grandparents, or the whole family for that "good old summertime" taste.

Insights: Delivering a product that is a conceptual/wishful idea to many people has the ability to produce purchase interest due to the hoped for excitement of an experience that may have occurred a long time ago or may just be a piece of memory never experienced.

It!s Convenient™, Crave It!™, Drink It!™ and Healthy You!™: Drink It! is an innovative integrated 30 category conjoint study that generates a database that can be used to understand the experience of beverages (from product, situation, emotions and brands/ benefits) in the marketplace.

Key attributes for an indulgent treat are: taste, creaminess (taste and texture), and packaging that supports the indulgence.

DPS/A&W took those insights, combined them heritage and nostalgia and played to a beverage world comfort food – the ice cream float.

Trends:

Key trends impacting the industry are convenience, indulgence, premiumness, and healthy.

Convenience: Manufactures are responding to consumers' hectic lifestyles by creating more convenient products that help the care giver have a chance to create something special yet not have to plan too long to achieve the specialness. If part of the convenience comes from the packaging, this becomes a bonus. Going to A&W can be a family tradition. But maybe there is no longer a "stand" near the minature golf course. So what better than to keep a pack of these in the pantry, toss a few glasses in the freezer and wow – make a small affordable event for the kids or grandkids.

Indulgence: Pleasure is important in this busy world. We have seen many products focused on healthy aspects, but the pleasure of an indulgent treat is still a strong behavior. Consumers use carbonated drinks and ice cream for enjoyment, moods, calories, and sometimes to deal with the stress and chaos in their world. Creating analogs of craveable foods allows consumers to have a luscious taste experience and satisfy a craving for the moment.

Premiumness: Premiumness in beverages has always come from the taste, package, and venue one can purchase the drink. Through the use of the bottle and beverage/food components, an experience that either requires a trip to a shop or the right combination of ingredients is made available. Combining nostalgia, ingredients, package, brand and situation – unique for this product – one can hope to cue premiumness.

Figure 3.2   (Continued)

Healthfulness: An overall industry focus on trying to help consumers reduce obesity combined with consumer behavior for treats equates to an intense desire to provide beverages that are healthier. Now this is a stretch but combining root beer which could have healthy herbs and provided a prepared version of a float – can shave off about 15-20% of the calories of a self-made or soda shop version – and you do have it highly portion controlled.

THE Experience: A & W Root Beer Float

There are two flavors of this product available. One is the Root Beer Float we are trying and the other is an Orange Sunkist Float. Both are available in a four pack at a fairly high register ring of $4.99. A single bottle at the C-Store is $1.69. The four pack carrier starts the message with a bold new at the top, the classic brown, orange, tan and white A&W colors and a lot of interesting messaging: "A bottle on the outside. A float on the inside." On an image which is like a large glass of root beer with the "handle" looking like neon lights and pointing to the A&W logo, there is another statement – "A creamy blend of rich A&W and Ice Cream flavor". When you turn the package you get directions for the "Perfect Float Experience" – using three circles they describe how you need to "twist it around', "pour upside down", and then "foam it up." The last direction on this side is, Drink it Down! Flipping to the back of the package you get the last of the story – "Pour in a glass for a Foamy Float Experience or enjoy straight from the Bottle. It's all the flavor of a real A&W Root Beer Float at the twist of a cap! Rich, Creamy and so Delicious you won't believe that it comes in a Bottle". This side of the story ends with "Serve Chilled." The fonts are classic A&W and evoke nostalgia from the first glance. And that was just the carrier!

The twist off cap and heavy (HEAVY), shapely bottle was kind of nostalgic. There were no "instructions" on the bottle itself, which can lead to a less than optimum use of the product. The bottles were clearly designed to go with the carrier. The bottle design echoed the carrier graphics with a full bottle shrink wrap label. Lots of images of bubbles and layers were depicted in the wrap. So here is what happened – some of the folks who tasted the product did not have the carrier and just drank the product from the bottle. Others chilled their glass in the freezer and when through the whole upside down experience. You get a lot more foam when you follow the directions. Without the idea of foam and fun, the folks who poured the product into a glass (NOT upside down) did not like the muddy water, slight opaque appearance. And they appreciated the full label to cover the color. It took one taster talking to the other taster to help them figure out how to use the product! Again – expectation and how you design the tasting is going to impact the rest of the taste experience. For those who knew what the directions were and followed them, they had a very pleasant and surprising tasting. The bubbles from the top of the soda in the glass gave off a cream soda smell. The product had a nice "head" and held it for a while. This made for a nice look – it made some start thinking . . . . it might be a float. The root beer flavor seemed A&Wish, some wanted a "better" vanilla ice cream but others thought it got the point across. It had creaminess like melted ice cream and is a nice mouthfeel. Some thought that the root beer tasted weaker than if you had made it yourself. Feelings about aftertaste varied with some finding pleasant aftertaste while others saw some strong flavor notes. We did collect a slightly different story for the non-label readers. They found the product to be very tangy in the first sip. This was unexpected from root beer or a root beer float. It was almost as if it was made with French Vanilla ice cream, which tasted odd in a float. After the first few sips that taste went away, but it was a little overly sweet, but since some of our tasters don't drink a lot of full-sugar sodas, this may just be a lack of experience. The biggest issue was the unpleasant aftertaste. It was either from the extreme sweetness or from the French Vanilla-ness of the ice cream. I had to go in search of water after I drank it to get rid of the taste, it just lingered . . . . In the glass there was more foaminess which was pleasant. Some found it a convenient product especially if they did not have ice cream and root beer at home. They felt it was handy to have on hand.

Figure 3.2   (Continued)

One comment was: "We drank the product on the hottest day (so far) of the year out on the back lawn swing because thinking about floats it seemed like we should be somewhere relaxing when we evaluated it."

This product definitely is a treat. At 260 calories for 11.5 oz, it has more than 50% more calories than a normal sugared CSD. The ingredient statement is fairly clean – carbonated water, sugar (yes – sugar), and skim milk represent the first three ingredients.

So does the product deliver the brand promise? For many, it did. Few brands could hope to create this type of experience, but the A&W concept really helps this one come together. But you needed to get into the experience. Drinking straight from the bottle gave a suboptimum product and did not do the creativity of the idea justice. Several thought it was a fairly adequate replacement for a root beer float.

How to make the Idea Bigger:

These are the same people who make what many diet drink people consider the BEST diet product on the market – Diet Dr. Pepper. We want them to channel their inner design skills to create the Diet A&W Float. All of what we had above with the experience, lose the lingering aftertaste, adjust that ice cream flavor and…every person in Weight Watchers will be drinking these non-stop. Additionally we can see further tweaking of the product design to help the product taste even creamier with less lingering sweetness as they figure out how to blend the cream/root beer taste together.

Thinking more broadly – the "healthy" aspect of root beer could probably driven further with beverages that are not yet mainstream like Indian lassi and Chai. We could think of many more, but time does not allow.

The clever crafting of product and package and messaging continues to intrigue. The bottle needs to communicate the experience better – the hold tilt and turn needs to be worked on so everyone who buys the product gets the fun.

Rating:

This product is good. But could be improved. It is a great start great.

Market Potential: OK. There will be moms and grandmoms that purchase this and enjoy it for the convenience that it brings to a very enjoyable experiential fountain product. The bigger impact will be what others (or Dr. Pepper Snapple) do to build upon this innovation with the bottle, the components, and the story. It will be fun to watch where this goes in a year or so.

Figure 3.2   (Continued)

lives his life, how the person perceives his world, and then how the person interacts with the product in the world in which the person lives.

These types of insights are powerful. With that knowledge and insight, the company researcher, developer, marketer will, in time and with some experimentation, figure out how to craft a product. The product will be developed after looking at the brand or the company as the foundational element that makes an implicit promise. That knowledge (or perhaps wisdom) recognizes that with every product development project the company is making a choice about what the ultimate goal should be.

Knowledge has a role here. Knowledge of the customer, of the *person* behind the mask of customer, behind their blog, the person removed from his Twitter or other new "today" tool, should lead to product development that is authentic both in terms of goals (*what is being produced*) and lives up to the brand promise (*what is being marketed*).

Economics comes in as well, economics that recognizes the reality of today, the demands of the company, the constraints of business, and most of all what's really important—"true affordability." The search to understand the real person underneath the customer/consumer demands a way of working, an approach, that is quick, that is affordable (so that it can be conducted all the time for every project and not be considered special), and that is really productive in terms of what it delivers. Gillette, H.J. Heinz, Kraft Foods, S.C. Johnson, General Mills, Avon, Altria, and P&G are known to have processes in place that allow for this type of activity to occur on a regular basis. These forward-looking companies have discussed their person-centric approach at a number of conferences, sharing their wisdom with colleagues from other groups, other companies, and other countries.

Companies don't stand still. We don't mean to say that the aforementioned companies are the only ones searching for the better way. It's clear that many companies are trying to figure out how to understand what to do better or differently because the current testing that has occurred has not led to success (Moskowitz, 2003; Noble & Kumar, 2010). From the sensory or consumer researcher perspective, it requires that one honestly look at the methods being employed today in order to assess consumer reaction, response, and feelings and ensure that the researcher understands the pros and cons of the methods and decisions used within the product development chain. Religiously adhering to metrics and methods that were highly relevant in bringing products to market in the 1960s might be outdated and even counterproductive in a world with one-to-one marketing. We're living in a world where the situation demands new thinking about how to achieve the end point, which has not changed—create and sell products that people want to buy, both now and later (recall Table 3.2).

Ipsative—the world of one

All of the aforementioned call for studying the person means data analysis at the individual level. Psychologists have a name for this—*ipsative,* or individual-level analysis, leading to *idiographic* rules (rules pertaining to the individual). Classical trained experimental psychologists know that this focus on lawful behavior of individuals, the ipsative approach, stands in direct contrast to the study of large groups, the *normative* or *nomothetic* approach. Each focus is valid—but only the ipsative approach digs into the soul of the individual as a customer, ferrets out needs and wants unique perhaps to that individual, and by so doing provides a necessary new key to successful product development efforts.

At the time of this writing, we know of few market researchers or sensory organizations that regularly analyze their qualitative or quantitative data on an individual-by-individual basis. From a glance at the range of advertisements, a form of competitive analysis, we find only a few companies today who have "experience understanding" programs in place. Yet, until one incorporates an individual's perceptions along with normative results, there will be a disconnect between delivering an experience—to an individual—and creating products that appear interesting but are not really compelling. And until one studies experiences on a regular, disciplined basis, it will continue to be difficult and disheartening to understand the extremely complex array of elements that weave together to create consequences relevant to groups of consumers.

The output from such a study of the individual can be daunting but worth the effort. Figure 3.3 and Table 3.4 show an example of both an idiographic profile obtained from a study of reactions to different experimentally varied ideas (conjoint analysis; Table 3.4),

Figure 3.3 An individual experiential thread.

and in turn the "threading" of that individual's qualitative experience through a discussion. Integrating the specific individual into the whole truly restructures one's understanding of the nature of an experience for an individual. Even more important is how entire thread for that person links from product or idea to emotions all the way to high-order values.

SOCIETAL CONSTRAINTS AND SOCIETAL DEMANDS

We don't live in a vacuum. We are influenced by the outside world in which we find ourselves. That's obvious. The implication is that decisions people make about items they purchase and use are influenced by societal norms. Examples of changing consumer attitude and behavior can be found in five of the many trends that emerged during the last 10 years:

(i) Water consumption increasing relative to brown cola beverages (Anonymous, 2001, 2003b; Bennett, 2004).

Table 3.4 Idiographic profile.

Female DB Personal constant = 70

Product			Added nutritionals			Emotions/structure function claims			Brands/ endorsements		
E_3	Indulgent, richly flavored foods with large visible pieces of real ingredients	36	E_18	100% Organic	33	E_19	A quick and easy meal	7	E_29	From a well known nationally available brand	45
E_4	Crisp and light texture... leaves a fresh feeling in your mouth	33	E_14	With inulin ... known to improve calcium absorption and improve digestion	30	E_21	Fillls you up with energy that lasts for 2–3 hours	7	E_35	Recommended by your doctor	31
E_1	Healthy eating that tastes great	32	E_12	Full of antioxidants and phytonutrients that help you maintain a healthy heart	26	E_22	Such pleasure ... knowing you're eating something healthy	7	E_36	Recommended by nutritionists and dieticians	31
E_5	A heartier crunch with layers of dense filling	12	E_10	Provides essential vitamins your body needs, including A, B12, C, and E	18	E_26	As part of a low fat, low-cholesterol diet may reduce the risk of some forms of cancer	5	E_34	Endorsed by the American Diabetes Association	31
E_6	A cripsy light texture on the outside with a warm gooey filling inside	12	E_11	Provides essential minerals your body needs, including potassium, magnesium, and zinc	17	E_24	Even better for you than you thought	4	E_32	Homemade like mom used to make	21
E_9	Irregular shapes, sizes, and pieces	-8	E_16	Contains essential Omega-3 fatty acids, which may reduce your risk of heart disease	13	E_25	Builds and maintains strong bones	-5	E_33	Endorsed by the American Heart Association	8
E_7	Great aroma	-15	E_15	Contains the essential nutrient lycopene ... which may reduce your risk of cancer	10	E_27	May reduce your risk of high blood pressure and stroke	-14	E_28	From a Store Brand	-5
E_2	A delicious, classic, familiar taste	-20	E_17	All natural...no artificial flavors, no preservatives	8	E_20	A food you feel good about feeding your family	-20	E_31	From a sit down restaurant like Applebees, TGI Friday's, or Olive Garden	-22
E_8	You can easily identify what individual real foods are in the product from looking at the pieces in it	-26	E_13	Made with whole grains	-23	E_23	Calms you down...just what you need when you're feeling stressed	-22	E_30	From a fast food restaurant like McDonalds, Wendy's or Burger King	-52

(ii) The popularity of low-fat foods (Gorman, 2003; Sloan, 2003; Theodore & Popp, 2004; Toops, 2004; Wade, 2004; Yager, 2010).

(iii) The popularity of low-carbohydrate foods (Goldman, 2003; Gorman, 2003; Sloan, 2003, 2004; Theodore & Popp, 2004; Toops, 2004; Urbanski, 2003a, 2003b; Wade, 2004; Wansink, 2006).

(iv) The changing perspective of good and bad fats (Shell, 2002; Theodore & Popp, 2004).

(v) Interest in organic foods (Krall, 2003; Murphy, 2004; Scheel, 2003; Sloan, 2003; Toops, 2004; Fromartz, 2006).

There's a practical application to all of this; look at the person and look beyond the person to society. You "can't have one without the other." When dealing with consumers' needs and wants, it is important at the same time to get a "fix" on how deeply the particular individual is anchored into a specific structure that guides beliefs.

A sense of long-term as well as short-term history is also necessary. That history informs; knowing it can put the right coloration on one's findings. Research suggests a number of "top of mind" factors that consumers "pull out" when they are interviewed:

When a large-scale study on anxiety-provoking factors ("Deal With It!TM") was executed, the second Iraq war had started in April of 2003 (Ashman, Teich, & Moskowitz, 2004). The United States had not yet begun to emphasize obesity as a major health issue. The study was executed well before concerns arose regarding the nature of an economic return without equal increase in job creation (the so-called "jobless recovery").

The research implemented an approach to measuring "emotional power" of phrases and subject areas. The authors categorized beliefs into three groups: situations that are part of the external world, interpersonal problems, and self and wellness issues (Table 3.5). Top-of-mind topics that surfaced highest for individuals in her research were obesity, terrorism, and war. *Whereas these three factors were the highest for top-of-mind (what two people might tell each other in casual conversation), in actuality only "self and wellness" issues had the most profound impact on the individuals themselves. Self and wellness issues of income loss, obesity, and loss of assets were of significantly higher import and strongest for people, compared to the more talked-about but really less important issues of terrorism, phobias, and war.*

The implications are that what people will say in surveys or more superficial interviews may differ extremely from what is really important to them, what really ends up directing their behavior. The key lesson for research: it is important in needs and wants research to understand when one is measuring top-of-mind versus when one is understanding consumers more deeply. The answers differ, and product development may move in different directions as well.

So what's the lesson here? "Where's the beef," in the immortal words of the late Clara Peller of Wendy's advertising fame? Current events and society trends must be understood by

Table 3.5   The 15 topic areas dealing with anxiety in the "Deal With It!TM" study.

The external world
Franken foods (genetically modified foods)
Environment degradation
War
Terrorism
Infectious diseases

Interpersonal problems
Unpleasant social interactions
Breakdown of interpersonal relations
Sexual failure

Self and wellness
Breakdown of health care systems
Loss of income
Loss of assets
Aging
Obesity
Loss of health
Phobias

the development teams and taken into consideration as they plan their work with individuals. Why? Simply, close interaction with the consumer exposes the team to more of the belief sets and cultural norms of a consumer group, and colors the information exchange. When one begins to work closely with the consumer, it's hard to anticipate what topics will turn out to be triggers or touch points for the individual. The only way to understand is to research more, investigate, enter into discussions with the consumer, and at the same time, be aware of these influences and how these affect the data collected (Kelly, 2001; Weiss, 2003):

> Historically, incorporating the potential influence of current events on the findings of consumer insights has not really been a consideration for the product development team. Until this is done, consumer interactions will continue to under-deliver on their potential because a large number of weak signals, emanating from the consumer participant, will simply be ignored, misinterpreted, or simply thrown out.

Approaches to take to "understand" the consumer

> The difference between the right word and the almost right word is the difference between lightning and a lightning bug.
>
> Mark Twain

In Davis's article about the role of market research in the development of new consumer products (Davis, 1993), he states that to understand the consumer successfully turns out to be an intrinsically complex task with elusive goals. This may well have been the case in 1993, but time has a way of adding to our arsenal of tools when the opportunities present themselves. Davis does point out that in this discovery phase there are some commonalities of structure: identify and understand the problem, identify possible technical solutions, and choose the best solution. IDEO, a highly successful design firm, suggests the steps are four: Learn, Look, Ask, and Try (Anonymous, 2003a). This is just one advance, a simple, structured system, rather than just tests.

Another practitioner from the commercial world, David Bakken, cautions that standard market research (and sensory) practices actually decrease the likelihood of finding emerging customer needs. Bakken points out that sampling and analytical methods can obscure emerging needs (Bakken, 2001). Bakken's concerns reaffirm the strictures on insights emerging out of market research and sensory analysis (see Table 3.2).

In a sense, nothing really changes in the world of consumer research, that is, in terms of objectives. Tools change, but objectives remain. The notion and goal of understanding consumer wants and needs is not new. Looking back at one of the early editions of Kotler's seminal book on *Marketing Management* (Kotler, 1972), the reader is informed that a marketing concept would suggest that customer needs and wants should be the *starting point* in the search for new product ideas. Kotler goes on to say that companies can identify customer needs and wants in several ways: (1) direct customer surveys, (2) projective tests, (3) focused group discussions, (4) suggestion systems and letters received from customers, and (5) perceptual and preference mapping of the current product space to discern new opportunities.

We see from these two professionals (Kotler and Bakken), spanning three decades, that there are different approaches to take with the consumer. It's a matter of what's in style to answer the same question. Depending on the mind-set of the "research broker" (Mahajan & Wind, 1999) or belief set of the business leader, one approach may be arbitrarily designated to be superior to another. In actuality, it boils down to the same question but differences in the mind-sets and experiences of the decision makers who author the different forms of research to address the same, never-changing problem.

Practical steps to successful consumer understanding

During the past decades, researchers have spent countless hours developing approaches by which to create a good understanding of the consumer. We use the word "create" very deliberately here, for the effort has not been to discover but rather to develop a method. These approaches, while more unfamiliar to food product developers, have been the mainstay of compelling design by engineering and design companies for a long period of time (Leonard & Rayport, 1997). There are many methods that emerged.

We proposed one sequence, not the only sequence of course, but one that has worked in practice and with which we are familiar. The key phases break down to (i) Focus, (ii) Frame, (iii) Find Out, and (iv) Quantify (Ashman, 2003; Karlsson & Ahlstrom, 1997). The steps are listed and explicated in Table 3.6:

(i)   *Focus*: Focus is about understanding what one knows and doesn't know. It can be called "learn" or early planning. This topic was discussed previously in Chapter 2. Mindful reuse of existing information plays an important role in this Focus phase. The information may be explicit and/or tacit. The opportunity to use past successes and disappointments by the company is a continuing failure by companies to use resources that are readily available. Those who are ignorant of the past are condemned to pay to learn it again, or so might be the operative phrase here.

(ii)   *Frame*: Frame is the phase where one develops strategies or hypotheses regarding what consumer linkage is believed to exist. The Frame stage addresses the business or question that came into focus during Phase 1. Framing is a highly critical phase for establishing the team/interviewer mind-set for the next or Find Out phase. The Frame activity generates a so-called value diagram, though other terms used for the same result are "hierarchical value map" or "linkage map." Figure 3.4 shows an example of this value diagram.

The Frame stage can be the most crucial in new product development. It's worth more detail, which follows:

(ii-a)   *Establish the value framework prior to interacting with the consumer.* This effort is important. Personal beliefs and bias will become a factor when dealing with people on a one-on-one basis if paths for consumer understanding are not

Table 3.6   Steps to follow to achieve strong consumer understanding.

Phase	Stage	Activity
1	**Focus**	*Knowledge map*: Understand what you know.
2	**Frame**	*Value diagram*: Develop a theory about what your product or topic really is and then develop a theoretical model of consumer behavior that you can review.
3	**Find Out**	*Qualitative in-context interviews*: Conduct a series of listening sessions to fully understand the roles in current/potential users' lives.
		Can expand to other observation activities if necessary.
4	**Quantify**	*Quantitative survey typically using trade-off methodology*: Expand the small-scale *listenings* to a larger scale audience to confirm theories and develop initial strategic product platforms that grow and expand the proposition for product design.

Value diagram: theoretical for coffee created by training group

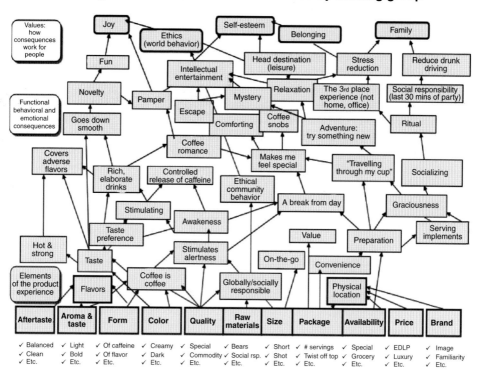

Figure 3.4 Example of a "value diagram" for coffee, established at the Frame stage. Note that linkages are simplified for illustrative purposes (the U&I Group Archives). Used with permission of Massachusetts Institute of Technology.

managed prior to the interaction. When not managed properly, the sought-after rapport between interviewer and interviewee may conceal something less desirable in the research process. It's likely that certain clouding effects will occur. One is that what the interviewee says may be adapted, albeit unconsciously, by the interviewer so that what's said accords with the theoretical (and ideological) preunderstanding and possible presuppositions of the researcher (Jarvinen, 2000). In simple terms, we see what we are ready to see, hear what we are ready to hear, and believe what we want to believe. Understanding the interactions between the product, the person, and the purpose is absolutely critical (Ratneshwar et al., 1999).

(ii-b) *Get to what's actionable; features and benefits provide a structure on which to erect a value framework.* There is extensive literature regarding relations between features and benefits (Hauser & Clausing, 1988; Holbrook & Hirschman, 1982; Olson & Reynolds, 1983; Reynolds & Gutman, 1988). Gutman's organizing concept of means–end chain offers food developers a way to position products by associating means (the physical aspects of products) with ends (consumption of products) to achieve a desired end (value state). To be central to marketing and product development and consumer research, the nature of the elements comprising means–end chains and the nature of the connections between these elements must be specified and the model operationalized. To be concrete is to be more productive and more insightful (Gutman, 1982). Unfortunately, traditional food development and marketing practices obscure the difference between means and ends. As a consequence, the clear, hoped-for operationalization does not occur, or if it manages to occur, the linkage remains unclear. The specific defined linkage between the product features (elements) and the consequence, and therefore values, often fails to be clearly delineated. All too often, food developers and sensory and consumer research at the products research phase may isolate potentially key variables, preventing their influence on the consumer experience (e.g., elements like package, price, and brand). Furthermore, market research may limit their discovery of what product elements link to the consequence and value assessment; they do so by the fact that in their job roles they are cut off from much of the relevant "physical" product information (Garber, Hyatt, & Starr, 2003a).

(ii-c) *Develop what you need and what allows you to work in a productive, easy manner.* All knowledge is not created equal. Make that knowledge usable. What should you do? Collapse all the learning and understanding, generated and clarified in the Frame step, into salient "sound bites," which represent what the company knows or does not. Make these silos or "buckets." These buckets are:

(a) *Elements*: Elements are the product factors that a company can control. They represent attributes of the product-relevant factors such as flavor, texture, color, quality, and package and business-relevant factors such as price, location, brand, and image. Elements represent the initial factors related to how a consumer first experiences a product or service.

(b) *Consequences*: Consequences constitute the functional, behavioral, and emotional consequences of the product experience (Vriens & Ter Hafstede, 2000).

(c) *Values*: Values are defined as the end-stage phase for a consumer's experience with a product (Gutman, 1991; Maslow, 1968–1999; Ratneshwar *et al.*, 1999; Whitlark & Allred, 2003).

In the Framing phase (Phase 2), participants develop a theoretical framework *before* field testing is commissioned (Find Out, Phase 3). This theoretical framework guides the participant. The participant is cued in and becomes aware of what is going to be looked for or listened for when dealing with people. This heightened awareness is critically important.

When a mental model has not been created and visualized for the Find Out phase (Phase 3), much of what one will observe or hear can be missed. Each person in the group needs to be aware of this framework so that no one misses anything when talking to people. Even senior management should be made aware of the theoretical framework so that they can appreciate what is happening.

Visualizing the mental model and then sharing one's own "take" on this model are not only reasonable, but also necessary. The information from people will be coming from a variety of modes (functional, emotional, behavioral). Sorting out this abundance of information and knowing ahead of time one's biases and "prisms" for understanding are critical. Each participant must become aware of, and clear about, his/her individual biases and mental "filters" *before* entering the dialog with the person (i.e., the end consumer).

A good Frame phase will allow one to construct far better Find Out sessions (Phase 3). A good Frame phase makes the participant receptive to new paths of thinking when the new information emerges in the Find Out phase (3). Finally, this systematic, open, sensitization approach fosters a scientific approach to observation and listening. The reason for such science is that the assumptions become public; hypotheses can be established prior to the field and tested against once in the field experience.

(ii-d) *When should one build the framework?* Processes have a way of working when executed correctly, and delivering either the wrong information or incomprehensible information when executed incorrectly, or in the wrong order. The framework-building activity should be scheduled right after the construction of the knowledge summary on a project (see Chapter 2). Frameworks can be created either during a team meeting or as part of a subgroup activity. Whereas one can build these frameworks the "old-fashioned way," with paper and pencil (Post-it® Notes and flipchart paper), technology has come to the fore with a variety of specific and general software (e.g., Inspiration, Visio, PowerPoint).

The framework-building activity should include all of the information the company knows about this topic. By including all this information, the so-called theoretical basis is really not "theory" but rather merely untested information about the consumer. Yet the untested information is still resident within the company.

Frameworks lead to hypotheses. Now what emerges turns out to be company-based collections of information plus hypotheses. These working hypotheses need confirmation, disconfirmation, or enlightened expansion. At this stage it's important to construct clear, cogent linkage diagrams. These diagrams portray the mental model of the team. That mental model makes it easier to move on with consumer testing. The mental model defines what might be working, and in fact can be used by those who want to stop at that point and not proceed further into the discovery phase, the Find Out phase.

(ii-e) *At the end of the day, what then is the real value of the Frame stage, prior to dealing with people?* When interacting with consumers it is very important to develop a model such as the one described. That mental model will move the professional from being an uninvolved observer to someone who is more ready to learn from the observations or exchanges.

There is another key benefit: the chance to reflect rather than the temptation to make snap judgments that color one's thinking later on and may turn out to be counterproductive. Behavioral research suggests that over 50% of the American adult population are task-oriented and enjoy making speedy decisions. Even more American adults, 60–80%, prefer to understand things sequentially and in an orderly manner. The problem comes from the contrast between the desire for speed and decisions versus the reality of what one observes. We are dealing with different universes. Consumers' lives are really not orderly; events that one watches or hears definitely are not as they appear to be. When the developer, sensory scientist, marketer, or almost anyone in the corporation attempts deep consumer learning without tools to prevent snap decisions based on little information, there is a very good chance that the result will turn out to be erroneous conclusions, despite all good intentions. These conclusions and the decisions emanating from them may lead one to an incorrect development path. Even more important, these errors "poison the well." They may turn out to generate the long-term counterproductive feeling that working with the consumer is unreliable and complicated (Kroeger & Thuesen, 1988; Martin, 1985).

(ii-f) *Who does what?* The activities in the Frame phase work best with a couple of team members who are very familiar with the knowledge base provided by the data. It is best when the team members come from different functional groups. Often one of the members will be a product/brand specialist (technical or marketing) and another member will be a consumer/sensory specialist (consumer insights, sensory scientist). The complementary talents ensure a broader experience base will be considered as the value diagram is built.

(ii-g) *How is it done?* The team members typically build the value diagram or linkage map using Post-it® Notes. The process develops a hierarchical structure. The specifics, the elements are at the bottom of the base (there are more of them than values), the consequences lie in the middle of the structure, and the values are placed on the top of the structure. When completed, the map typically looks like a triangle. (Other structures have been used; however, we suggest this approach since it tends to drive people from lower order attributes and consequences at the bottom to higher order consequences and overarching values at the top of the diagram). This activity can be a part of a team meeting or it can be done with a couple of team members who work separately and then bring the map into the group for review and editing.

(ii-h) *Why is it done?* Why is this important? The theoretical value diagram ensures that knowledge and insights that the company has or will get in Phase 3 (Find Out) are connected from the parts that a consumer experiences (flavor, temperature, package, price, etc.), through the various ways the product can interact with an individual's life (the consequences), all the way to the highest level value states for an individual. Not planning out what this set of connections might be from the large base of explicit or tacit data before the listening will lead to missed listening and building during consumer conversations. The listening step becomes much more productive, more insightful with the value diagram as a framework for understanding. Without the prework leading to the value diagram, a consequence might be inadequate field execution, perhaps leading to questionable findings (Stewart, 1998).

(ii-i) *What are the outcomes?* Useful outcomes result from the preparation. The investigator goes into the fieldwork with consumers understanding the potential boundaries are for the consumer experience. There are at least three of these outcomes:

(a) *Veracity*: Veracity is about validity. We are looking for verisimilitude of depiction. When veracity is not sought as a goal, there will be limits to the learning during the field portion of the study.

(b) *Objectivity*: Objectivity is about reliability. We look for a transcendence of perspectives. When objectivity is not sought as a goal, then later on issues may arise in the in-context interviews, reactivity can occur, and finally some unexplainable but potentially important outcomes may occur that cannot be understood and therefore may be dismissed.

(c) *Perspicacity*: Perspicacity is about generalizability. We are looking for broad applicability of insights. When perspicacity is not sought as a goal, there can be issues related to creation of insights and broader application of the knowledge.

(ii-j) *What are the various outputs of value diagramming, and to what use can these outputs be made?* Table 3.7 lists value diagramming and a number of other

Table 3.7 Different forms, analyses, and diagrams that can be created with company information to further clarify information in the knowledge maps.

Type of map	Purpose	Value
Activity analysis	Details all tasks, actions, objects, performers, and interactions involved in a process.	Provides organizing principles for who needs to be included in the Find Out phase.
Character profiles	Articulates potential archetypes.	Begin to think about who the potential end user might be.
Error analysis	Lists all things that could go wrong when using product or service.	Understand potential areas of product or service failure. Specifically focus on the negative.
Flow analysis, sequence mapping	Represents flow of information through phases or system or process.	Provides structured understanding of process.
Historical analysis	Compares factors across a category from a historical perspective.	Identifies trends and cycles of behaviors and habits.
Value diagramming	Organizes in a hierarchical format the elements, consequences (functional, behavioral, emotional), and value links for a particular product or service.	Provides a visualization of the relationship of elements (triggers) for a consumer experience and the subsequent events that are discussed by individuals. Is a framework for effective consumer listening?

Source: The Understanding & Insight Group (U&I) methods pack; Anonymous, 2003a.

    forms of analyses that can be used in the Framing phase. The table shows that there are a fair number of different types of outputs. The key here to remember is that the exercise uncovers and then structures existing information, generally available to the participants, but perhaps not really structured in such a way that the information can be used.

(iii) *Find Out*: In this third phase, one begins to interact with the consumer. The interactions can take place in one of the two ways: observation of actual behavior or interview, where the consumer reports to the interview. There are no right or wrong methods here; all forms of observation techniques and interview techniques can be used, and are used, in this Find Out phase. Table 3.8 lists just a few of the many techniques that can be used to interact with consumers to understand their behaviors, emotions, and belief states relative to a product or a service. The approaches come from direct observation, interviewing, and simulation, respectively.

The Find Out phase—five considerations

As in every search for knowledge, one doesn't know what one will find. The Find Out phase is exactly what it says—discovery. Although one doesn't know what one will discover, it's certainly possible to make the proper preparations. Here are five considerations, which should not be considered as exhaustive but rather things to keep in mind. Executing the Find Out phase in light of these considerations increases the chances of discovering something major:

(i) Does the interviewer understand the question(s) that need to be answered in the Find Out phase, from the specific research exercise?

Table 3.8 Techniques that have been used during the "Find Out" phase of consumer understanding.

Observation "Look"	Interviewing "Ask, Hear"	Simulation "Try"
A day in the life	Camera journal	Adaptive creations
Behavioral archeology	Card sort	Behavior sampling
Big brother	Cognitive map	Be your customer
Bumbling shopper	Collage	Bodystorming
Competitive product survey, benchmarking	Conceptual landscape	Costuming
Fly on the wall	Cultural probes	Experience and insight groups
Expeditions	Draw the experience	Interactive conjoint
Experiences	Extreme user interviews (lead users, creative consumers can be a part of this)	Predict next year's headlines
Getting lost	Five whys?	Prototyping, modeling
Going shopping	Foreign correspondent	Role playing
Guided tours	Game show	Simulated shopping
Personal inventory	Metaphors	Situation evaluations
Photo shoot	Narration	Scenarios
Powertasting^{TM}	Past experiences	Take home; try it yourself
Psychodrama	Repertory grid	
Rapid ethnography	Talk show	
Topic immersion	Storytelling	
Shadowing	Surveys & questionnaires	
Social network mapping	Unfocus group	
Video tracking	Word-concept associations	
	ZMET^{TM}	

Source: The U&I methods pack; Anonymous, 2003a.

(ii) Can this phase be executed in the consumer's "real-world" environment to provide more opportunity for the context to stimulate unexpressed thoughts and activities (Marshall & Bell, 2003)?

(iii) Can this phase be done in a one-on-one or two-on-one situation in order to deepen the level of conversation, depth of learning, and ease of connection (Griffin, 1992)?

(iv) Does the interviewer have enough knowledge of himself or herself to know what part of the interaction needs to be managed for the interviewer?

(v) Can the interviewer use empathic interviewing techniques and apply a laddering approach to the discussion (Wansink, 2000)?

The group of techniques known as "observation" are emerging as the most popular approach to use in much of the social and natural sciences when one wants to develop new ideas or new products (Coopersmith, 1966; Kelly, 2001; Martino, 2003). Observation is the foundation undergirding scientific inquiry. Yet the pursuit of what might be called "rigor" observation has been cheated, treated as a step-child. As mentioned earlier in this chapter, both sensory and consumer researchers often ignored observations. The rationale, often expressed, was that a business needs to have all aspects of the product and market inquiry conducted in such a way that the results, and thus the conclusions, would be perceived to be valid and reliable. One outcome was overfocus on quantitative methods and often a deliberate snub of observational methods. The pursuit of the appearance of science dictated that the softer, nonstatistical, qualitative methods be relegated to second-class citizenship (Erickson & Stull, 1998). If it didn't have the trappings of mathematics and rigor, the procedure was deemed to be old fashioned, not appropriate for the "new age of scientific business."

Nothing stands still. Times change, fashion changes, and even what is considered to be valid changes. At the end of the day, ethnography is enjoying popularity as never before. Lately, observation research, along with other approaches (commonly referred to as "ethnographic methods"), has become trendy. Although "new" to food development, most of these methods are an important part of other sciences (e.g., social, behavioral) and other industries (e.g., design, architecture; Abrams, 2000; Kelly, 2001; Leonard & Rayport, 1997; Stewart, 1998).

While watching people shop, or eat, or enjoy a piece of fruit or a meal, and having them talk about the situation seems so obvious, one might wonder what consumer food researchers were thinking by placing people in booths with tiny sinks behind guillotined doors, or handing out bags of white packaged snacks with plain labels and specific directions about how to eat a cookie! Those overly clinical approaches now seem so strange. They deliberately lose sight of the many aspects of consumer understanding in the quest for purity of data and simplicity of situation; people simply do not behave normally in these booths, with less-than-meaningful test stimuli. It is probably fair to say that these newly "discovered" (or better, newly accepted) approaches, subsumed under the rubric of "ethnography," work because they allow researchers to see the world outside of their cubicle, lab bench, and conference room, and let these researchers appreciate first-hand the world in which their products are used.

Additional considerations in the Find Out phase when using ethnographic methods

How many interviews, observations, or consumer interactions are necessary? Base size, for that's what we are talking about, remains a commonly asked question among researchers, often a favorite since it's a matter of opinion. Base size is very important in the Find Out phase. Since Find Out is not passive and does require contact or time for field observation, it is commonly believed to be expensive. The Find Out phase can become expensive if one loses sight of what is needed in the discovery phase.

A little preparation here can save the day and reduce the cost. When the investigators use an approach that frames the issues (Phase 1) and presents the theoretical behavioral model (Phase 2), Find Out simply turns into one more standard, integrated step in the process, typically far less expensive than one might think. The homework, up front, provides a framework in which the output from a relatively limited number of people can generate important observations and hypotheses. Psychologists are well aware of the richness of the so-called ipsative world—just a few observations can provide massive insights:

> From author Beckley's point of view, about 16–24 observations/interviews during this phase generate the necessary information and constitute a satisfactory sample. The base of 16–24 allows for approximately 4–6 interviews/observations per relevant consumer segment, sufficient to establish qualitative patterns of behavior. The goal is to do this work rapidly and small scale and then move rapidly to the fourth and final phase, Research. The base sizes previously reported by others support this number (Griffin & Hauser, 1993). Figures 3.5 and 3.6 present some supporting evidence as well.

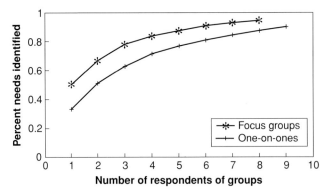

Figure 3.5 Research showing focus groups versus one-on-one interview. Reproduced from Silver & Thompson, 1991. Used with permission of Massachusetts Institute of Technology.

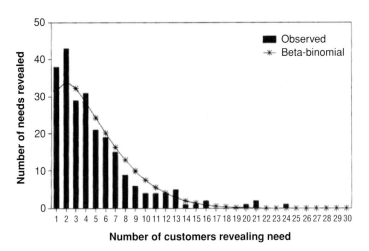

Figure 3.6 Research showing number of needs identified by customer number. Reproduced from Silver & Thompson, 1991. Used with permission of Massachusetts Institute of Technology.

Focus groups and their issues in the context of the Find Out phase

In his book *The Art of Innovation*, Tom Kelly, writing about one of the "secrets" of IDEO (a design firm) success, says quite bluntly:

> We're not big fans of focus groups. We don't much care for traditional market research either. We go to the source ... actual people who use the product or something similar ... plenty of well-meaning clients duly inform us what a new product needs to do ... Of course we listen to these concerns. Then we get in the operating room ... and see for ourselves.
>
> Kelly, 2001

For the past 60 years, focus groups have been used to get close to consumers and understand them. Review of sensory and marketing research journals, food trade publications, and the popular press attest to the popularity of this tool. This part of the consumer research industry bills about $1.1 billion in market research per year (Wasserman, 2003). In the last 30 years, this approach to getting close to the consumer has expanded due to increasing familiarity with groups and the availability of practitioners and facilities designed to create the opportunity for consumer interactions. Additionally, most marketing programs educate individuals about the benefits of focus groups so they are part of the tool set that a good businessperson should use (Kotler, 2002).

Any technique or tool that becomes extremely popular and is used to obtain answers for the multitude of questions raised by business soon generates an equal and opposite reaction. The wide acceptance generates questions. In this spirit of opposing forces, focus groups have come into question, perhaps in some ways quite deservedly. There are many anecdotal stories about how a participant in a focus group states one thing, but his or her observed behavior or that of his or her child suggests a totally different perspective or understanding of the information that was heard during the focus group (Fellman, 1999).

Concerns related to focus groups range from inappropriate use of groups, to group think, leading (i.e., biasing) moderators or discussion guides, poor listening habits of the back room, and lack of delving deeply into a subject due either to interview constraints or researcher inadequacy. In a word, focus groups are subject to the vagaries of motivation and hidden agenda, typically on the part of those who commission them, occasionally on the part of those who execute them.

Dumaine is blunter, arguing that the fear of risk among marketers leads them to follow a formula and not a feeling, a strategy that puts at risk the effectiveness of marketing. It was marketing that ordered the research in the first place in order to increase knowledge and reduce risk (Dumaine, 2003). When this risk-aversive behavior is juxtaposed against anecdotal reports, which cite successful product launches and the importance of the observation research for success (Wasserman, 2003), one can almost see the pendulum swing from one type of world view and methods to another world view and its methods. What might be better is a serious look at the merits of a mixed selection of approaches and a balance between information and judgment. That mix comprises an entirely different topic, indeed a very controversial one, a veritable business minefield.

Current thinking suggests that despite its widely proclaimed problems, the focus group is not dead and its demise is not going to be soon. Recent research suggests high levels of relevance (90%) and limited feelings of dissatisfaction (Langer & Last, 2003).

The current criticism has generated many good suggestions for continued improvement in the method (e.g., take more time; ask fewer questions; recruit groups, not groupings; Rook, 2003). Dissatisfaction with focus group is continuing to drive the search for better methods. One consequence is that we see increasing interest in, and use of, in-context and expanded observation/anthropological approaches. These methods, at first given a chance to showcase themselves as a counterpoint to focus groups, will inevitably become even more popular because they will become streamlined. They will eventually become less dependent on the person's skill and more intimately linked with people and analytics by technologies and mind-sets that empower this type of research (Beckley & Ramsey, 2009).

What then should the reader make of this discussion about focus groups? The discussion should not be about the use or nonuse of focus groups. Rather, the discussion should be what one could achieve when groups of people gather in a room with a moderator. For the world of food, this means that the focus group should allow the developer and marketer to get closer to the person so that a deeper understanding of his or her life can be incorporated into the food development. Malefyt summarizes the perspective well: "In the end, the real challenge of using any qualitative tool, albeit focus groups or ethnography, is to skillfully understand what exactly is being observed, and what exactly it means" (Malefyt, 2003).

Moving closer in to our topic of sensory and consumer research, it will be important to remember that there is a *person* in the consumer or customer. All too often this has been forgotten when developing sound test designs with human subjects, rather than people (Garber, Hyatt, & Starr, 2003a). This humanistic point of view needs to be maintained whether one is using traditional focus group methods or hybrids (online, community panels, etc.) that are the new age version of the traditional method. That point of view will demand from the researcher a disciplined response to the potential that electronic interactions with consumers can turn into large-scale, superficial interactions, rather than in-depth interactions.

Does the "who" tell us enough: demographics and their limits in the Find Out phase

Demographics are easy to measure. It makes some intuitive sense to believe that "birds of a feather flock together," or that people who seem similar to each other in terms of demographics should be similar in their responses to products. Demographics, the "who" of a person at the most basic level, should somehow co-vary with the pattern of likes and dislikes. Indeed, we often talk about different generations, different groups of people, like the boomers, the milennials, and so forth, basing our classification on demographics but implicitly thinking that these groups are "similar" to each other in lifestyle and preferences.

Increasingly, in the past decade or two, questions have been raised about flawed strategies incorporating demographics, and segmenting/selecting consumers on the basis of such factors as age, gender, race, and income (Clancy & Krieg, 2000; Larsen, 2004; Schwartz, 2004; Silverstein & Fiske, 2003; Stengel, 2004; Costa & Jongen, 2006). Despite these concerns, however, when most companies conduct consumer research, they still require respondent screeners that categorize people by age breaks, gender, and so forth. To some extent, sensory research has been less concerned with pure demographics, whereas market research has been more concerned. Enlightened researchers in both fields have moved away from pure demographics and focused on more relevant criterion for a product, such as product use and familiarity.

Beyond demographics, there are some additional factors that help us in the Find Out phase. When we look at consumers as individuals, there are some opportunities to further understand factors that might play a role in choice. One of these factors is life stage. For example, consider people who have just finished college and find themselves in the stage of creating a home, doing so in the face of limited resources. These individuals will certainly have a very different cluster of needs and wants than the young mother, just a few years older than the college student, but who finds herself having to care for a child, who has to buy baby clothes, and in general is now dealing with guiding and helping a new life, not her own. The foregoing suggests to us that life stage might be more important than age or zip code, or indeed most easy to measure demographics (Beckley *et al.*, 2004).

However, life stage isn't enough. A person's response to products may vary somewhat (not strongly) with demographics and may be far more strongly affected by life stage, but yet there is something deeper, some individual variation that transcends the "who" a person is, and the "when in the life" the measurement is made. The differences among people in terms of preferences, what's important, may be far more traceable to psychological differences between people. At one level, these differences lead to attitudinal segmentation. Psychographic profiles of these segments have become popular as a result, but again may generalize the type of person rather than understand the individual. The profiles are given cute names to describe how they differ (e.g., the carefree graduate, the serious mom, the empty nester, and so forth). At an even deeper level, author Moskowitz has demonstrated through a series of projects that traditional demographic breaks and life stage breaks have little to do with choice. Rather, it is the "mind-set" of the individual that is far more significant (Beckley & Ashman, 2004; Moskowitz & Ashman, 2003; Moskowitz, Jacobs, & Lazar, 1985).

New vistas in the Find Out phase: lead users and mind-sets

Once we move away from conventional approaches into observations, a new world of techniques opens up. Recently, Von Hippel's approach to mind-set, *lead user methodology,* has found some popularity in the discovery phase of product development (Herstatt & Von Hippel, 1992; Von Hippel, 1988). The lead user approach has been reported to work well in Europe and in certain low-technology fields.

Von Hippel is very clear about the requirements to be classified a lead user:

(i) Lead users are people who face needs that will be general in a marketplace, but they face them months or years before the bulk of that marketplace encounters them.
(ii) Lead users are positioned to benefit significantly by obtaining a solution to those needs.

There are four necessary steps to work with the lead user method:

(i) *Know the people.* Specify the characteristics to which lead users will attend. This will more precisely identify the trends on which these users lead the market. This will also specify indicators signaling what high benefits the lead users will gain.
(ii) *Recruit the people.* Identify a sample that meets the criteria in (i).
(iii) *Do the interview.* Bring these people together with engineering and marketing for problem-solving sessions.
(iv) *Test actual ideas.* Test concepts that are relevant to the lead users among more typical users as well (Urban & Von Hippel, 1988).

There doesn't seem to be much in the way of lead user work in the world of food product development. If the work has been done, it has not been reported in journals. Research that author Beckley conducted (Beckley, 2003) shows that individuals who were potentially lead users represented about 8% the population. These potential lead users showed food concept preference that clearly differed from the preference pattern of the remaining test sample. The preference of the most *conventional* group (45% of the population) was clearly different from the lead users. The difference was so great that we surmised that it would be extremely difficult to predict when, if ever, the lead user interest would work with the more general population. Still, the idea has supporters and may offer some opportunities if utilized by sensory and consumer researcher applying a strategic framework of thinking.

The fact that lead users may not work well in the world of food product design may suggest the difficulties in creating new "standardized" ways to look at consumers. The sensory community cannot drive this change since they are perceived as too far away from consumers. Until market research and advertising agencies drive to a new understanding of how to set "targets" for consumers, the disconnect between what is seen in the marketplace and what consumer and sensory researchers do will continue to exist (Trump, 2004).

GETTING EVERYTHING TO WORK—THE NEED FOR COHERENCE IN RESEARCH

The three parts (Focus, Frame, and Find Out), along with the standard requirement to Quantify, work in tandem, complementing each other like the parts of an automobile. When you are missing one of the following, gas, wheels, oil, or a driver, the car will not really run. Now, there is a lot more complexity, but you really have to have those four things to drive a car that is manned.

Many times an anxious, motivated product development team will want to jump to Phase 2 or Phase 3 before they execute Phase 1 (Focusing). It's tempting to skip steps; after all, the goal is reached that much quicker. Yet, avoiding Focus is one of the most common reasons why Phases 2 and 3 are less valuable than they could be, and why those two phases are all too often criticized as being the weak link in the PDP.

As we begin discussing the different types of tasks for Phases 2 and 3, this work assumes that a robust Phase 1 has been engaged in prior to these steps (Figure 3.7).

HAVING MEANINGFUL CONSUMER INTERACTIONS—THE RULES OF "HOW"

Suspend assumptions—avoid taking a position and defending it. Act as colleagues—leave your position at the door. Spirit of inquiry—explore thinking behind views.
Peter Senge, The Fifth Discipline (1994), in discussing what is necessary to have more meaningful outcomes when interacting with others

Integrating the experience

- Use a variety of research techniques to understand the experience from the consumers perspective.
- Associations in the form of images, usage, and emotional connections provide clearer understanding.

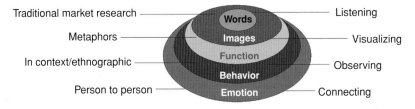

Figure 3.7 Graphic representation of what measuring experience entails.

It's all about a multidimensional, multidisciplinary, profound "sense of the consumer," a feel, or what is now known as "empathic design." Leonard and Rayport (1997) suggested in their *Harvard Business Review* article that:

> few companies are set up to employ (empathic design) since the techniques require unusual collaborative skills that many organizations have not developed. Market researchers generally use text or numbers to spark ideas for new products, but empathic designers use visual information as well. Traditional researchers are generally trained to gather data in relative isolation from other disciples; empathic design demands creative interactions among members of an interdisciplinary team.
>
> Leonard and Rayport, 1997

Garber raised a number of questions about taste tests and their failure to produce predictable market performance (Garber *et al.*, 2003b). The editor of the journal that published the paper opened up the journal to rebuttals. Both called into question the ability of today's research to spark really new developments.

In today's sensory and consumer research we are finding an increased awareness of and acceptance that new strategies need to be applied. However, the adoption of new structures is moving slowly. Indeed, the world of "sensory" is the slower of the two fields to adopt new methods, perhaps due to an innate conservatism and risk aversion (Ashman, 2003; Beckley & Ashman, 2004).

However, there are pockets of hope:

> We know of one group that embraces many of these attributes. They are called the 'Insight Panel' and work as a group of people to help Heinz, North America develop products. The people on the Insight Panel comprise a self-directed group designed to assist the discovery process. They have a manager who is extremely skilled in the field of product design and through her leadership they help create products that can be breakthrough for that company.
>
> Charles, 2003

AT THE END OF THE DAY, CONSUMERS ARE STILL PEOPLE

One of the hardest behaviors for developers, sensory scientists, and consumer researchers to understand when adopting this "new" concept of the consumer is that they are people. The entire market research industry is coming around to this after decades of treating consumers as "subjects," with the goal of understanding them as a scientist understands a "preparation." Kathman (2003) summarizes the process of consumer interactions quite clearly and cogently:

> Constantly talking to consumers, visiting their homes around the world, understanding the mindset they bring to the shopping experience (which varies by channel and region) are all part of acquiring empathy. Leadership comes through the heart, supported by the mind.
>
> Kathman, 2003

Many sensory and market research practices have enabled product development through the mind, much less through the heart. The IDEO Methods Card sums it up best (see Table 3.9).

Table 3.9   The IDEO Method Card.

Respect your participants!

Some methods depend upon other people, often strangers, sharing generously their time, thoughts, and feelings. Have consideration for their health, safety, privacy, and dignity at all times. Some of the principles that IDEO uses are as follows:

- Approach people with courtesy.
- Identify yourself, your intent, and what you are looking for.
- Offer to compensate participants.
- Describe how you will use this information and why it's valuable.
- Get permission to use the information and any photos or video you take.
- Keep all the information you gather confidential.
- Let people know they can decline to answer questions or stop participating at any time.
- Maintain a nonjudgmental, relaxed, and enjoyable atmosphere.

Source: IDEO Method Cards instruction pamphlet; Anonymous, 2003a.

INCORPORATING CONSUMER LEARNING DURING SUBSEQUENT PHASES OF PRODUCT DEVELOPMENT

It's one thing to know your consumers—it's a far greater challenge to develop your technology and balance sheet around them.

Rushkoff, 2004

When consumers are viewed as being an integral component of the development and marketing process, it becomes easier to include their initial voice and then subsequent comments during the full product development cycle. Having perspective is instructive and corrective. It's important to realize just how important food has been in the story of human existence. It is critical to recognize that no one product or project exists in isolation of the forty thousand others in the marketplace.

The world is dynamic, and product development has to recognize these dynamics and connections. Meiselman reminds us that food researchers continue to forget or at least ignore the fact that attitudes, opinions, and habits that an individual brings to the eating situation might be part of the reference event or part of the context. That is, it is very hard to know where the start and finish happen to be for a specific food situation (Meiselman & MacFie, 1996). This comment calls for a deeper, more mindful recognition of an integrated approach to consumer and sensory testing, rather than the episodic activities that characterize most product development projects.

Rethinking and then renovating the role of the consumer in product development is what is required now. A systematic approach for consumer interaction, such as we have presented in this chapter, is called for. Such systematic interaction must dovetail with the established product development system, complement it, learn from it, and then inform it. The consumer interaction that can occur during the discovery stage needs to remain more alive, memorable, and vital through all phases of the development cycle. Keeping the spirit of the consumer alive is important so that if a product development team member believes during a back-end stage that the need/want prescription has been lost, then there is the call and the opportunity to question and revisit the situation. The benefit can be process, but also a commercial. McWatters *et al.*, for example, noted that involving consumers early on in the PDP of a peanut butter tart actually increased the success of the product in the market (McWatters *et al.*, 2006).

Yet all is not easy. There's work involved. Keeping the voice of the consumer or customer alive during a product development project is hard. It is even harder to keep this learning alive in the corporate mind and corporate memory. At least eight approaches are being tried by marketing research professionals. The approaches may seem silly, corny, trite, but they work, and that's what's important to the business:

   (i) Shadow boxes that contain collages that represent specific consumers or consumer groups.

  (ii) "Albums" with pictures and video that can be called up to review by current or new project teams.

 (iii) Small pocket books that are visuals of the experience, and capture in "sound bites" the essence of the consumer listening.

 (iv) Special "books" that are prepared for the project teams and are stored in both a library and electronically to remind and recall the event, the people, and the implications. When the budget permits, then these "books" are created with special binding and in special sizes so that they become significant memorabilia.

  (v) Map books to store the value diagrams or the maps created by the team for a project. They are organized by category and are available in both a physical copy and electronically.

 (vi) The consumer insighting process is "ritualized" so there are certain "must do" and "must remember" factors. Some people have adapted this to be bolder and more memorable using the model of network TV. For example, on the TV show *Survivor*, there is tribal council and people are voted off the island, and the words "the tribe has spoken" are given. In a similar vein, in *The Apprentice,* Donald Trump says, "You're fired" in the boardroom. These rituals remind us that the benefit is a process. The rituals are easily remembered, and most important, the rituals add a bit of fun to a serious process.

 (vii) The reporting of results is done with a number of different interest groups. It's important to involve those functional groups who need to make use of these data early on in this front-end reporting. This deliberate involvement of groups, even perhaps before they are absolutely necessary in the process, can be important in larger companies, where people only get involved in the process for short periods of time. Reporting to groups gets these tangentially involved individuals more deeply involved.

(viii) Creating groups of consumers who are not company employees but do work for the company and who follow projects. In most companies they are known as the Consumer Design Panel, the Insight Panel, the Consumer Powered Design, or the Experience Panel. One company named the group POP for Perception and Opinion of People.

By changing its product development strategies to match more closely the wants and needs of the marketplace, a firm can transform product development into a formidable competitive weapon. Just as formidable, however, is the effort that this transformation requires. Established organizational structures and corporate politics present significant barriers to achieving fundamental changes in product development strategy (Karlsson & Ahlstrom, 1997). New structures, new practices have to take the place of old ones. Table 3.10 lists new paradigms that must be embraced for change to occur.

Table 3.10   Advances that are enabling adaptation of untraditional market research tools.

New paradigm	Implication
Understand consumers as people.	Individual experience provides specific triggers.
Get face to face with the insight.	Must deal with anomalies.
Generate lots of relevant data.	Feeling of "too much" data.
Focus on underlying business need.	Solve specific questions.
Discover deeper knowledge that sells.	Strategies of the past might not fit this model.
Capture information using multimedia.	Even more data are gathered.
Silos are part of the problem.	Collaborate to survive.
Focused resources.	Spend appropriately.
Communication-savvy customers.	Past successes may not work well anymore.

Sensory and consumer research need to reconsider certain fundamentals that were brought forward as helpful suggestions forty some years ago. It is time to relook at and reevaluate the value of the concepts that constitute our legacy from people who shaped the work of the late 1960s and early 1970s. We today, 30–40 years later, must reshape the methods and practices to recognize that good scientific work has been done, but maybe some changes need to be made. Indeed, when individuals who are not part of the "conditioned" professions of sensory and consumer research see issues and call them into question (e.g., Garber *et al.*, 2003b; Simester, 2004), it is time for consumer and sensory research to stop negotiating and to start taking action and change the paradigm.

Let's end the chapter by one final restatement. Getting closer to the consumer is not going to replace sensory or market research. Rather, it will contribute to the entire product system to make things that have an opportunity to be more relevant, rather than additive. Zander and Zander (2000) in *The Art of Possibility* suggest two of twelve practices of possibilities: 1: "It's All Invented" and 6: "Rule Number 6—don't take yourself so goddamn serious." We should take counsel in these two very simple but profound pieces of advice.

REFERENCES

Abrams, B. 2000. *The Observational Research Handbook*. Chicago: NTC Business Books.

AMA. 1961. American Marketing Association. www.ama.org.

AMA. 1995. American Marketing Association. www.ama.org.

Amerine, M.A., R.M. Pangborn, & E.B. Roessler. 1965. *Principles of Sensory Evaluation of Food*. New York: Academic Press.

Anonymous. 2001. The 2001 soft drink report. *Beverage Industry* 92(3):18–26.

Anonymous. 2003a. IDEO Method Cards. IDEO Incorporated. Palo Alto, CA.

Anonymous. 2003b. Water: growing by leaps and bounds. *Beverage Industry* 94(7):32–34.

Anonymous. 2004a. The Internet Encyclopaedia of Personal Construct Psychology. www.pcp-net.de/encyclopaedia/constr-alt.html (March 1, 2004).

Anonymous. 2004b. Personal Construction Theory and Repertory Grid. www.Brint.com (March 1, 2004).

Ashman, H. 2003. Understanding the impact current events have on consumer behavior and how to use your knowledge of "today" to make better business decisions "tomorrow". Future Trends Conference (IIRUSA), November, Miami, FL.

Ashman, H. & J. Beckley. 2003. *Sensory & Professionalism II*. Available from the author and also at sensory.org.

Ashman, H., I. Teich, & H.R. Moskowitz. 2004. Migrating consumer research to public policy: Beyond attitudinal surveys to conjoint measurement for sensitive and personal issues. ESOMAR Conference: Public Sector Research, May, Berlin, Germany.

Bakken, D. 2001. The quest for emerging customer needs. *Marketing Research* Winter:30–34.

Beckley, J. 2003. Hot, hot buttons. Institute of Food Technologist Annual Meeting and Food Exposition, July, Chicago, IL.

Beckley, J. 2007. How have things changed? Results collected over six years to understand the changes in mindset of sensory professionals. 7th Pangborn Sensory Science Symposium, August, Minneapolis, MN.

Beckley, J. 2008. Linking consumer insight with sensory, material and formulation science to develop consumer preferred dairy textures. Institute of Food Technologist Annual Meeting, June.

Beckley, J., H. Ashman, A. Maier, & H.R. Moskowitz. 2004. What features drive rated burger craveability at the concept level. *Journal of Sensory Studies* 19(1):27–48.

Beckley, J. & C. Ramsey. 2009. Observing the consumer in-context. In: *An Integrated Approach to New Food Product Development*, eds M. Moskowitz *et al.*, Chapter 12. Boca Raton, FL: CRC Press, Taylor & Francis Group.

Bennett, B. 2004. Water redefined. *Stagnito's New Products Magazine* 4(1):32–33.

Birdwhistell, R. 1970. *Kinesics & Context*. Philadelphia: University of Pennsylvania.

Bogue, J. & C. Ritson. 2005. Integrating consumer information with the new product development process: the development of lighter dairy products. *International Journal of Consumer Studies* 30(1):44–54.

Bucklew, J., Jr. 1969. Features of conventional behavior. In: *Panorama of Psychology*, ed. N.H. Pronko, pp. 253–254. Belmont, CA: Wadsworth Publishing Company.

Casale, R. 2003. It's time to retool consumer research. *Brandweek* 44(40):24.

Charles, S. 2003. A system for capturing the whole product experience—Getting IT!®. The Voice of the Customer—Co-Creation of Demand, Listening and Looking at Innovative Ways for New Product Opportunities (IIR & PDMA), December 8–10, Miami, FL.

Clancy, K. & P. Krieg. 2000. *Counter-Intuitive Marketing*. New York: Free Press.

Cooper, R.G. 1988. The new product process: a decision guide for managers. *Journal of Marketing Management* 3(3):238–255.

Cooper, R.G. 2001. *Winning at New Products*. Cambridge, MA: Perseus Publishing.

Cooper, R.G. 2002. *Product Leadership*. Cambridge, MA: Perseus Publishing.

Coopersmith, S. 1966. *Frontiers of Psychological Research*. San Francisco and London: W.H. Freeman.

Costa, A. & W. Jongen. 2006. New insights into consumer-led product development. *Trends in Food Science and Technology* 17(8):457–465.

Crow, S. & J. Trott. 2002. Verse in "Soak Up the Sun" from *C'mon, C'mon*. Warner-Tamberline Publishing Corp.

Davis, R.E. 1993. From experience: the role of market research in the development of new consumer products. *Journal of Product Innovation Management* 10(4):309–317.

Dijksterhuis, G. 2004. The cognition-emotion controversy and its implications for marketing and consumer research. Unpublished document available through author.

Dumaine, A. 2003. Where's the intuition? *Ad Week* May 5.

Einstein, M.A. 1991. Descriptive techniques and their hybridization. In: *Sensory Science Theory and Applications in Foods*, ed. H.T. Lawless and B.P. Klein, pp. 317–338. New York: Marcel Dekker.

Erickson, K. & D. Stull. 1998. *Doing Team Ethnography Warnings and Advice*. Thousand Oaks: Sage Publications.

Fellman, M.W. 1999. Breaking tradition. *Marketing Research* Fall:21–24.

Fromartz, S. 2006. *Organic, Inc*. New York: Harcourt, Inc.

Garber, L.L., Jr., E.M. Hyatt, & R.G. Starr, Jr. 2003a. Measuring consumer response to food products. *Food Quality and Preference* 14:3–15.

Garber, L.L., Jr., E.M. Hyatt, & R.G. Starr, Jr. 2003b. Reply to commentaries on Garber, Hyatt, and Starr, "Placing food color experimentation in a valid consumer context." *Food Quality and Preference* 14:41–42.

Gladwell, M. 1991. On second thought, skip the explanation. *Dallas Times Herald* March 18.

Goldman, D. 2003. The body politic. *Adweek* July 21:16–18.

Gorman, C. 2003. How to eat healthier. *Time* October 20:49–59.

Griffin, A. 1992. Evaluating QFD's use in US firms as a process for developing products. *Journal of Product Innovation Management* 9:171–187.

Griffin, A. & J. Hauser. 1993. The voice of the customer. *Marketing Science* 12(1):1–27.

Gutman, J. 1982. A means–end chain model based on consumer categorization processes. *Journal of Marketing* Spring:60–72.

Gutman, J. 1991. Exploring the nature of linkages between consequences and values. *Journal of Business Research* 22:143–148.

Gutman, J. 1997. Means–end chains as global hierarchies. *Psychology and Marketing* 14(6):545–560.

Hauser, J.R. & D. Clausing. 1988. The house of quality. *Harvard Business Review* May–June:63–73.

Herstatt, C. & E. von Hippel. 1992. From experience: developing new product concepts via the lead user method: a case study in a "low tech" field. *Journal Product Innovation Management* 9:213–221.

Holbrook, M. & E. Hirschman. 1982. The experiential aspects of consumption: consumer fantasies, feelings and fun. *Journal of Consumer Research* 9:132–140.

Hunt, M. 1994. *The Story of Psychology*. New York: Anchor Books.

Institute of Food Technologists. 2004. Undergraduate Education Standards for Degrees in Food Science. www.ift.org (March 1, 2004).

Jarvinen, M. 2000. The biographical illusion: constructing meaning in qualitative interviews. *Qualitative Inquiry* 6(3):370–391.

Jensen, R. 1999. *The Dream Society: How the Coming Shift from Information to Imagination Will Transform Your Business*. New York: McGraw Hill.

Karlsson, C. & P. Ahlstrom. 1997. Perspective: Changing product development strategy—a managerial challenge. *Journal of Product Innovation Management* 14:473–484.

Kash, R. 2002. *The New Law of Demand and Supply: The Revolutionary New Demand Strategy for Faster Growth and Higher Profits*. New York: Doubleday.

Kathman, J. 2003. Building leadership brands by design. *Brandweek* December 1:20.

Kelly, G.A. 1955. *The Psychology of Personal Constructs*. New York: Norton Publishing.

Kelly, T. 2001. *The Art of Innovation*. New York: Doubleday.

Klemmer, E.T. 1968. Psychological principles of subjective evaluation. In: ASTM Committee E-18. 1968. *Basic Principles of Sensory Testing Methods STP 433*. ASTM Special Technical Publication, pp. 51–57. Philadelphia: ASTM.

Kotler, P. 1972. *Marketing Management: Analysis, Planning and Control*. Englewood Cliffs, NJ: Prentice-Hall.

Kotler, P. 2002. *Marketing Management*. Englewood Cliffs, NJ: Prentice-Hall.

Krall, J. 2003. Big-brand logos pop up in organic-foods aisle. *Wall Street Journal* July 29.

Kroeger, O. & J. Thuesen. 1988. *Type Talk*. New York: Dell Trade Paperback.

Langer, J. & J. Last. 2003. Commentary on "A fresh look at focus groups." *Brandweek* November 24:14.

Laros, F. & J.B. Steenkamp. 2005. Emotions in consumer research: a hierarchical approach. *Journal of Business Research* 58(10):1437–1445.

Larsen, P.T. 2004. The company that hogs the TV control in the US. *Financial Times* February:8.

Lawless, H. 1991. Bridging the gap between sensory science and product evaluation. In: *Sensory Science Theory and Applications in Foods*, eds H. Lawless and B. Klein. New York: Marcel Dekker.

Leonard, D. & J. Rayport. 1997. Spark innovation through empathic design. *Harvard Business Review* November–December:102–113.

Longinotti-Buitoni, G.L. 1999. *Selling Dreams: How to Make Any Product Irresistible*. New York: Simon & Schuster.

Mahajan, V. & J. Wind. 1999. Rx for marketing research. *Marketing Research* 11(3):7–13.

Malefyt, T. 2003. Ethnography and focus groups: mutually exclusive? *Brandweek* 44(43):14.

Marshall, D. & R. Bell. 2003. Meal construction: exploring the relationship between eating occasion and location. *Food Quality and Preference* 14:53–64.

Martin, C. 1985. *Estimated Frequencies of the Types in the General Population*. Gainesville, FL: Center for Applications of Psychological Type, Inc.

Martino, G. 2003. Does the sensory profession provide business with results that grow businesses? Presented at Pangborn Sensory Symposium, Boston, MA.

Maslow, A.H. 1968–1999. *Toward a Psychology of Being*. New York: John Wiley & Sons.

McMath, R.M. 1998. *What Were They Thinking?* New York: Time Books, Random House.

McWatters, K.H., M.S. Chinnan, R.D. Phillips, S.L. Walker, S.E. McCullough, I.B. Hashim, & F.K. Saalia. 2006. Consumer-guided development of a peanut butter tart: implications for successful product development. *Food Quality and Preference* 17(6):505–512.

Mehrabian, A. 1971. Nonverbal betrayal of feelings. *Journal of Experimental Research in Personality* 5:64–73.

Meiselman, H.L. 1993. Critical evaluation of sensory techniques. *Food Quality and Preference* 4:33–40.

Meiselman, H.L. & H.J.M. MacFie. 1996. *Food Choice Acceptance and Consumption*. London: Blackie Academic & Professional.

Moskowitz, H.R. 2002. The intertwining of psychophysics and sensory analysis: historical perspectives and future opportunities—a personal view. *Food Quality and Preference* 14:87–98.

Moskowitz, H.R. 2003. When bad data happen to good researchers: a contrarian's point of view regarding measuring consumer response to food products. *Food Quality and Preference* 14:33–36.

Moskowitz, H.R. 2004. Personal communication to the author (Beckley).

Moskowitz, H.R. & H. Ashman. 2003. The mind of the consumer shopper: creating a database to formalize and facilitate the acquisition and use of insights. Proceedings of the ESOMAR Conference "Excellence in Consumer Insights," March, Madrid.

Moskowitz, H.R., B. Jacobs, & N. Lazar. 1985. Product response segmentation and the analysis of individual differences in liking. *Journal of Food Quality* 8:168–191.

Moskowitz, H.R., S. Porretta, & M. Silcher. 2005. *Concept Research in Food Product Design and Development*. Ames, IA: Blackwell Publishing Ltd.

Murphy, R. 2004. Truth or scare. *American Demographics* 24(2):26–32.

Noble, C. & M. Kumar. 2010. Exploring the appeal of product design: a grounded, value-based model of key design elements and relationships. *Journal of Product Innovation Management*. 27(5):640–657.

Olson, J. & T. Reynolds. 1983. Understanding consumers' cognitive structures: implications for advertising strategy. In: *Advertising and Consumer Psychology*, vol. 1, ed. L. Perry and A. Woodside. Lexington, MA: Lexington Books.

Pangborn, R.M. 1989. Sensory science, an exciting and gratifying career. In: *Sensory Evaluation: In Celebration of Our Beginnings*, eds ASTM Committee E-18 on Sensory Evaluation of Materials and Products, pp. 14–17. Philadelphia: ASTM.

Paredes, D., J. Beckley, & H. Moskowitz. 2008. Bridging hedonic and cognitive performance in food and Health and Beauty Aide (HBA) products. Presented at the First Society for Sensory Professionals Meeting, November 3–5, Cincinnati, OH.

Pine, B.J., II & J. Gilmore. 1999. *The Experience Economy: Work Is Theater & Every Business a Stage*. Boston: Harvard Business School Press.

Pruden, D.R. & T.G. Vavra. 2000. The needed evolution in marketing research is respecting relationships with customers. *Marketing Research* Summer:15–19.

Qiu, T. 2004. The role of marketing managers in shaping proactive organizational culture—report on 2003 survey results. University of Illinois, Urbana–Champaign.

Rainey, B.A. 1986. Importance of reference standards in training panels. *Journal of Sensory Studies* 1(2):149–154.

Ratneshwar, S., A.D. Shocker, J. Cotte, & R.K. Srivastava. 1999. Product, person and purpose: putting the consumer back into theories of dynamic market behavior. *Journal of Strategic Marketing* 7(3):191–208.

Reynolds, T.J. & J. Gutman. 1988. Laddering theory, method, analysis and interpretation. *Journal of Advertising Research* February/March:11–31.

Rokeach, M. 1968. *Beliefs, Attitudes, and Values*. San Francisco: Jossey-Bass.

Rook, D. 2003. Out of focus: groups. *Marketing Research* 15(2):11–15.

Rushkoff, D. 2004. The Power of Three: When Technology, Business and Marketing Converge—Part III. www.thefeature.com.

Scheel, J. 2003. The health of organic foods. *Prepared Foods* May:25–30.

Schwartz, B. 2004. *The Paradox of Choice: Why More Is Less*. New York: Harper Collins.

Senge, P. 1994. *The Fifth Discipline: The Art & Practice of the Learning Organization*. New York: Currency Doubleday.

Sharma, J.R. & A.M. Rawani. 2009. Quality function development: a new paradigm for involving customers in product development process. *International Journal of Quality and Innovation* 1:16–36.

Shell, E. 2002. *The Hungry Gene*. New York: Atlantic Books.

Silver, J.A. & J.C. Thompson, Jr. 1991. Understanding customer needs: a systematic approach to the "voice of the customer." Sloan School of Management in partial fulfillment of a Master of Science in Management, Massachusetts Institute of Technology.

Silverstein, M. & N. Fiske. 2003. *Trading Up: The New American Luxury*. New York: Penguin.

Simester, D. 2004. Finally, market research you can use. *Harvard Business Review* 82(2):20.

Sloan, A.E. 2003. What, when and where Americans eat: 2003. *Food Technology* 57(8):48–66.

Sloan, A.E. 2004. The low-carb diet craze. *Food Technology* 58(1):16.

Sopheon, PLC. 2002. Improving the business impact of product development. White paper.

Stengel, J. 2004. Stengel's call to arms. *Advertising Age* February 16:front page.

Stewart, A. 1998. *The Ethnographer's Method*. Thousand Oaks, CA: Sage Publications.

Struse, D. 1999/2000. Marketing Research's top 25 influences. *Marketing Research* Winter/Spring:5–9.

Theodore, S. & J. Popp. 2004. Obesity: weighing the options. *Beverage Industry* 95(2):16–24.

Toops, D. 2004. Food Processing A.M. Update, March 8.

Trump, D. 2004. Interviewed on *Larry King Live* March 15.

Tuorila, H. & E. Monteleone. 2009. Sensory food science in the changing society: opportunities, needs and challenges. *Trends in Food Science and Technology* 20(2):24–52.

Urban, G. & E. Von Hippel. 1988. Lead user analyses for the development of new industrial products. *Management Science* 34(5):569–582.

Urbanski, A. 2003a. Getting the carbs out. *Frozen Food Age* 52(4):20.

Urbanski, A. 2003b. A passing fad or a permanent lifestyle? *Frozen Food Age* November:38.

Von Hippel, E. 1988. *The Sources of Innovation*. New York: Oxford University Press.

Wade, M. 2004. Surfing the trends of weight control formulations. *Prepared Foods* 173(2):34–40.

Wansink, B. 2000. New techniques to generate key marketing insights. *Marketing Research* 12(2):28–36.

Wasserman, T. 2003. Sharpening the focus. *Brandweek* 44(40):28–32.

Weiss, M. 2003. Inside consumer confidence surveys. *American Demographics* 25(1):22–29.

Whitlark, D. & C. Allred. 2003. Driving your market. *Marketing Research* 15(4):33–38.

Yager, S. 2010. *The Hundred Year Diet*. New York: Rodale, Inc.

Zaltman, G. & L. Schuck. 1998. Seeing through the customer's eyes with computer imaging. In: *Sense and Respond: Capturing Value in the Network Era*, pp. 145–172. Boston: Harvard Business Press.

Zander, R.S. & B. Zander. 2000. *The Art of Possibility: Transforming Professional and Personal Life*. Boston: Harvard Business School Press.

Zikmund, W.G. 2000. *Business Research Methods*. Fort Worth, TX: Dryden Press.

4 Innovation's friend: integrated market and sensory input for food product design and development

Many people working in the Sensory area in the food industry still think that we need to be more objective and clinical because that's what we did in school (and besides, if we act scientifically, people will respect us more). There is the impression that their ideal testing environment would be a laboratory setting where people tasted the product in separate booths, and never saw each other or the servers, where no distracting noises or smells could possibly get in, and where even the way they eat the products would be controlled. Of course, the subject matter in school was how people smell or taste. The subject matter in industry is "which course of action is the right one?" "Should I use 10% or 12% sugar in my product?" Whenever the occasion permits, I (author JB) just ask the consumers what they think I should do, and they invariably tell me.

> Personal communication to author, 2004, by an established PhD reflecting
> on issues in sensory in the food industry today

Whereas consumers have changed beyond recognition, marketing has not. Everything else has been reinvented—distribution, new product development, the supply chain. But marketing is stuck in the past.

> Elliott Ettenberg, then CEO of Customer Strategies Worldwide, quoted by Kevin Clancy in
> *Surviving Death Wish Research* (Clancy & Krieg, 2001)

INTRODUCTION

Being consumer-centric is a path to success. Resources need to be managed well in a highly value-driven marketplace. There is no room for the continuation of work silos that separate people who need to understand the consumer. There needs to be extensive flow of information and thinking between workers and suppliers whose function it is to understand consumers, find out what exactly they want, make sure that the offer being created meets that need or demand, and then make sure that the product design delivers on the promise of that design. The original product and market research model provided for tests conducted by different organizations with output coming together in a highly linear fashion. Technology, science, and experience allow for a more flexible and time-conscious approach to the utilization of these inputs earlier and with greater integration. There is a demand for more problem-solving than tactics today. There are, however, great issues in this area that are related to power and control.

Sensory and Consumer Research in Food Product Design and Development, Second Edition.
Howard R. Moskowitz, Jacqueline H. Beckley, and Anna V.A. Resurreccion.
© 2012 Blackwell Publishing Ltd. Published 2012 by Blackwell Publishing Ltd.

This chapter addresses the reasons why the traditional model of product development with sensory and consumer research is broken and needs to be fixed. The chapter presents approaches that effectively integrate consumer and sensory research activities. Approaches include specific actions that need to take place during the discovery phase of the product development process and at other points along the path. A method will be demonstrated that more effectively blends qualitative and quantitative information, enabling more effective value creation during food product design and development.

UNDERSTANDING YOUR MARKET AND PRODUCT FROM THE CUSTOMER/CONSUMER PERSPECTIVE

Homo economicus has little to do with his half brother *Homo sapiens.*
> Eric Roston in "Inside Business" (*Time* magazine), April 2004. Headline from an
> article about the lack of rational behavior really associated with money and purchases

Customers purchase products that they want or need. When they are standing in the aisle or sorting through a catalog or hunting on a Web site, they are looking for products to buy that satisfy a desire. Whether it is a flat-out hunt or a casual stroll, shopping and buying is fundamental to the human animal today. The rapidly emerging field of "neuroeconomics" continues to show us something we already knew but economists didn't allow for in some of their theories—that the perfectly reasonable (human) being who maximizes his utility and gains and is always clear about seeking the right thing to do with his money is probably a myth. This "economic man" concept (*H. economicus*) pervades much of what we think about in the sales and marketing of products and services, even though this model is now openly challenged (Fox, 2009). But why should that matter in food product design and development (Rousseau, Fauchereau, & Dumont, 2001; Marshall, 2003)?

For the business reader, choice matters a lot. The topic of choice—what I buy, why I buy it, what motivates me to buy more—is a subject in sensory, market research, and general business literature (Garber, Hyatt, & Starr, 2003; Grunert, 2003; Chernev, 2006). Much is beginning to be discussed, evaluated, and published regarding these areas (Paredes *et al.*, 2008; Teratanavat *et al.*, 2008).

The current product model within food companies today, while evolving, still does not regularly look at the product as the consumer looks at it. The immediate act of picking something off the shelf, or the menu, or from a Web site is an active process for people, as is sticking the piece of food in one's mouth. That moment is very emotional and fairly transient. There are those behaviors that might or might not lead to the selection and then to reflect on what was selected. But one way or the other, it is a whole product that is selected by a specific individual. Drucker calls this "whole product utility"—what the product or service does for the buyer (Drucker, 2001).

At the practical level, how do you create a product having this or any specified utility? In our desire to try to make this often chaotic swirl of events appear more rational than it is, we design sequential processes to subdivide the product experience (Miller, 1998; Longinotti-Buitoni, 1999; Pine & Gilmore, 1999; Twitchell, 1999; Underhill, 1999; Ashman *et al.*, 2003; Hine, 2002; LaSalle & Britton, 2003). We thereby make the process, and in turn, consumer choice, *appear* logical and sequential and manageable. It is not surprising that the relationship of consumer testing and sensory testing has a rather disjointed design, which works as often against the product in the consumer world as working with it.

Table 4.1   Department/group organization. (From Beckley, 2004.)

How is your department or group organized?	N	%
Market research/consumer insight department within the marketing department	18	10
Sensory department within the R&D function	88	47
Blend of consumer insight and sensory reporting into a consumer insight person	6	3
Blend of consumer insight and sensory reporting into a sensory person	8	4
Blend of consumer insight and sensory reporting into a product development person	11	6
Other	58	31

Source: 189 respondents surveyed over the Internet from both market research and sensory fields.
N, number.

The model that continues to be most prevalent in food product design and development today is a sensory group within an R&D function and the market research (consumer) function in marketing.

This is a relatively classic model, fairly similar to models that have existed for the last 30 years. What we see are two different functions not necessarily working together, nor uniformly on the same product. Within the survey reported in Table 4.1, there are emerging a small number of groups, comprising blends of consumer and sensory researchers. Market research group appear to be more uniform in their identification of what that particular group of researchers does, whereas sensory professionals reflect much more difference of opinion related to what they call themselves or their departments.

What is evolving is the emerging discipline of "products research." This group focuses on the entire product-consumer interaction. The products research group comprises individuals who may have had classic sensory or market research training but focus on the product function as a pivotal, "ownable" aspect of the brand or company. For decades now, Procter & Gamble has used this designation of a product/consumer research function. Groups of researchers who understand the essential role of a disciplined approach to product design and development have come to realize that this is a function separate and distinct from sensory testing or market research. (Cooper, 2008; Switzer, 2005). Tables 4.2a, 4.2b, 4.2c, 4.2d, and 4.2e provide examples of these evolving mind-sets.

Table 4.2a   Job description. (From Beckley, 2004.)

When you describe yourself to others, which one of the phrases best represents what you say about what you do for a living?	N	%
I am a scientist	41	22
I am a researcher	26	14
I am a businessperson	8	4
I am a teacher or professor	6	3
I am a psychologist	0	0
I am a sensory professional	39	21
I am a consumer scientist	16	8
I am a trend tracker	0	0
I am a business executive	1	1
I am a statistician	6	3
I am a market researcher	32	1
None of the above	14	7

Source: 189 respondents surveyed over the Internet from both market research and sensory fields.
N, number.

Table 4.2b Preferred title. (From Beckley, 2004.)

If you had to make a choice, which phrase best describes you?	N	%
Consumer insight person	21	11
Market researcher	22	12
Sensory analyst	42	22
Sensory professional	61	32
General manager	6	3
Other	37	20

Source: 189 respondents surveyed over the Internet from both market research and sensory fields.
N, number.

Table 4.2c Job tasks. (From Beckley, 2008.)

Does your sensory group work on both sensory science (i.e., drivers of liking, overall liking, etc.) and consumer understanding (i.e., consumer behavior, usage, consumption)?	
Sensory science only	11%
Consumer understanding only	2%
Both sensory science and consumer understanding	86%

Source: Sensory professionalism study, July 2007. Sample size *N* = 88. Data available from author or through sensory@yahoo.com.

Table 4.2d Methods used most by sensory professionals. (From Beckley, 2008.)

Which of the following methods do you use most often?	
Discrimination testing	59%
Descriptive testing	77%
Consumer (affective) testing	88%
Qualitative testing	62%
Observational testing	24%
Human factors testing	6%
Web-based structured questionnaires	32%
Test markets	6%
Trade-off analysis/conjoint	26%
Focus groups	51%
Gamma testing	0%
Customer site visits	9%
Beta testing	5%
Pre test markets	8%
Ethnography	19%
In context testing	22%
Lead users	12%
Voice of customer	15%
Creativity sessions	17%
Concept engineering	6%
Fit to concept test	35%
Alpha testing	7%
Other	12%

Source: Sensory professionalism study, July 2007. Sample size, *N* = 88. Data available from author or through sensory@yahoo.com.

Table 4.2e Importance criteria for achieving workplace success. (From Beckley, 2008.)

Please check all that you feel are of high *importance* for a sensory professional to achieve success in the workplace.

Length of experience/time in business	40%
Breadth of experience in a specific segment of business	32%
High level of flexibility	69%
Champion of extremely high standards	45%
Capacity to understand and make sense of complexity in business	69%
Extraordinary interpersonal/social relations	58%
Ability to recognize employees properly for their contributions	28%
Ability to support business partners to target research to solve the right problem	77%
Usage of most effective methods to find the solution to a product development problem	82%

Source: Sensory professionalism study, July 2007. Sample size, $N = 88$. Data available from author or through sensory@yahoo.com.

If this understanding is actually evolving, then what causes the problems with product research that many who read this book struggle with on a continuing basis?

(i) *Poor business process design*: The design of businesses with separate functions for marketing and product development reinforces organizational barriers to integration (Khurana & Rosenthal, 1998; Bond & Houston, 2003; Kahn *et al.*, 2003; Plummer, 2003; Rein, 2004). Maintaining roles that are no longer useful can provide sensory and consumer research with labels that reinforce behaviors. These roles may be outdated, but they still create mind-sets that lead to poor collaborative consumer and product understanding. The outcome: competition for a share of resources and management attention. Looking deeper into the research, the indication is that not enough dialog has occurred between members of the respective disciplines and as a result, alignment of focus is missing (Gladwell, 2005; Ashman & Beckley, 2001, 2003).

(ii) *Disjointed activities*: Integrating consumer and product research has not occurred in food design in a large way. The unhappy result: disjointed understanding about consumers, products, and the levers one uses with consumers. (Meiselman, 1992; Cardello, 2003; Wansink, 2003). New ideas and thus new activities promoting "user-oriented innovation" have not been heavily applied to the food industry (Grunert *et al.*, 2008). The result is there are still many more missed opportunities than are necessary, given the technologies and understanding we have today in this marketplace. Testing a fragment of a concept or not being able to look at pricing as a variable, when it matters, overlooks knowledge and produces an incomplete, perhaps even erroneous product design. Needless to say, all too often market share follows, but the loss is less dramatic, because the rate is slower.

(iii) *Different mind-sets*: On average, the mind-set of the sensory professional and the market researcher are different (Moskowitz, 2003). There are individuals within the sensory profession whose thinking is very much like other business associates, that is, business focused, but this is not the normative thinking or behavior for the sensory professional (Table 4.3). For sensory professionals the normative focus seems to be on tactical approaches (Ashman & Beckley, 2004).

(iv) *Incorrect or nonproductive research methods when assessing product performance*: Market research, with all its benchmarks and metrics, has not been delivering on the

Table 4.3 Impact sensory professionals feel they have on meeting consumers' needs and wants and how they feel other members of the organization view their impact.

The field of sensory science has been one of the key areas that has allowed companies to develop better products that meet consumers needs and wants. How much do you agree with this statement?

Agree completely	55%
Agree somewhat	38%
Neither agree or disagree	4%
Disagree somewhat	2%
Disagree completely	0%
Do not know	1%

Thinking about the statement, "The field of sensory science has been one of the key areas that has allowed companies to develop better products that meet consumers needs and wants." How much do you think the senior management of your company would agree with this statement?

Agree completely	22%
Agree somewhat	49%
Neither agree or disagree	18%
Disagree somewhat	6%
Disagree completely	1%
Do not know	4%

Thinking about the statement, "The field of sensory science has been one of the key areas that has allowed companies to develop better products that meet consumers needs and wants." How much would your peers in other areas of your company or organization agree with this statement?

Agree completely	17%
Agree somewhat	54%
Neither agree or disagree	17%
Disagree somewhat	8%
Disagree completely	1%
Do not know	2%

Source: Sensory professionalism study, June/July 2003. Data available from author or through sensory@yahoo.com.

needs of businesses (Simester, 2004). It is interesting to note that, since about 1985, there has been a very public conversation regarding an inability to deliver reliable and significant business results in the market research field for almost 25 years (Mahajan & Wind, 1999; Liefeld, 2003)! To understand this situation, let's look at a popular method in both market research and sensory, preference testing. Here is one example. Every time a preference test is conducted with two specific products, one brand of popular cheese crackers versus another, one of the products on a blind basis usually wins. However, the market leader, and the one that consumers say by name, whether they have that product in the pantry or not, is usually the loser in the blind preference study! Anomalous results? No. It often happens with products. And, with these types of results, one should begin to suspect the method of paired comparisons to measure preference because it just doesn't track with the market reality. Yet, two-sample preference tests and the single sample product concept are loved in all parts of business since they are simple to understand, appear to have compelling results, and have the benefit of years of use (Cialdini & Rhoads, 2001).

Recommendation 1: Goals that are shared and linked between consumer, sensory, and products research groups to foster and reward collaborative behavior.

Recommendation 2: Building upon Recommendation 1, the two groups, sensory and market research, identify their common and unique ground, respectively. Management acknowledges these similarities and differences and provides mutual recognition and reward.

Recommendation 3: Sensory, product research and consumer research become a greater contributor at all phases of the project planning. All groups work to understand their specific benchmarking or grading biases and work to build consensus in what goals will be used and what is the rational point of view (from all perspectives) for those choices.

Whereas Recommendations 1, 2, and 3 may sound overly optimistic and simplistic, they are founded on the principles of confronting the brutal facts but not losing faith, as explained in Collins's book *Good to Great* (2001).

MISSING THE IMPORTANT THINGS

In the drive by consumer and sensory researchers to be credible and believable, much of the product experience and understanding of it have been missed. Today, an experience can be something very special or an everyday event, like showering. Many individuals never tire of their morning shower and for many, it brings them joy or calm or bliss. Yet some look at the product and say, "but it is just a shower!" Well, the market numbers show us that it is not just a shower, with less than 50% of the soap sold in its traditional form and market share continuing to be taken by liquid soap and soap gels.

It's not the reality that's elusive, it's the methods that are weak, or downright misleading. What we find is that methodologies and process, designed to be helpful in understanding product questions, become the destination, the goal of the research. The reality, the good stuff from people, users, and our customers can get lost. No consumer looks at your product as a spider graph. Consumers look at their kids when they give them a slice of bread and peanut butter and see them giggling, happy, and fed. How often does a driver of liking research measure the impact of the giggle plus the belly fill and the smell and taste of the peanut butter as it is spread on the bread? As we explore the use of more emotion-based research and more specific neurological understanding, are we challenging ourselves to truly understand the WHY of the delight or merely training one form of data collection for another? We hope the former rather than the latter. (Frazier, 2007; Young, 2010).

Who loses when we don't put the whole product understanding experience together? Everyone. The consumer, the company (and their stockholders), and individuals in the consumer/products/sensory professions are all big losers. Consumers lose because a great product they want might never make it to market so that they can purchase it. The company loses because it is not selling truly compelling products even though funds have been commissioned in a number of areas. Business professionals lose because they have diminished self-esteem, which can be reflected in perceptions of them by others as being negative, defensive, abrasive, or hard to work with. Analysis by top R&D executives suggested personal passion and job excitement were of key importance in delivering successful business results (Mermelstein, 2001).

In the work that author Beckley has done in the last 5 years, she has had the privilege to look at the overall business files for a broad range of industries and across several hundred projects or subject areas. The results show consistently that companies, large and small, have much of the data and information they need to solve most of the problems troubling them. What companies, or better the relevant groups, lack is some specific knowledge in specific categories regarding specific consumer behavior, beliefs, attitudes, or specific product acceptance. Yet, many of these companies have taken a repetitive, not productive path to commission more research (*How many segmentation studies are required? One, two, or fifteen!*). Again and again it turns out that the commissioned research was not particularly additive to what has been collected. Again and again market, product, and sensory professionals reported being frustrated by the lack of cohesiveness and disheartened, because they feel they are not being listened to by management and peers. And peers and management continued to say that the market, product, or sensory person is not listening to the need of the customer or consumer!

All of this leads to a waste of many company assets. The wastes are in the form of the following:

(i) *Projects*: Unsuccessful projects that fail in the market. And this is in spite of talent and corporate commitment. Critical gaps appear to be articulating, delivering, and identifying. Articulating is figuring out how to articulate the product benefit so consumers with expectations of immediate understanding (most of them, everywhere). Delivering is delivering on the salient, communicated benefit to most. Identifying is properly identifying the core audience for the product and the communication.

(ii) *People*: Less enthusiastic employees. In the different research areas—market, product, and sensory—the staff gets "burned out" doing the same work over and over again, without really attracting the attention of management in food product design and development or marketing. Sensory product and consumer researchers continue to report a lack of acknowledgement of their efforts. Desire or curiosity to take on new/different work can be reduced due to overall company recognition behavior.

(iii) *Money*: Appropriation of funds to gather more information in areas already known—monies are spent on "posterior protection" rather than increase of knowledge that builds understanding (Van der Bij, Song, & Weggeman, 2003).

(iv) *Time*: Loss of time—while the data are being collected for issues that have been addressed before, the marketplace continues to change. That loss in time and in turn in motivation, can be devastating to projects or program profitability.

(v) *Overgeneralization*: Whether it is in people, belief systems, and product trends. For example, there has been a tendency for the food industry to categorize all sensory practitioners as similar. Data from surveys (Ashman & Beckley, 2001, 2003) suggest that although there may be several consistent patterns, there is a very wide range of perspectives that is not well understood by the client base (Moskowitz, 2003).

(vi) *Execution*: If research by sensory and market research professionals for the last 30 years has all been leading to positive outcomes and the promise of product success, then why is the success rate with new products in the food sector around one in seven (Trout, 2000)? Is it time to listen to critical voices about discrimination testing, descriptive evaluations, segmentation and mapping research, and model building, and consider change (Moskowitz, 1995; Garber, Hyatt, & Starr, 2003; Wansink, 2003)?

If food product design and development is looking for productivity, there is a need to start looking at market and product understanding and where these areas can be helped by collaboration of consumer and sensory evaluation groups. For food product design and development today to be more innovative, it needs to take the leadership in removing barriers separating those whose job it is to understand the market, to understand the consumer, and to understand the product. It requires a little more emphasis on design (thinking, planning, and modeling) and a little less emphasis on development. We explore in the later sections some of the proactive measures that need to take place (or have started to be implemented) so that there is an open flow of information running between business and research and inclusive of the voice of the customer.

Since product development is an acknowledged way for firms to build and strengthen their competitive advantage (Karlsson & Ahlstrom, 1997), it is important for food product design and development to have more complete knowledge. Furthermore, it's important for product design and development to exert more influence on functions originally considered exclusively marketing. Moskowitz suggested that part of the trouble was that top management was unwilling to get more involved in product issues, and that they should demonstrate this through behaviors that suggested their life depended on it (Moskowitz, 2003).

What does the foregoing mean in a practical sense? It's one thing to take responsibility, but it's another to specify what one must do. At the level of deliverables, there is the need to have product differentiation that is more compelling and more ownable. That is, the products that emerge must be different, better, and specific to the company's brand. We believe that one of the only ways to have this type of product "insulation" is to drive the product through the specific attributes a company can control and to integrate the attributes and elements so that they deliver the consequences that the customer wants or needs (Figure 4.1). As Figure 4.1 illustrates, those factors are shared elements *between* product development and marketing and therefore must be shared by the testing

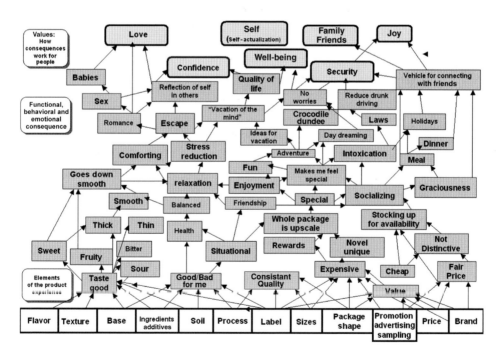

Figure 4.1 Value diagram describing the attributes of a product experience and its consequence and values to a consumer.

groups—consumer research and sensory science . If not, the end results are products that are not close, and lose or lack stickiness in the marketplace (Barabba & Zaltman, 1991; Stanton, 1997; Bradley & Nolan, 1998; Pine & Gilmore, 1999; Rogers, 2000; Clancy & Krieg, 2001; Zaltman, 2003). And that undifferentiated, nonsticky product is a way to fail in the food marketplace today.

There's another factor playing a role. This is the FUD—fear, uncertainty, and doubt—but mainly fear (Speece, 2002). A company's fear of change is often reinforced by the market's apparent fear that the brand will change. The result, a false sense that the right move is to stick with what's been working. "Do nothing, so no one will get hurt." The expression of fear can be misinterpreted and paralyzing. Change is risky. Change is scary. Change is expensive. A quick look at brands such as Smith Corona, Zenith, TWA, and Amtrak, however, reminds us there are riskier, scarier, and more expensive things than change. Failure to change, embracing the status quo, is a decision as well (Speece, 2002). But change can get you fired, whereas the brands just mentioned died slow deaths that preserved some individual jobs for a period of time. Recognizing this realpolitik and need for market-driven change, Al Clausi, a former top R&D executive at General Foods (Kraft), said it very well:

> It is simply that as members of a consumer products company, we are all in "marketing."
>
> Mermelstein, 2001

SILOS GET IN THE WAY OF COMPELLING PRODUCTS FOR THE CONSUMER: NOW WHAT TO DO?

> Often, consumer research is a dead weight rather than a springboard to connecting with our markets. The problem is inadequate thinking, not inadequate data.
>
> CMO of a major consumer products company on
> an Advertising Research Foundation (ARF) Web site, March 2004

Barabba and Zaltman (1991) many years ago commented that it was a flawed concept that the only customers of market research were the marketers. Advocates from the sensory community (Pearce, 1980; Stone & Sidel, 1985; Moskowitz, 1995; Kindle, 2000) have been making a similar point for a long time. Our recent research suggests that some interleaving of consumer and sensory is going on; however, for the most part the silo for market research (or consumer insight or consumer strategy and insight) is located with marketing on the business side and sensory or product evaluation located on the product development side. These groups have trouble informing each other, trusting motives of each other, or feeling that insight found by one group can benefit all. In a word, there is mutual suspicion, often unspoken antagonism, and all too often undeclared warfare.

A review of market research books, periodicals, and journals, comparing them to similar sensory media, suggests that both groups share many of the same core issues of what it means to be a researcher. Shared beliefs and practices include the following:

(i)   Emphasis on rigor in what data are used.
(ii) Concern about quality of collection of data.
(iii) Need to be a good communicator.
(iv) Being creative in test design, tools used.
(v)   Cost of method, approach.

Yet, there are clear differences across the groups:

(i) Access to decision makers within a company typically is better for market research on average.
(ii) Size of budget is larger for market research on average.
(iii) Perceived "sexiness" of the work done is greater for market research.
(iv) Academic training.
(v) Fundamental skill level of professional.
(vi) Expectations regarding performance metrics.
(vii) Assumption of what the job really is. (For points iii–vi.)
(viii) Open or more public discussion of issues and challenges for the field is much more extensive in market research.

Whereas there are many similarities, the differences appear to be in the greater business access and performance accountability that market research enjoys in the corporation. A theme that is shared by both groups is knowledge that their work is important but oftentimes not understood or acknowledged by the organization. Table 4.4 shows a difference of almost 40% between the impact sensory professionals feel they have had on products in their businesses versus what they believe their peers feel their contribution is to the products. The gap reflects a significant issue in how the sensory professionals feel they are acknowledged.

Surveys and job postings show that the most prevalent observable model for sensory today is a highly tactical approach. Consumer or market research surveys and job boards show demonstrably more emphasis on thinking and strategy, along with knowing how to look at data. Review of ongoing training courses in both sensory and consumer research produced similar results (Figure 4.2). With more training programs and an awakening around the disciplines, the area that should grow because of need and relevance is that of the food product design specialist. This specialist bridges the gap between the two disciplines.

MAKING THE MOST OF THE SPECIALTIES FOR A BETTER BUSINESS

In today's food business, the hallmark needs are speed, less cost, and excellence in intelligence; we believe a new model for consumer understanding for food product design and development is required. A new organization is needed. Figure 4.3 illustrates this new organization. This new model is based on several general facts:

(i) Food choice is complicated and plays on a variety of levels for individual consumers (Grunert, 2003).
(ii) Food R&D executives have competing needs to deliver, a demand that occasionally reduces their ability to support long-term consumer understanding and sensory research (Mermelstein, 2001; Cardello, 2003)
(iii) To outpace the market consistently, companies must not only create fluid organizational structures but also must provide for unyielding rigor in measurement and decision-making (Aufreiter, Lawver, & Lun, 2000).

Table 4.4 Content analysis to determine similarity/difference in language used to message to individuals who are sensory or consumer insight/market researchers.

	Professional courses offered—key phrases	Job listings—key words
Sensory field	Issues of behavioral measurement.	**Very senior level:**
	Design your own test.	Advance knowledge of product.
	Do it yourself rather than blindly follow others.	Interpret.
		Conduct.
	Understanding the senses.	Support.
	What goes on in the head when a judge is given a numerical estimate?	Design.
		Implement new methods.
	Explore methods used in the sensory evaluation of consumer products.	Enhance.
	Application of sensory science and consumer testing principles.	Working knowledge.
		Lead.
	Consumer test methods.	Experienced with automated data collection systems.
	Explore how to use your senses to evaluate and test food, beverages, and nonfoods.	**Senior level:**
		Advance knowledge.
	Participate in various basic test designs.	Working knowledge.
	Be more aware of physiological and psychological influences on sensory judgment.	Apply.
		Interpret.
		Have interpersonal skills.
	Experience analyzing and interpreting results.	Be able to communicate.
		Work with designs.
	Learn how to train judges.	Conduct.
		Coordinate.
		Recommend.
		Analyze.
		Lead.
		Manage data collection.
Market research/ consumer insight field	Rapid rate of change in customers' values and the diversity of response available.	**Senior level:**
		Leadership.
		Integration.
		Identify.
	Customer demand.	Information needs.
	Competitive environment.	Generalist.
	SWOT analysis.	Strong interpretive skills.
	Importance of understanding customer perspective.	Communication.
		Resourceful.
	Understanding why individuals act as they do.	Must develop strong relationships.
	How to be relevant in these pursuits.	Adept at building relationships.
	Presenting results: How to speak marketing research with a management accent.	Coordinate.
		Work.
	Setting marketing research objectives.	Roll up sleeves.
	Problem-solving with market research.	Involved.

Table 4.4   (Continued)

	Professional courses offered—key phrases	Job listings—key words
Market research/ consumer insight field	Leverage your brand for more effective fund-raising. Maintain your unique positioning and stay relevant to constituents. Protect your brand in times of turbulence. Decode industry forces. Learn how to frame critical competitive issues. Apply early warning systems and learn how to avoid long-term surprises. Improve your execution: What are the critical questions to ask and not ask; how can you improve planning and execution?	**Associate level:** Leadership. Integration. Strategic consultation. Identification. Identify information needs. Provide guidance. Well-grounded. Know techniques/methods. Working knowledge. Skilled in application. Coordination. Assist.

Source: Summarization by author conducted in spring 2004 across brochures, Web sites, and courses offered by professional organizations.

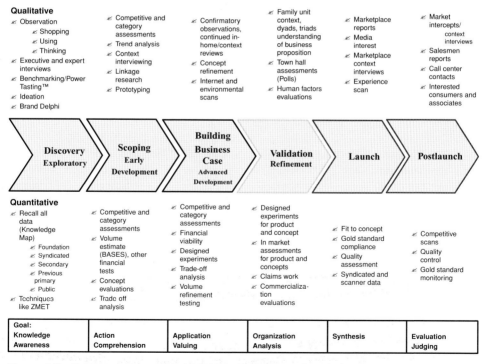

Figure 4.2   Designing qualitative and quantitative components from the beginning.

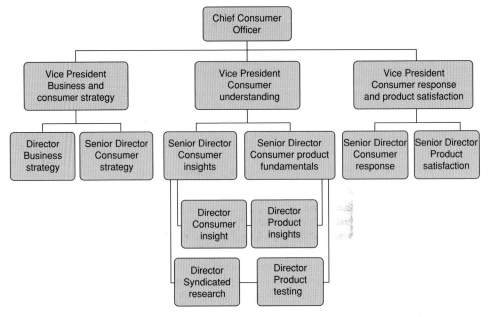

Figure 4.3   Recommended new organizational model to address the need for integration of the consumer and their needs.

(iv) As hopeful as the field of sensory evaluation has been about becoming a full partner in the research process, due to limits on product development resources, past program emphasis, and ongoing academic emphasis and training, sensory evaluation cannot drive the understanding of the whole product alone and must collaborate extensively with the consumer business focus and needs (Tassen, 1980; Holden, 1989; Mermelstein, 2001; Delwiche, 2004; Moskowitz, 2010).

(v) Hierarchy is more than nature's way of helping us to process complexity. Powerful psychological forces come into play. Hierarchies provide clear markets that let people know how fast and far they have come (Leavitt, 2003).

(vi) Creation of a new structure begins to clearly define a new way of thinking for an organization. Key in this structure is the general role of consumer understanding. Consumer understanding is the merging of market research (consumer insight), products research, (consumer and product fundaments) and sensory evaluation/sensory research (sensory measurement and its implications for monitoring design).

THE "NEW MODEL"—THE CHIEF CONSUMER OFFICER AND THE STAFF

At the time of this writing (2010), the evolution of corporations in the package goods industry suggests that it's time to take seriously the call of the consumer. We realize that a lot of companies have taken some steps, with certain of these steps fairly profound, other steps simply window dressing, and going with the "Zeitgeist."

In the creation of a chief consumer officer (CCO), a company begins to think about the *consumer*, much the same as the company thinks about marketing, product development, manufacturing, or sales. It's not a question of voice of the consumer, or insights into the consumer, or even a retitled consumer research function. In the model we propose, we underscore the need to have aspects of the organization focused on tactical issues whether they are needs of a product or trade deal. In the CCO model, this person would have responsibility for a variety of topic areas, including strategic market research, competitive intelligence, knowledge management, and methods development. The CCO model should drive core and consolidated understanding by managing all of these issues through his or her executive team. Duplication of similar fact-finding would be eliminated and a consistent voice of the consumer could be heard through this new organization.

This general idea is really not new. Moskowitz (1976), in personal communication with this author (Beckley), indicated that Erik Von Sydow, then head of SIK (Swedish Institute for Food Preservation Research), felt that this was the only way for high-caliber understanding to be created in companies. The studies that Tables 4.1, 4.2a, 4.2b, 4.2c, 4.2d, and 4.2e come from show us that there is movement in this direction.

Here is what we envision are the benefits of the structure, which in its fundamental design embraces integration of consumer and sensory research completely:

    (i) Creates a common "community" of individuals whose job it is to make the consumer real within the organization (Leavitt, 2003).

    (ii) Enables a *learning strategy* rather than a tactical strategy (Senge, 1994, 2006).

    (iii) Explores the entire product experience through the eyes of the consumer, whether that exploration entails empathic interviews, novel measurement approaches like neuroimaging, a small- or large-scale quantitative study, or a customer satisfaction intervention.

    (iv) Acknowledges the multidisciplinary needs of this area.

    (v) Hires people for roles, not jobs, allowing the organization to have enhanced flexibility as market or business needs shift (Aufreiter, Lawver, & Lun, 2000). The pool of candidates can become as broad or narrow as a company culture requires.

    (vi) More strategic use can be made of the consumer research budget when all of these groups work to cooperate for both short-term and long-term consumer understanding goals.

    (vii) Common leadership fosters common goals and knowledge sharing. The approach leads to more balanced project implementation rather than duplicated or unacknowledged recognition.

    (viii) Doer and vendor mind-sets (see Table 4.3) can now be mingled to produce a better-balanced use of resources within and outside the company.

    (ix) The hot project or program can shift as is necessary for the business need, rather than having endowed status due to one's access to a specific power broker within the organization.

    (x) Balances the needs of product development, marketing, manufacturing, and sales can all be balanced.

    (xi) Leadership would now be expected to have a generalized knowledge of all areas of consumer understanding, rather than biased or filtered knowledge based on exposure to one methodology or thinking process. Leadership would have responsibility to identify opportunities, instill accountabilities for failure to deliver promised results, create access to a broad base of inputs from the consumer marketplace, and designate a mixture of individuals who can play the most productive roles at the time.

Why do we recommend a new model? While reviewing the data and looking at the failure of change within both consumer and sensory research it became increasingly clear that it is not about an individual group of people or a profession. The models for consumer and sensory research need to change for the demand economy in which we live in today. The underlying issue for lack of change goes very deep. The issues are more political and organizational than they are functional (Joni, 2004). Furthermore, in today's marketplace no executive can be sufficiently skilled in all areas of the market mix. We know of no CTOs or CMOs who have the broad base of consumer knowledge to adequately manage this combined group. Some amount of reward and responsibility must be given to this consumer function, thereby providing that position with a focus on actionability of work. Outside of the scope of this chapter and this book, the organization in Figure 4.3 also addresses needs that arise in both the quality and customer response areas.

The excitement about this organization is that it truly gives the consumer and company a fair shot at understanding the product and the promise from all sides, not from the necessary filters required by marketing or product development, the most common "owners" of these organizations today. There is the opportunity to create a true, necessary diversity of skills, perspectives, and mind-sets. Budgets can be established to provide for the present and the future. Since this group is responsible for the management of trends and "futuring," it's easier to apply the right level of technology solutions. Furthermore, those people who are best suited for the more holistic view of product and consumers will be vested with the role of integrators, whereas there will be a number of as-needed participating specialists who might have much narrower but occasionally very relevant, specific skills (Aufreiter, Lawver, & Lun, 2000).

What about emotions? That is, in light of this new vision of the corporation, what will the feelings turn out to be among those who are already doing some of the job? Should current knowledge workers in marketing or food product design and development be afraid? We do not believe so. The new organization is not going to replace people, but rather it is going to access knowledge in a more direct way, related to end-use, rather than to being constrained by, and functions within power-sensitive corporate boundaries.

What are examples of this new knowledge accession? Now the needs that both areas require for their specific goals can be accessed in a very direct way. When purchase behavior needs to be studied early, it can be by this group, without worrying about which department will be asked to do the study, and who, as a consequence, actually will not be involved because of the existing corporate silos. When the problem that needs to be solved is best understood by testing the stimulus unbranded, then that can be done. The impeding, artificial silos of companies will be broken down. The notion that sensory evaluates everything about the stimulus but purchase intent or price, and then hands the responsibility and job to consumer research will be a thing of the past.

All too often in today's world, sensory tests everything about the product; consumer research fighting for its place at the table ends up grabbing the rest of the testing job, but not permitted to probe the product attributes because "that's sensory's job." We have here a territorial department issue, not a business, marketing, or product development issue. The new organization recognizes the need for the consumer voice throughout the product development process and product life cycle.

The new organization does not specify that one aspect of this mix is more significant or important than another. Casale (2003) explains: "It is difficult enough to disaggregate a stand-alone research objective from interconnected business dynamics. It is myopic to expect real solutions—motivating insights and ideas that can sell more products—without this holistic understanding of how acquired knowledge can impact the total value chain." (Casale, 2003).

A new world opens up after we break down the silos, and work on the business problem that increases corporate sales, not corporate egos. This new organization model allows companies to move rapidly. But it's not just companies; it's education as well. The curriculum to educate consumer, product, and sensory researchers must be restructured to address the business reality and need. As we look across food science, business, and master of business curriculums, none of these programs prepare an individual for the ability to conduct consumer, product, and sensory research today. Proper training at the university level and could help further this sector (Tuorila & Monteleone, 2008).

There is an intermediate strategy. One need not redo the entire edifice all at once. Forward-looking companies could use the same approach that industries like aluminum and electronics did for strategy; hire talent and then cultivate the remainder of the skills within the organization. We know of a few progressive companies who are doing something similar with their innovation/sensory/ product research/consumer/product development groups. The way they are "moving the needle" is through purposeful training programs that embrace the company's culture. The training forces the development teams to live the new approaches and customized the thinking and tools to achieve deeper consumer connection to product development. The development is no longer just the particular product developer but rather the developer, their product researcher/sensory/consumer insight leads, and innovation managers. With good leadership, supportive business culture, and talent, these hybrid groups are showing progress and better success with market products. It must be recognized that these are new frontiers and the models are still being built that will show us the nature of the successful ones.

This new organization tries to operate in accordance with the directive of Mahajan and Wind. These very respected academics, grounded in the discipline of marketing, recommend an integrative approach. The integrated approach gets rid of artificial walls that are so counterproductive. For example, marketing research, like many other management disciplines, have created its own silos. By arbitrarily distinguishing among types of marketing research approaches (such as qualitative vs. quantitative), and in separating themselves from other information, marketing researchers miss a huge opportunity to produce more effective results and have a bigger impact on the firm.

These divided processes need to be brought together: quantitative and qualitative; marketing research and modeling; marketing research and decision support systems; marketing research and adaptive experimentation; market research and databases; market research and other information sources—they all must expand the strategic impact (Mahajan & Wind, 1999).

THE OLD SAWHORSE: QUALITATIVE AND QUANTITATIVE

Let's face it; people like to talk to people. Stories, interviews, the human touch works again and again. We're social creatures; we learn from what others are saying, often more than we do from impersonal surveys. But there are problems. Witness some of the pundits in the field talking about focus groups, and the qualitative "feel" for the consumer:

> This is a one-trick pony. It's necessary but not sufficient to employ one qualitative methodology
> to gain deeper insights, since different perspectives add understanding to explanation.
>> Ric Casale (2003), on his concerns about focus groups

It's not my objective to eliminate qualitative exploration. What is done has value. But, I submit, it is not research. It's not the application of science to decision making, and it shouldn't be given the imprimatur of research. I'm all for qualitative explorations when they're used appropriately. But to the degree they're misused for making decisions—and almost all of them are—they are an insidious force in the marketing of goods and services that could just as readily mislead a company as give it good direction.

Achenbaum, 2001

As exciting as these tools are (referring to new online tools), they are only as good as the qualitative research consultant (QRC) who is using them.

Savin, 2009

Since 2000, one of the primary tools of qualitative research—focus groups—has come under attack (Clancy & Krieg, 2000; Letelier, Spinosa, & Calder, 2000; Pigott, 2002; Rook, 2003). In their book, *Counterintuitive Marketing*, Clancy and Krieg identify some areas that disturb them, including fuzzy focus groups, and asking people what is important. By fuzzy groups, Clancy and Krieg mean groups that are conducted without focus or with agendas that shift. Clancy and Krieger also believe that when you ask people direct questions, they provide you with top-of-mind responses that are virtually always obvious. Fuzzy can also come when different interests are dueling it out in the back room and the agreed-upon discussion topics are superseded by individual interests and needs of the moment (Clancy & Krieg, 2000).

The advice proffered by Mahajan and Wind (1999) to use both qualitative and quantitative together is the path with the greatest future. The way to integrate the two first requires that "qualitative types" ("the devil is in the details") and "quantitative types" ("if it is not a number, I do not want to see it, I am a scientist!") try to build a way to trust each other.

PERSPECTIVES ON QUALITATIVE AND QUANTITATIVE #1— SIMPLY STATED, NOT SO SIMPLY DONE

Laudable is not the same as easy to do. The merge of qualitative and quantitative is not easy, and indeed the gap between the two forms of research can be rather large. Practitioners often face the battle of "n" (numbers or base size) and "i" (insight) (Feig, 1989).

There is a way to bring these two worlds together, at least somewhat. We refer right now the process of knowledge mapping a category or subject area (see Chapter 2). The union of the two methods comes not from agreeing to agree, but from doing the work together. It is very easy to have respect for both snapshots of a subject, since they are reviewed at the same point in time and can be very useful to explaining anomalies in an area or understanding why a gap exists in a category or in knowledge.

Once respect and acceptance for each piece of the data are achieved, it is critical to view any interaction with consumers as requiring specific, *well thought out needs* for the research. The emphasis is on well thought out and on needs. Thinking means that the individuals on the team really think about what output is necessary and why, and what the "size" of that input needs to be. By size, does it really require three focus groups in each of four cities *or* an optimization study or can that need be satisfied with a much smaller piece of research or fact-finding?

PERSPECTIVES ON QUALITATIVE AND QUANTITATIVE #2— WHY ARE FOCUS GROUPS SO POPULAR?

Discussion guides for focus groups are another place for problems. Rarely does the discussion guide bring to bear much depth of knowledge about the circumstances around the research. Depending on the mind-set of the moderator, the research may be viewed as tactical (ask questions, get answers) or strategic. The issue of quality of the person conducting the interviews or involved in the interviews is also very critical. As much as it frustrates people like Clancy and Krieg, the root reasons for focus groups being used as much as they have been is that they are:

  (i) *Convenient*: Easy to convene.
  (ii) *Transparent*: Everyone thinks he or she "knows" what focus groups are and what the groups do.
  (iii) *Useful*: Professionals who are trained in marketing, marketing research, sensory, and food science think that focus groups are useful.
  (iv) *Cost effective*: There is a huge market to support these efforts. The vendor segment has a lot of capital and resources, which depend upon this business, producing an entrenched group who want to maintain the status quo.
  (v) *"Learnable"*: Due to the format for focus groups, it is reasonable that one can learn to administer a group with efficiency.
  (vi) *Busy*: They look like one is doing something, with the report as a tangible work product.
  (vii) *Face valid*: They appear to be "listening to the consumer."
  (viii) *Productive*: They appear to solve the problem of "get input from the consumer."
  (ix) *Cost effective*: They appear to be cheaper than some market research or other forms of qualitative.
  (x) *History*: They have helped successful marketing people achieve success in the past.
  (xi) *Luck of the draw*: There are ideas and insights you are bound to get from the groups.
  (xii) *Friendships*: Some of the moderators are trusted friends, business associates of the requestor, and they have the ability to help the client solve problems.
  (xiii) *Shared business intimacy*: There is a community built between the moderator and the clients; it makes business less hostile.
  (xiv) *Red carpet treatment*: People are treated special at the facilities.
  (xv) *A welcome break*: It is a way to get out of the office.
  (xvi) *Food*: They have M&M's® and sometimes chefs with great choices in menus.
  (xvii) *Entertainment*: The task of observer is fairly easy and not as hard as a lot of the other work-related tasks a marketer or product development person has to do.

When we explore this list there are some very good aspects to focus groups that help them maintain their loyal users, even when they have been identified as flawed. The most powerful elements on this list are actually the teaming aspect of getting out of the office, the ability to loosen up a bit and think, having a little fun, and meeting with people that are not part of your company. Whereas none of these seem to be on the actual task, they are very important factors for helping executives make better decisions.

So focus groups are not going away anytime soon (Last & Langer, 2003). How do we get more out of them and make them work better for quantitative? Table 4.5 is very prescriptive and designed to help foster better integration of qualitative and quantitative.

Table 4.5 Summary of an approach that designs qualitative and quantitative components from the beginning to assist in answering the same questions from different perspectives.

Research question needing to be answered	Approaches to answer questions	What is different with this approach?
1. What do we know, not know about this initiative?	**Qualitative:** Shopping experience Executive interviews Trend plotting **Quantitative:** Knowledge mapping (see Chapter 2), map review. This assumes that you have pulled together all relevant information on the product, the market, the marketplace, the consumer, and your company and that all of this information will be coming into the mapping.	Allows for collection of all information company has on the subject. Allows for consideration of information *collectively*, thereby allowing review of what barriers or filters the client/company/consultant is imposing. Although there may be one team leader, this must be a shared event. Allows team to figure out what the real question that needs to be answered is *before* too much has been done. Forces company to consider what has been done before. Focuses everyone, attempts to get everyone on same page. Highlights areas that lack consensus and will provide for flawed execution later if not dealt with now.
2. Given the knowns and gaps we have identified, why do we not know this information?	**Qualitative:** Focus on probing, not measuring. Use observation, theoretical model building, and real world constructs (qualitatively map the product space, qualitatively benchmark). Work iteratively, keep building and refining. Can use consumers or insight/experience panels to assist in this effort. Look to confirm or disconfirm findings in the knowledge map. Have category/subject matter experts review map and theoretical models to confirm or disconfirm assumptions.	There is often a rush to get quantitative data or numerical measures (concept scores, focus groups, descriptive analysis data). Rushing substitutes activity with thinking and reframing the questions.
3. Do we understand what we are trying to make and sell? Do we understand the fundamentals that will drive excitement or lack thereof for our product? Do the ideas work with the product protocepts? If not, why not? If so, why so? Look from consumer perspective and from product/brand perspective.	**Qualitative:** Start laying out the sequences of beliefs now. Not before. Do context interviews, do a lot of small-scale work that *builds* on the knowledge. Work with protocepts *and* concepts. **Quantitative:** Do Internet screens or small conjoints to understand the fundamental concept triggers, words, and images. Refine, iterate rapidly.	You are still open to adjusting the product, the concept, and the questions/needs. This is shared responsibility between marketing, development/research, and finance. *Shared* is the operative word.

Table 4.5 (*Continued*)

Research question needing to be answered	Approaches to answer questions	What is different with this approach?
4. What will it take to sell this? What do people want to pay/will they pay for this?	**Quantitative:** Start running the financials and have these folks work through your map and your learning so they are connected to the concept and can adjust financial filters to be more or less conservative.	Provides a context for the financial numbers and ability for the whole team to dialog with finance on the trade-offs and assumptions working against the entire knowledge base, not a few assumptions.
5. Given this background, have we made this "product" idea before, and if so where, when, why? Why are we making it again?	**Quantitative:** Begin to do some of your traditional work, quantitative category appraisal, if necessary and not covered by work done in #2 through qualitative measures. Early stage designed experimentation should begin for PD and manufacturing. More assessments of the marketplace and concept/price/product fit will continue.	Focuses on reflecting on what has been discovered and requires that the same past is not repeated. Requires focus on the product/concept interface before a lot of product development effort is expended.
6. Can refinement begin? Are we ready to start refining the idea and developing the product to fit the purpose, concept, and need?	**Qualitative:** Where necessary, continue contact with consumers. During this phase "town hall" or "game show" formats work well and allows for interactive polling and feedback and efficiently allows interactive surveying. **Quantitative:** Use what is part of your regular PDP process but look to see that the tools are grounded in consumer fundamentals and not hybrids that disconnect the product from the marketing promise. Can be implemented through contextual location testing, traditional CLTs, and in-homes.	This step is generally what companies do for their product development efforts. Some of the qualitative techniques allow for more efficient and direct contact with consumers.
7. We are starting to scale-up, commercialize; are there any deal breakers appearing? Is the product and package delivering on the brand promise?	**Qualitative:** This is where the focus can shift from PD to sales. Now should begin a dialog with sales force, brokers, and field to make sure that product fit/concept works for marketplace. Run an experience screen to see if key proposition points continue to be valid and have met the business criteria **Quantitative:** Consumer defined specifications, quality assurance specifications, and product documentation.	Invites the sales component in to provide insight and understanding of what they are perceiving in the field and to confirm the behavioral/observed/evaluated components of the product and the business proposition.

Table 4.5   (Continued)

Research question needing to be answered	Approaches to answer questions	What is different with this approach?
8. Launch has started; are we making what we said we would make? How do people feel about the product?	**Qualitative:** Can repeat the efforts in #2 if goals/budget work for this part of the effort. Use the consumer response area to collect important feedback and be ready to respond and confirm or disconfirm issues via more consumer contact. Observe the marketplace and see if the product/market plan is working given the assumptions made during the early phases. **Quantitative:** Quality audits, scanner and marketplace data, and strength data.	Direct consumer contact is designed in to supplement the regular quantitative assessments. This provides the important early feedback loop that the product and proposition are working or need to be tweaked.

CLT, central location test; PD, product development; PDP, product development process.

In addition to respecting the perspectives of both types of thinking, a critical need is to recognize that consumers can articulate very well a lot about what they need and want. The big issue is being ready to hear what they say and realizing some of the messages we have heard about inarticulate consumers and untrained panelists does not square with the research (Griffin & Hauser, 1993; Moskowitz, 1996; Hughson & Boakes, 2001; Zaltman, 2003).

Let's clarify a few points. Research shows us that "qualitative" means focus groups to most people (Brandweek, 2003; Cardinal *et al.*, 2003; Greenbaum, 2003; Wasserman, 2003). For the purposes of this section, *this is not* what we mean:

(i) By qualitative, we mean specifically any set of tools that focus on the *quality* of the interaction between the researcher and the person (consumer). These tools are designed to collect information that is not completely duplicated by any measure that we have today, which we would label "quantitative."

(ii) By quantitative, we mean specifically an interaction with people where the quality of the action is less important and is traded off with collection of quantity of responses. The conclusions from quantitative are perceived to have greater validity because they are numerical and in greater *quantity.*

PERSPECTIVES ON QUALITATIVE AND QUANTITATIVE #3— MARRYING THEM OR LEAST JOINING THEM TOGETHER

Starting out with a base of respect discussed earlier, plan ahead for any project for how the two methods will support each other. Decide ahead of time what one believes will be the individual contributions of the qualitative and the quantitative portions. Plan on doing both, rather than turning the planning into a face-off or a confrontation. (Some call this approach "mixed method" testing.)

Knowing ahead of time what one will do with the data is always helpful, and here with qualitative and quantitative even more so. Develop a *plan that integrates* both parts to solve the business and development questions that must be understood. When clear understanding of the purpose of the research cannot be articulated and therefore implemented in a clear, rational way, then *no* testing should take place. It is unfair to people you will invite into your testing situation to conduct research with them that is essentially an ill-planned fishing expedition when you do not even have a clue as to what the real question (Why?) happens to be that motivates the research.

At this point, many who are reading this are saying, "Wait, I never do research that is not well thought out." Well, a lot of people do, and many of these fishing expeditions are responsible for poor qualitative reactions and projects that end up with lower success than is necessary (i.e., some of the seven out of ten products that fail in market).

Viewing an entire project as a science experiment that builds in an interactive fashion is very healthy for the outcomes. Table 4.5 is an example of a plan that integrates qualitative and quantitative from the initial stages. Notice that the outcomes are anticipated and are viewed like a science project with hypotheses so that the entire process looks to confirm or disconfirm changing hypotheses as the process continues.

PERSPECTIVES ON QUALITATIVE AND QUANTITATIVE #4— SOME GUIDELINES

For food product design and development today to be more innovative and truly include the voice of the customer, it is important for food product design to have more complete knowledge and greater impact with functions that were originally considered exclusively marketing. The reason is the need to have product differentiation that is more compelling and more ownable than anyone else. This is only achieved by an active, engaged product development area that knows the limitations and benefits of different consumer and product strategies.

With the criticality of understanding again underscored, here then are 17 guidelines, or at least thinking points. Consider these to give you the knowledge you need when you link together qualitative and quantitative in your next research project:

   (i) Know where your research is starting from—really new, new, established, confirmatory, etc.
   (ii) Provide context where possible; this facilitates more naturalistic responses and thinking patterns.
   (iii) Think about amount of depth you should have for the qualitative.
   (iv) Be balanced.
   (v) Know the reason for qualitative and what forms of quantitative naturally emerge from the specific qualitative selected. Not everything is connected. There should be a flow.
   (vi) Understand consumer's point of view—group consensus versus individual.
   (vii) Work all components of the product experience. Identify and work with as many relevant avenues as is possible.
   (viii) Go to the story. It's not just factoids that you see and observe, but strive for coherence—recency, memorability (metaphor is memory), stimulus response.
   (ix) Focus on questions to ask.

(x)   Design impeccably, in order to control or eliminate, or perhaps even just balance out context, bias, psychological factors.

(xi)   Start with a hypothesis in mind—you're going on a journey, not a meandering walk.

(xii)   Use tools to understand potential, belief path. Find out the tools that you can use, and try them out.

(xiii)   Incorporate stimulus response. Look for relations between and among variables.

(xiv)   Analyze your own data and thinking. Insert a check on your judgment, checks along the way for your filtering, teams. Confirm/disconfirm/check/balance/recheck.

(xv)   Confirm, clarify, expand with trade-off analysis and individual subject analysis.

(xvi)   Use value diagrams as a tool to understand relationship among all aspects as you continue through phases of research. Write it down and formalize. It will force you to be disciplined.

(xvii)   Be aware of impact of context, understand implications of sensory factors.

WORKING EFFECTIVELY IN A FUTURE OF SCARCE RESOURCES

There are many factors that are driving value issues within food product design and development. Key market factors are: a high number of value-oriented consumers and retailers (Haden, Sibony, & Sneader, 2005; Hale, 2005), raising costs for many items that one needs to support a household today (Weston, 2004; US Department of Labor, Bureau of Labor Statistics, 2005), rapid changes in the marketplace (Levy, 1999), and fierce competition. These events have led to business process hurdles that need to be considered and addressed. The hurdles affect learning (knowledge), people, and organizations, respectively.

Getting knowledge, staying up to date, being effective

In a hypercompetitive business world that is constantly "on," it is very hard to stay current and contemporary. Most editors of professional journals periodically quietly ask themselves, is anyone reading what we publish? Senge addresses this impatience with learning by suggesting that if calculus had to be introduced today, it would be a failure since it could not be learned in a 3-day training course! (Senge, 1994).

Discipline is required for professionals who try to stay current and contemporary. A number of systems, or at least "rules of thumb" are called for. They instill this discipline through their regularity:

Daily

(i)   Set aside 30 minutes to review all newsletters, trend letters, and other professional materials you subscribe to or your company provides.

Weekly

(ii)   Identify four newspapers that you feel have broad range of perspectives and scan them three times per week.

(iii)   Choose a set of periodicals from different sources (electronic, print, DVD, etc.) that you must review.

Monthly
  (iv) Use Amazon or another bookseller who will send you ideas about books that are related to topics you find interesting. Define a budget for book purchases and buy at least five per year. Some professional memberships offer this as a premier benefit and it is another way to stay informed.
  (v) Identify three individuals with whom you can interact. The three individuals should provide opinions and thoughts that expand, differ, or complement you and your company's thinking. Look for individuals who bring diversity to your thoughts. If you have the budget, figure out how to "fund" this activity through meals with these people, small consulting activities, or some other way to get the interaction.

Quarterly
  (vi) Have a periodical and reference material stack or section in your office, study, or home, and go through it quarterly. Read every cover and figure out which articles must get into your reading stack.

Half yearly
  (vii) Twice per year go outside of the range of periodicals and professional journals you read and spend 4 hours seeing how other fields deal with similar topics.
  (viii) Look at courses and seminars offered both in and outside of your core interest area and note what is continuing as a theme and what are new approaches to thinking that surprise or startle you.

Yearly
  (ix) Develop a program to get a broad range of feedback from a collection of advisors. When funds allow, bring them into your office and meet with your staff for the gathering. When funds are scarce (Sharp, 2001), stay connected and have this organized as a phone conference with an agenda, and so forth. Both you and your advisors will benefit from this interaction and feedback. Depending on your budget, you may or may not be able to pay them, but giving people permission to tell you things about your business that they observe can be highly productive for ongoing quality improvement.

Lifetime
  (x) Embrace a mind-set that welcomes change, embrace a personal way of thinking that ensures that you continue to broaden your perspective about subjects and your world. This mind-set of welcoming change is absolutely critical to keeping ideas and programs fresh. This is your job. This takes work and planning. We observe that about 10% of the professionals engage in this behavior. Those who do appear to enjoy a much broader perspective regarding the opportunities and issues with which they must deal as professionals.

MANAGING KNOWLEDGE IN A WORLD OF EXPONENTIAL GROWTH OF FACTOIDS, INFORMATION, OPINIONS, WORLD VIEWS

The Sisyphean problem of trying to understand what is the "best" knowledge today is complex. So much more information is available than in the past, in so many channels, promoted, hawked, even screamed by voices, some more believable, some less. Managing the knowledge stream and understanding that human behavior will direct each of us to try to filter information. Reducing complexity is important. It shows what can be truly breakthrough contemporary knowledge. Just because one does not understand an idea immediately is no reason to dismiss the idea as unworthy.

The answer is really simple; many roads lead to knowledge, one road just might lead to ignorance and error. Donald Fiske devoted his life to trying to help people conduct better research. His classic paper on methods for social science advocated use of multiple methods in research (Campbell & Fiske, 1959), with a need to assure convergent and discriminant validation. There's more to the story. There's the angle of reality. Talking specifically about short courses and best practices, but really aiming at the world of the knowledge worker, Preston Smith summarizes knowledge acquisition well:

> To me, such (executive) training provides value only when the participant activity initiates improvement in their approach to product development as a result of the training. That is, 'compelling concepts,' 'best-practice tools' and such are of no value until the participant applies them.
>
> Rosenau, 2002

MIND-SETS OF PROFESSIONALS

The sensory function in a company is typically allocated smaller budgets than the sister discipline, market research/consumer insight side of business. Sensory's work, and indeed its self-promoted world view, is that its work is tactical, driven by a mentality of do it yourself. That world view did not come about by accident. It came about due to evolution and deliberate strategic thinking early on in the game, in the 1950s–1970s. Sensory is about the heroes search, the validation of oneself, earning a seat at the table while bringing good value to the company. The desire to do things oneself (for economic and/or personal reasons) comes from the reactions to and by other professionals in the sensory community, close associates, and teachers.

Market research is about consultation, needed budgets to "purchase" knowledge or assistance, and seeing oneself as the vehicle for achieving results, not doing the task exclusively, that is, not proclaiming oneself as the "low cost supplier" competing with outside vendors. As a result, we see that there is a bigger arena of people who educate and practice in the market research side of the business as compared to sensory, predominately because of the economic resources that are made available.

There is a story that is told by sensory scientists, perhaps one that constitutes the field's own "urban legend." The story is that sensory scientists are individuals who have a broad range of educational and career backgrounds. They are told that they should market themselves as an interdisciplinary function with applications throughout the product development and evaluation cycle. Added to that, they are encouraged to emphasize how sensory evaluation principles are applied by market researchers in their product tests, home economists in the product showings, and sensory scientists in all their work (Westerman, 1989). And this author (Beckley) agrees. Yet fundamentally, the field is defined by some test methods and procedures. (Lawless, 2009). In an era when CEOs of businesses are needful of more strategic thinking and better problem-solving, few sensory scientists are sitting at the senior management table. What's wrong with this story, this picture?

In Westerman's well-written article in *Sensory Forum*, written more than 20 years ago, it was suggested that the best place to start to market sensory was within senior management of R&D. The promise of that suggestion at the beginning of the 1990s had really not occurred. Of course there has been some progress. There are more members of the sensory science field that sit as directors, vice presidents, and senior executives today than at the beginning of 1990. There are several members of this professional group who own and operate successful firms that market sensory, market research, or business strategy services. But the hoped-for evolution of the sensory professional into a status equal to the market researcher just has not happened:

> We are seeing an evolution in market research, where greater participation in strategy discussions has begun. We are seeing the evolution of the product researcher who is a valued component of the discussion about successful product design that may win hearts and minds if it is communicated properly. It has occurred in market research by bringing a large number of nonmarket research people into business organizations (Vence, 2004). Is this a model that should be adapted for the product development/sensory evaluation community?

THINKING, DOING, SUCCEEDING: WHAT HAS TO BE DONE TO "GET IT RIGHT?"

Market, product, and sensory research ends up be a thinking person's game. There are tactical aspects—the tests, the surveys, the data summarization. All too often they overwhelm the more quiet aspect of thinking. Also in the dance that is service versus line, the businesspeople or food product design and development, people feel they have already thought about the subject. Now, or so they feel, it's time to just *go do*. This is a bad idea. It helps perpetuate a problem of thinking action is what to look for within sensory or market research functions. When the measure of a sensory or market research group is based on number of tests per year or number of subjects per test at a given price point, companies that practice that approach are on the path of lower product quality, poor business thinking, and lower benefit from new and innovative product launches.

Currently there are programs in place in most major companies that are starting to train people about diversity—in ethnicity, gender, people choices—and the way people think. As companies try to do more with less, it is critical to move from a mind-set of command and control (Go Do!) to one of empowered individual thinking. Individuals such as Marcus Buckingham and Donald Clifton and the Center for Creative Leadership advocate diversity and empowerment of thinking (Buckingham and Clifton, 2001; see also www.ccl.org). Yet, again and again, the available data show that as humans we tend to gravitate toward what is familiar (tasks and people who we believe think and act like us) and those people who we feel we can trust.

When resources are scarce, a company benefits from a process that allows a group of thinkers with mixed viewpoints and skills to solve problems. Better solutions arise from that approach. In "The Geography of Trust," Joni presents a scalar approach for managing different types of "trustees" that allows a manager to approach all relationships (business, family, and friends) with a more realistic and grounded strategy. In her article, Joni explores six classes of advisors and breaks the structure down on the basis of the following:

(i) What each group offers.
(ii) What the relationship needs.
(iii) What structural level of trust is achievable, for it is trust, as we will see, that is the ultimate enabling or disabling agent.

Table 4.6   Market research: Past and future.

Old paradigm (not gone)	New paradigm (exists to a small extent)
Define research budgets	Define strategic plans
Establish benchmarks	Understand consumers as people
Say that you are getting face-to-face with the consumer	Get face-to-face with implications
Generate lots of data	Generate appropriate amount of relevant data
Focus on research objective	Focus on underlying business need
Work to build brand image and brand/product preference	Discover the knowledge that will sell and sustain sales and consumer relationships
Insights by the pound	Motivating insights

Source: Beckley and Casale, 2003.

Joni's approach appears to have great merit since it allows a much larger, yet structured approach to managing different people or companies. When the goal is to have a broad base of advisors to allow more diversity of thinking, implementation of a systemic approach described by Joni has merit (Joni, 2004). Furthermore, Joni's approach increases the opportunity for concerned managers to understand how arbitrary their advisory base might or might not be in areas such as preferred suppliers and priority vendors.

ORGANIZATIONS: INTELLECTUAL HONESTY

Many of the factors that provided the illusion or perhaps reality of a constant, reliable marketplace disappeared toward the end of the twentieth century. The consequence has been significant, often major, and occasionally disturbing changes in the marketplace (Kash, 2002). Choice (too much) and value (craving for) are issues that are topical (McDonough & Braungart, 2002; Silverstein & Fiske, 2003; Hurd & Nyberg, 2004; Levy, 2010). However, new approaches and new thinking are being introduced, not much appears to have changed with respect to how products are evaluated in the end or how we assess success prior to the numbers in the market. How can this be?

What is the role of sensory in this change? In fact, does sensory even figure in this change, or does sensory simply remain a service organization, happily providing tests as a "Tests R Us" capability. Some suggest that the stakes are fairly low when it comes to the work sensory does, whereas at the same time they suggest the stakes could be fairly high today for market researchers (Table 4.6).

TRUST—A NEW BOTTOM LINE

Whereas preferred supplier relationships have become valuable assets for suppliers and effective resource management tools for companies, inability to commit honesty (inside and outside a firm) will eventually impact the quality of products, employees, and external relationships. It is trust that is the new bottom line that will drive innovation.

Today there is a lack of trust, compounding the aforementioned problems in organizations. It's getting increasingly more difficult for people within organizations to feel they work in environments that are trustworthy (Buckingham & Coffman, 1999; Joni, 2004). The stress of the business environment today is believed to have a large impact on how much an individual can maintain his or her integrity (Shea & LeBourveau, 2000), the consequence of the stress is the disappearance of truly honest and often profoundly productive relationships with bosses, peers, subordinates, and suppliers. Williams (2004) suggests that areas that undermine relationships over time are:

(i) *Breach of trust*: Sharing competitors' plans, failing to live up to commitments, generally not meeting expectations for fairness and straightforwardness.
(ii) *Poor communication*: Late or incoherent information, poor listening, failure to build understanding.
(iii) *Mixed messages*: Lack of functional alignment, cherry-picking projects.
(iv) *Commoditization*: Disincentive to add value, works against loyalty.
(v) *Win–lose thinking*: Dialog must be two-way or at some point relationships fail.

When all is said and done, therefore, and at the end of the day, innovation turns out to be about people working in organizations, giving of their talents, and being open to new ideas and new opportunities. Sensory has a long way to go in this regard. The voice of the consumer will be heard much better in an environment that appreciates its workers, understands the diversity of talents, and is ready to move from the status quo to the next profitable opportunity. In succeeding chapters, we will show the methods, now that we have dealt with the structure, the people, the egos, and the preconceptions.

REFERENCES

Achenbaum, A. 2001. When good research goes bad. *Marketing Research* 13(4):13–15.
Ashman, H. & J. Beckley. 2001. *Sensory and Professionalism.* Available from the author and also at www.sensory.org.
Ashman, H. & J. Beckley. 2003. *Sensory and Professionalism II.* Available from the author and also at www.sensory.org.
Ashman, H., S. Rabino, D. Minkus-McKenna, & H.R. Moskowitz. 2003. The shopper's mind: What communications are needed to create a "destination shopping" experience? Proceedings of the 57th ESOMAR Conference "Retailing/Category Management—Linking Consumer Insights to In-Store Implementation." Dublin, Ireland.
Aufreiter, N., T. Lawver, & C. Lun. 2000. A new way to market. *The McKinsey Quarterly* (located on March 19, 2004, at http://www.marketingpower.com/live).
Barabba, V. & G. Zaltman. 1991. *Hearing the Voice of the Market: Competitive Advantage through Creative Use of Market Information.* Boston: Harvard Business School Press.
Beckley, J. 2004. *Integrating Sensory and Marketing Research Impact.* Available from the author or at sensory@yahoo.com.
Beckley, J. 2008. Linking consumer insight with sensory, material, and formulation science to develop consumer preferred dairy textures. Presented at the Annual Meeting & Food Expo for Institute of Food Technologist. New Orleans, June–July, 2008. Available from the author or at sensory@yahoo.com.
Bond, E.U., III & M. Houston. 2003. Barriers to matching new technologies and market opportunities in established firms. *Journal of Product Innovation Management* 20(2):120–135.
Bradley, S.P. & R.L. Nolan. 1998. *Sense and Respond.* Boston: Harvard Business School Press.
Brandweek. 2003. Letters to the Editor. vol. 44, 43. November 24:14.

Buckingham, M. & D. Clifton. 2001. *Now, Discover Your Strengths.* New York: Simon & Schuster.

Buckingham, M. & C. Coffman. 1999. *First, Break All the Rules.* New York: Simon & Schuster.

Campbell, D.T. & D. Fiske. 1959. *Convergent and discriminant validations by the Multi-trait-multi-matrix method. Psychological Bulletin* 56:81–105.

Cardello, A. 2003. Idiographic sensory testing vs. nomothetic sensory research for marketing guidance: Comments on Garber *et al. Food Quality and Preference* 14:27–30.

Cardinal, P., A. Flores, A. Contarini, & G. Hough. 2003. Focus group research on sensory language used by consumers to describe mayonnaise. *Journal of Sensory Studies* 18:47–59.

Casale, R. 2003. It's time to retool consumer research. *Brandweek* 44(40):24.

Cialdini, R. & K. Rhoads. 2001. Human behavior and the marketplace. *Marketing Research* 13(3):8–13.

Clancy, K.J. & P.C. Krieg. 2000. *Counter-Intuitive Marketing.* New York: Free Press.

Clancy, K.J. & P.C. Krieg. 2001. Surviving death wish research. *Marketing Research* 13(4):8–12.

Collins, J. 2001. *Good to Great: Why Some Companies Make the Leap and Others Don't.* New York: Harper-Collins.

Cooper, R. 2008. The Stage-Gate idea to launch process-update, What's new and NexGen systems. *J. Product Innovation Management* 25(3):213–232.

Delwiche, J. 2004. Some sensory. *Sensory Evaluation Division of IFT Newsletter* March:1.

Drucker, P. 2001. *The Essential Drucker.* New York: HarperCollins.

Feig, B. 1989. Wrestling with research. *Food and Beverage Marketing* June 14.

Frazier, M. 2007. Hidden Persuasion or Junk Science. *Advertising Age.* September 10:1.

Fox, J. 2009. *The Myth of the Rational Market.* New York: Harper Business.

Garber, L.L., Jr., E.M. Hyatt, & R.G. Starr, Jr. 2003. Measuring consumer response to food products. *Food Quality and Preference* 14:3–15.

Gladwell, M. 2005. *Blink. The power of thinking without thinking.* New York: Little, Brown.

Greenbaum, T. 2003. The gold standard. *Quirk's Marketing Research Review* 17(6):22–27.

Griffin, A. & J. Hauser. 1993. The voice of the customer. *Marketing Science* 12(1):1–27.

Grunert, K.G. 2003. Purchase and consumption: The interdisciplinary nature of analyzing food choice. *Food Quality and Preference* 14; 39–40.

Grunert, K.G., B.B. Jensen, A.M. Sonne, K. Brunso, D.V. Byrne, C. Clausen, A. Friis, L. Holm, G. Hyldig, N.H. Kristensen, C. Lettl, & J. Scholderer. 2008. User-oriented innovation in the food sector: relevant streams of research and an agenda for future work. *Trends in Food Science and Technology* 19(11):590–602.

Haden, P.D., O. Sibony, & K.D. Sneader. 2005. New strategies for consumer goods. *McKinsey Quarterly,* January 10.

Hale, T. 2005. Winning retail strategies start with high value consumers. *AC Nielsen,* Spring.

Hine, T. 2002. *I Want That! How We All Became Shoppers.* New York: Harper Collins.

Holden, D. 1989. The 1990's: A decade of growth or decline for sensory evaluation? *Sensory Forum* (A Newsletter of the Sensory Division of IFT) 46.

Hughson, A.L. & R.A. Boakes. 2001. Perceptual and cognitive aspects of wine expertise. *Australian Journal of Psychology* 53(2):103–108.

Hurd, M. & L. Nyberg. 2004. *The Value Factor.* New York: Bloomberg Press.

Joni, S.A. 2004. The geography of trust. *Harvard Business Review* March:82–88.

Kahn, K., F. Franzak, A. Griffin, S. Kohn, & C.W. Mill. 2003. Editorial: Identification and consideration of emerging research questions. *Journal Product Innovation Management* 20:193–201.

Karlsson, C. & P. Ahlstrom. 1997. Perspective: Changing product development strategy—a managerial challenge. *Journal of Product Innovation Management* 14:473–484.

Kash, R. 2002. *The New Law of Supply and Demand: The Revolutionary New Demand Strategy for Faster Growth and Higher Profits.* New York: Doubleday.

Khurana, A. & S.R. Rosenthal. 1998. Towards holistic "front ends" in new product development. *Journal of Product Innovation Management* 15(1):57–74.

Kindle, L. 2000. Liking sensory patterns. *Food Processing Magazine* November:69–70.

LaSalle, D. & T.A. Britton. 2003. *Priceless: Turning Ordinary Products into Extraordinary Experiences.* Boston: Harvard Business School Press.

Last, J. & J. Langer. 2003. Still a viable tool. *Quirk's Marketing Research Review* 17(11):30–39.

Lawless, H. 2009. Sensory Evaluation from the perspective of a textbook author. Presented at 8th Pangborn Sensory Science Symposium, July 30. PL5.1

Leavitt, H.J. 2003. Why hierarchies thrive. *Harvard Business Review* March:96–102.

Letelier, M.F., C. Spinosa, & B.J. Calder. 2000. Taking an expanded view of customer's needs: Qualitative research for aiding innovation. *Marketing Research* 12(4):4–11.

Levy, P. 2010. How to reach the new consumer. *Marketing News*. February 28. 16–20.

Levy, W. 1999. Beware, the genie is out of the bottle. *Retailing Issues Newsletter* 6(6):1–4.

Liefeld, J. 2003. Consumer research in the Land of Oz. *Marketing Research* 15(1):10–15.

Longinotti-Buitoni, G.L. 1999. *Selling Dreams: How to Make Any Product Irresistible*. New York: Simon & Schuster.

Mahajan, V. & J. Wind. 1999. Rx for marketing research. *Marketing Research* 11(3):7–13.

Marshall, D. 2003. Commentary on Garber *et al.*, measuring consumer response to food products. *Food Quality and Preference* 14:17–21.

McDonough, W. & M. Braungart. 2002. *Cradle to Cradle*. New York: North Point Press.

Meiselman, H.L. 1992. Methodology and theory in human eating research. *Appetite* 19:49–55.

Mermelstein, N. 2001. Top executives analyze food R&D in 2001 and beyond. *Food Technology* 55(9):36–58.

Miller, D. 1998. *A Theory of Shopping*. Ithaca, NY: Cornell University Press.

Moskowitz, H.R. 1976. Personal communication to the author.

Moskowitz, H.R. 1995. Advice to young researcher. In: *Consumer Testing of Personal Care Products*. New York: Marcel Dekker.

Moskowitz, H.R. 1996. Experts versus consumers: A comparison. *Journal of Sensory Studies* 11:19–37.

Moskowitz, H.R. 2003. When bad data happen to good researchers: A contrarian's point of view regarding measuring consumer response to food products. *Food Quality and Preference* 14:33–36.

Moskowitz, H.R. 2010. *You! What you MUST know to start your career as a professional*. S. Charleston, SC: CreateSpace.

Paredes., D., J. Beckley, & H. Moskowitz. 2008. Bridging Hedonic and Cognitive Performance in Food and Health and Beauty Aide (HBA) Products. *Society of Sensory Professionals*. Cincinnati, OH. November.

Pearce, J. 1980. Sensory evaluation in marketing. *Food Technology* 34(11):60–62.

Pigott, M. 2002. Looking beyond traditional methods. *Marketing Research* 14(3):8–11.

Pine, B.J., II & J.H. Gilmore. 1999. *The Experience Economy*. Boston: Harvard Business School Press.

Plummer, J. 2003. The co-creation of demand. Presentation at Product Development & Management Association (on behalf of Gerald Zaltman). December 9.

Rein, G.L. 2004. From experience: Creating synergy between marketing and research and development. *Journal of Product Innovation Management* 21(1):33–43.

Rogers, P. 2000. Getting to know you. *Prepared Foods* June:15–19.

Rook, D.W. 2003. Out-of-focus groups. *Marketing Research* 15(2):10–15.

Rosenau, M. 2002. From experience: Teaching new product development to employed adults. *Journal of Product Innovation Management* 19:81–94.

Roston, E. 2004. The why of buy. *Time*, April.

Rousseau, F., K. Fauchereau, & J.-P. Dumont. 2001. People's food liking does not warrant consumer choice. Delivered at 4th Pangborn Sensory Science Symposium, 22–26 July; Dijon, France.

Savin, A. Have technology, will investigate. *Quirks' Marketing Research Review* December, 2009. Volume XXIII Number 12:24–27.

Senge, P. 1994. *The Fifth Discipline: The Art & Practice of the Learning Organization*. New York: Currency Doubleday.

Senge, P. 2006. *The Fifth Discipline: The Art & Practice of the Learning Organization*. New York: Currency Doubleday. Revised and Updated.

Sharp, S. 2001. Truth or Consequences: 10 Myths That Cripple Competitive Intelligence. www .marketingpower.com.

Shea, C. & C. LeBourveau. 2000. Jumping the Hurdles of Marketing Research. www.marketingpower.com.

Silverstein, M. & N. Fiske. 2003. *Trading Up: The New American Luxury*. New York: Penguin.

Simester, D. 2004. Finally, marketing research you can use. *Harvard Business Review* 82(2):20–21.

Speece, M. 2002. Top of mind marketer's malady: Fear of change. *Brandweek* August 19:34.

Stanton, J.L. 1997. Who's no. 1? Consumers! *Food Processing* December:55–57.

Stone, H. & J. Sidel. 1985. *Sensory Evaluation Practices*. San Diego, CA: Academic Press.

Switzer, L. 2005. *Radically Accelerate and Improve New Product Development for Maximum Impact*. PDMA Washington, DC. November 15.

Tassen, C. 1980. Sensory evaluation in research and development. *Food Technology* 34(11):57–59.

Teratanavat, R. & M. Jeltema. 2008. An approach to integrate long-term behavioral measures to identify opportunities for new products. *Society of Sensory Professionals*. Cincinnati, OH. November.

Trout, J. 2000. *Differentiate or Die: Survival in Our Era of Killer Competition*. New York: John Wiley & Sons.

Tuorila, H. & E. Monteleone. 2009. Sensory food science in the changing society: Opportunities, needs and challenges. *Trends in Food Science and Technology* 20(2):54–62.

Twitchell, J.B. 1999. *Lead Us into Temptation: The Triumph of American Materialism*. New York: Columbia University Press.

Underhill, P. 1999. *Why We Buy: The Science of Shopping*. New York: Simon & Schuster.

US Department of Labor, Bureau of Labor Statistics. 2005. *Consumer Price Index*. July 18.

Van der Bij, H., M. Song, & M. Weggeman. 2003. An empirical investigation into the antecedents of knowledge dissemination at the strategic business unit level. *Journal of Product Innovation Management* 20(2):163–179.

Vence, D. 2004. Better! Faster! Cheaper! Pick any three. That's not a joke. *Marketing News* February 1:1.

Wansink, B. 2003. Response to measuring consumer response to food products: Sensory tests that predict consumer acceptance. *Food Quality and Preference* 14:23–26.

Wasserman, T. 2003. Sharpening the focus. *Brandweek* 44(40):28–32.

Westerman, K. 1989. Sensory professionals, sensory resources—beyond the laboratory. *Sensory Forum* (A Newsletter of the Sensory Division of IFT) 44.

Weston, L.P. 2004. 25 Items—Butter to Bikes—Soaring in Cost. www.msn.money.com.

Williams, D. 2004. Partnering for successful product development. *Food Technology* 56(11):28–32.

Young, C. & A. Shea. 2010. The fundamental things apply here, too. *Quirks Marketing Research Review*. 19(4):38–43.

Zaltman, G. 2003. *How Customers Think*. Boston: Harvard Business School Press.

5 A process to bring consumer mind-sets into a corporation

While social networking has continued to drive the consumer and personal lives deeper into the mind-set of a company, there continues to be a "*disconnect*" in today's business practice. Although companies and employees say they want to interface with consumers and customers, often the relationship continues to be static. Creating and using a seamless *connection* with the marketplace is a laudable goal. This chapter shows how to connect with the mind of the consumer using approaches that build on both personal and technology-based advances.

The following quote suggests that the way things work in business occurs more often than one wishes:

> Work is organized a little like the court of Louis XIV, very complicated and very ritualized so that people feel they are working effectively when they are not . . . corporate culture is nothing more than the crystallization of the stupidity of a group of people at a given moment.
>
> Translated excerpt from Corinne Maier's book *Bonjour Paresse*, from Smith (2004), discussing her runaway success with a book that discusses the state of work in France

Historians feel one does not have a good idea of what is really going on with a situation until it is 50 or so years in the past. The world has changed. Was it Wal-Mart, the Internet bubble, the last Great Depression, or technology? The economic downturn after 9/11, 2008, or global terrorism? Moore's law being applied everywhere? Probably all of these and then some.

With economic changes and challenges in recent years, we are still trying to figure out how to function in a demand economy as compared to a supply economy. From a consumer perspective this economy looks like every child's dream come true—along with the scary stuff under the bed. Lots of choices. But with lots of choices come lots of trade-offs to consider. For the manufacturer, there are no guarantees, and if one doesn't differentiate, one might well die. So along with managing the entire supply chain, rapidly chasing "new news" for differentiation, the enhanced needs of the value-seeking consumer/customer, survival becomes a basic company need. This is the reality, raw and unpleasant, but vital for the professional who wants to understand the consumer, the product, and the development opportunities.

Sensory and Consumer Research in Food Product Design and Development, Second Edition.
Howard R. Moskowitz, Jacqueline H. Beckley, and Anna V.A. Resurreccion.
© 2012 Blackwell Publishing Ltd. Published 2012 by Blackwell Publishing Ltd.

Business does not, and in fact cannot, respond particularly rapidly because few people recognize all that quickly what has actually happened. For every example of fluid information flow (like the phenomenon of mobile devices in Iran in 2009), we have examples like the financial tsunami of 2008, which was ignored for 2–3 years before the calamity. Additionally, business and people processes take time to adjust. There was, of course, a lot of information to presage that tsunami, but no one seemed to be paying attention, or perhaps no one cared. In the words of Voltaire, everyone was "tending his garden."

The interconnected world is helping to push this faster—yet business practices inherently slow things down. In Evans and Wurster's book *Blown to Bits* (2000), they very clearly suggest that the "ten rules for succeeding in the information economy" will disappoint, since the task of rethinking strategy is specific to each business and cannot be short-circuited by simplistic formulas. We found that out on Wall Street, USA!

Technology tools abound. But let's look at what might be going on daily in a major food company and see exactly what the much-heralded technology is actually doing. A tool like Microsoft Outlook® is fairly ubiquitous today. To most people, Outlook® is a familiar scheduler and calendar. Outlook® will be used to schedule a much-needed meeting to accomplish certain business objectives. In order to have a meeting, everyone needs to meet. So what else is happening? Or not happening, after we factor in Outlook®?

Everyone acknowledges that things get done better and faster when they divide the work among different people who are capable of doing the specific jobs. Adam Smith had it right in his *Wealth of Nations*.

The product developer can be apprised of new thinking, new technologies, and new tools. The marketer can get a sense of the timing changes and may meet new deadlines more easily. The joint venture partner can be "brought up to speed." This is getting the "team" together.

Think of an important problem that comes up, requiring a real, concrete, meaningful, and business-relevant solution. The meeting being organized is high priority, to deal with a brand that is one of the "jewels" for the organization. The brand competes in a category that is absolutely "red hot." Internal and external partners or potential partners are to be invited. It's already the end of May, so we are fast approaching the Memorial Day holiday. A date is found in early June when everyone can meet. Unfortunately 2 days before the meeting, a key new stakeholder makes a simple, personal telephone call demanding a meeting in another city with the joint venture partner, who was supposed to attend the team meeting. This second meeting is scheduled, taking priority over the big team meeting. No one is available for a 2-hour team meeting until the end of the month. Our 2-hour team meeting simply never materializes. The schedule shifts, the window of opportunity closes, and the corporation moves on. Typically, in these situations, the product developers end up going back to their old approach of thinking through the problem and do the best they can. As a consequence, the new technologies and processes that were envisioned to help the business in fact are so reactive that they cancel themselves out.

What's just happened here? Many procedures (such as the simple telephone call to request a meeting) that have been in the company for a long time don't seem to be getting breakthrough thinking. Yet these procedures exist, continue to work, can be implemented without much thought, and so they are. It's a case of the business colloquial phrase *Fast to go slow*. New approaches trumped by what is. Sound familiar? (P.S.: In the aforementioned story, it would have been nice if someone had an administrative assistant to help coordinate the meeting planning, but that role has been eliminated in that company at the level of the product development business unit.)

Peter Drucker suggested that the biggest issue in the new millennium in information technologies is/will be the social impacts. These will turn out to be much more important for goods, services, and business than the impact on the material civilization. He suggested that we are currently in the fourth surge of entrepreneurship (entrepreneurship being one of the signals we can observe). The first occurred during the mid-seventeenth to early eighteenth centuries and was called the Commercial Revolution. The second occurred during the eighteenth to the mid-nineteenth centuries and was called the Industrial Revolution. The third started around 1870, triggered by new industries, which in turn had been triggered by advances in biology and information. What is critical is that this surge now has impacted a range of human behaviors, with politics, government, education, and economics being most noteworthy, since there are no national boundaries for information (Drucker, 2001).

Let's introduce the consumer into this maelstrom of change. The customer thrives in this whirlpool. Companies are also living in the world and trying to survive, but they cannot keep up with their buoyant and radiant customers. Companies attempt to stay upright and sell what they can. It's a buyer's market. For whoever lives in the world just described, keeping up to speed with the consumer mind-set is difficult at best. But in an era where perception is the center point, how does one keep up with percepts that lack even a tangible reality but that can drive a product to be successful one day and passé the next? Whereas we think that the material world may impact us on a day-to-day basis, it appears that the larger social impact of technology on consumer mind-set is the one we must understand in order to stay on the curve, or hopefully ahead of the curve.

A BALANCED THINKING PORTFOLIO IS REQUIRED AND YOU MUST REGULARLY REFRESH IT

> What is now obvious to me—but was not at the time I wrote The Tipping Point—is that we are about to enter the age of word of mouth, and that paradoxically, all of the sophistication and wizardry and limitless access to information of the New Economy is going to lead us to rely more and more on very primitive kinds of social contacts.
>
> Malcolm Gladwell, Afterword to *The Tipping Point*

If Drucker is right and we are in an era that looks more like biology, then the learning from that system is that the power of the past (size) will be trumped by the power of function. For the last 100 years, size has mattered a lot. Size matters in the worlds we are discussing—both market research and sensory research. The bigger the sample, the more reliable the results would be. The bigger the company, the more expensive their tool chest. Therefore, the better their knowledge of the market. The larger the university, the better their thinking. The bigger the market research or sensory department, the better it would be. Size matters a lot.

Big today provides one highly significant advantage—the ability to spend more before you need to or are forced to stop (Bowman, 2009; Eisenfeld, 2009; Beckley, 2010; Johnson, 2010). Size remains a significant factor, to be sure, but not necessarily a predictor of the best solutions or best results. Merely looking at one example tells much of the story about where big starts, stops, and how far it can pull the company. Most companies today have implemented one or another form of segmentation research in order to better understand their consumers. The segmentation research can be based on a number of different strategies:

behavioral, attitudinal, psychographic, moment-in-time, preconscious thoughts, and so on. Most segmentation suggests a nice story; between nine and thirteen segments and therefore delivers unique segments at the 4–12% of the population level size. An interesting analysis, and lots to think about. In the end, however, how does a multibillion dollar company craft a business with a goal of 50–75 million dollars out of small, nicely explained segments that are worth perhaps 2–4 million dollars each? That interesting analysis leads to business paralysis.

Given this seeming "law of consumer research," that stories from segments matter, and the more interesting the story the better, then what should the company do? Everyone in the consumer research department of a Fortune 500 company has to do segmentation research of this kind or use the results for something. Segmentation has created a nice ecology, a business, a group of "mavens," or segmentation experts. Many market research companies have found that segmentation and storytelling help grow their balance sheets. But in the end, how does one make these segments meaningful from a business perspective (Freeman, 2001; Unilever promotion materials, 2006; Byron, 2010)?

The bottom line from the foregoing story is that the methods that have been used before do not help in a world that is changing. The standard comfortable methods may not cause anxiety because they have been around for a while, but they may not lead to success either. Survival today demands continually unsticking a company from a behavior or belief set surrounded with too much comfort. And this is not easy, not for the company, a vendor, a department within a corporation, or an individual (Lerer, 2010).

In order to stay healthy and to keep moving the business forward, it is a good practice to have a mix of inputs into your business, provided to you through workers, suppliers, and consultants. To really stay on top of the knowledge game, one must be fairly ruthless about how he or she classifies companies and individuals. The company needs to realize that biases in judgment are fairly easy to fall into and that these biases can lead down the wrong path (think Enron and Arthur Andersen accounting, Lehman Brothers). Admit that you use certain consultants because they have a high-profile name and you need that to assure your board. Deal with the fact that you have worked with a specific sensory consultant for 25 years and even if the methods really don't work or cost more, you are comfortable with that source. Be honest with yourself.

A matrix to use for understanding where you are getting your inputs and how to evaluate these inputs appears in Table 5.1. We have divided the nature of the input into four groups: emerging, leading, parity status/quo, behind. This structure allows the reader to classify an external source in terms of how the company benefits.

Also relevant for the business professionals today are the sources of people collecting and disseminating the consumer mind-set information. Thirty years ago, there were ad agencies, market research agencies, a couple of sensory agencies, and a few very large consulting firms like McKinsey. Today, the mix is very different. Holding groups own whole areas of market research (e.g., Nielsen, as well as the different holding companies within the WPP family). There are large and small strategy companies, larger and smaller market research firms, and larger and smaller sensory consulting firms. The mix of people offering ability to understand the consumer today is much broader and significantly more complex.

When the corporation's stable of workers, consultants, partners, and suppliers all cluster in only one or two sections in Table 5.1, then it is important to explore the thinking about the choice of inputs and how they will provide only a certain limited, biased view of the world. This view might, or just as easily might not, allow the company to become and to remain competitive. The view will definitely structure the way in which the company connects with and "hears" the consumer mind-set.

Table 5.1   Mixture of inputs for strategy, planning, and research.

Emerging: Emerging capabilities are often too new to have proven return to investment (ROI), but they appear to provide a distinct competitive advantage. Companies that adopt emerging capabilities are risk takers and are looking for untested, breakthrough strategies. Where leading strategies cross boundaries within the company, emerging strategies integrate the extended enterprise, including outside vendors and partners.

Example: Business development firm that has a new take on how to do strategy.

Function: Take your business thinking to a new level.

Emerging capability: Allow you to look at a situation with a new set of eyes.

Case vignette: You want to innovate. You call in a number of firms who say they do this. Firm ABC provides you with some thoughtful approaches and while they scare you, they are within your budget and you do the project.

Parity/Status quo: Parity capabilities are common in many organizations and have proven ROI. In general, they focus on automating administrative tasks, but they do not transform the way the company performs its business processes, and only a single department or business function usually uses them. Companies justify their adoption of parity capabilities primarily based on ROI and usually adopt them under pressure from competitors that have adopted or are also in the process of adopting the same capabilities.

Example: A leading expert in the field of experimental design.

Function: Bring you a point of view that is practiced and well accepted.

Parity capability: Will get you to where you need to be to be competitive with others who have already put this strategy in place.

Case vignette: Leading beverage manufacturer, who is redoing its entire product development unit. While this approach has been in place for many years, the robust designs and flexibility of thinking have not been in place. This firm/individual helps resource and bring along the entire business unit so that they are functioning at a higher though not cutting edge level.

Leading: Leading capabilities are new enough that their proven ROI is limited, but they appear to provide a competitive advantage. They generally transform the way the company carries out its business processes, and they cross functional boundaries to integrate the company internally. Companies that adopt leading capabilities are generally looking for opportunities that can quickly separate them from their competitors.

Example: Package design firm that has demonstrated they can deliver success.

Function: Keep you moving at the speed of today.

Leading capability: They have the capacity and the capability to deliver the results on a repeatable basis.

Case vignette: You need to get new packaging for a lead brand. You don't have the time or resources to do a lot of the critical thinking to get there. You hire this firm to get to where you need to go and do most of the work for you.

Behind: Lagging capabilities are mature capabilities that companies have been using for some time. They generally include static, computerized, and manual processes. They have well-proven ROI with established industry standards. Companies that use lagging capabilities in their marketing research/sensory strategies usually adopt them after the capabilities are so widespread that they have become industry-wide standard practice and therefore are felt to be necessary for basic functions.

Example: A well-established firm in the field of sensory evaluation.

Function: Introduce methods that are over thirty years old to this firm.

Lagging capability: This is a well-recognized "brand" in the field and, therefore, while the methods are not at all cutting edge, they can be known to bring the best practice available.

Case vignette: Worldwide leader in a packaged good field changes senior management and also research management. The lead executives for the research function are comfortable with the output from this lagging function and want to have the expertise in place to be able to have that function covered within the company.

Source: Table adapted from Anon., 2001.

HAVING INTEGRITY ABOUT PLANS AND STRATEGIES IS CRITICAL

Table 5.1 was designed to help understand where a businessperson's comfort zone is with respect to those who supply information, and in turn the quality of the information, and the level of risk. Knowing one's sources of knowledge input allows the company to see what choices it makes so as to better understand the risk profile with respect to research. Why is that important? Why bring in risk based on the input source (the information supplier) when the whole idea of the input source is, in the first place, to reduce risk!

The events described earlier in the chapter make today's work with consumers exciting, but also filled with the unexpected, and occasionally filled with knowledge and insight that disturb. No amount of screening can keep you from having a very diverse consumer base that has experienced life in a way unheard of 50 years ago. But that is what the technology revolution has created. As a result, these consumers see through much of a company's marketing and advertising facades (Kuczynski, 2009; Hampp & Parekh, 2010; Helm, 2010). Even when one is not aware of issues and biases introduced by members of the company or their suppliers to the testing event, the results can be biased, and perhaps even the business can suffer. Mercifully, the suffering occurs quickly. The feedback loop for unpleasantness is shorter, making change perhaps a bit easier.

In "Looking beyond traditional models," Pigott (2002) suggests that the tyranny of the dominant paradigm freezes and fixes our ability to see what is available to us. Pigott's premise is that most businesspeople find that their best work has resulted from discoveries along a voyage, rather than a destination at the outset. *You don't know what you don't know.* Yet in both market research and sensory, we practitioners have inadvertently found ways to remove the anthropological context. The research culture defeats attempts at brave insights and courageous recommendations. In the end, it's the same thing, again and again, with the same patterns repeating. It's sort of like each of Vivaldi's 626 works, repetitions, albeit without Vivaldi's talent. At the end of the day, little exists surrounding the context of the research or the knowledge obtained. As one tries to incorporate the consumer mind-set into the company's knowledge base, it is critical to understand that often we simultaneously prevent the very same minds we want to hear from actually speaking to us. Following steps to consciously ensure research openness and integrity might help here (see Table 5.2).

Table 5.2 Checklist of questions to maintain one's research integrity.

(i)	Do I know why I want to engage in a dialog with the consumer?
(ii)	Do I care about how I appear to the consumer?
(iii)	Do I care about the consumer?
(iv)	Am I intending to use this information appropriately?
(v)	Am I collecting these data trying to prove a point or uncover a truth?
(vi)	Do I really know whether this research approach can provide ROI?
(vii)	Am I doing this research because everyone else does it?
(viii)	Do I care about the results that I am collecting?
(ix)	Am I using this consultant/firm/supplier because they are doing the best job, because they are a friend, because I like them, because I get a kickback, or because they are cheap and on the approved supplier list?
(x)	Do I look back at the work that I have done to see if it has left intended results or unintended results?

ESTABLISHING A FRAMEWORK TO INCORPORATE THE CONSUMER MIND-SET INTO YOUR MARKET OPPORTUNITIES AND YOUR PRODUCT

Frameworks are important. Frameworks are ways of doing things. They allow the corporation to work on "automatic pilot" and not to continue to fight battles. The right framework for a corporation, like the right habits for a person, can help move toward success even when one's mind is occupied with other matters. Incorporating the consumer mind-set on a continuing basis in the corporation needs this framework. It provides guidance about what to do, and it can incorporate metrics to show how successful the corporation has been in the past, and how successful the corporation continues to be.

Here are five favors that define a framework to include the consumer mind-set:

 (i) A blueprint approach to specifically describing consumers on whom the company wants to focus their marketing and products.
 (ii) Design that allows for databasing both quantitative and qualitative findings (Zahay, Griffin, & Fredericks, 2003).
(iii) Method to document the value set for the person and the product (value diagramming, Chapter 2).
(iv) Strategies that connect technologies to product features, to marketing, and to the consumer (Frankwick, Walker, & Ward, 1994).
 (v) An ability to value the impact of consumers/customers (Ambach & Hess, 2000; Gupta, Lehman, & Ames Stuart, 2004; Gerzema & Lebar, 2009).

A formalized, functioning, publicly available framework acknowledges and requires a cross-functional communication plan to allow effective incorporation of consumer mind-sets. One discipline alone simply cannot do all the work. No one in the company is omniscient; no single corporate silo can provide all the information, all the insight, all the guidance. For professionals marketing recognize that this acknowledgment is not easy. It is just very necessary. As Wittink suggested during his introductory paper when he became editor of the *Journal of Marketing Research*:

> I can only indicate my strong support for multidisciplinary research. Unfortunately, tenure and promotion criteria usually do not favor researchers who cross boundaries . . . The desire for specialization in academia is inconsistent with the need for business decisions to represent dependencies across areas . . . Inertia, but also active resistance from thought leaders, limits willingness to pursue multidisciplinary approaches.
>
> Wittink, 2004

Multiple points of understanding the consumer are required when one wants to succeed in the task of incorporating mind-sets. A complementary set of approaches will more likely achieve the hoped for complete view, rather than a provincial, but politically more acceptance single focus. No single approach does it all, no matter how ballyhooed the approach might be, no matter how much praise the business world lavishes upon it.

Later in the chapter, we have organized the approach we take into a series of seven steps. The steps may be expanded, contracted, varied, but the message to take away is that the approach is multifaceted, bringing in the talents, viewpoints, and even eccentricities of different people, different world views, different executions. The steps demand continual vigilance. The steps call for surveys, but in additional operationally clear, defined actions by which to introduce these new findings into the relevant business clusters. Jim Stengel, the CMO for Procter & Gamble, gave a talk to the American Association of Advertising Agencies in February of 2004. When one reviews the presentation he gave, it is clear that Jim understood and tried to integrate the consumer into the entire marketing plan. His comments such as, "we must accept the fact that there is no 'mass' in mass media" reflect an understanding of who the consumer is today and how they are thinking (Bianco, 2004; Stengel, 2004).

As we go into specifics, we ought to keep in mind that to a great extent what we have used to understand the consumer constitute "rearview mirror" approaches and techniques. Trend-spotting and cool hunting hopes to see the future via today (or the past). Custom surveys and semisyndicated work reported later can provide future vision, but unfortunately a lot of the consumer-based work is about the past, about tracking, about what happened. Much of market research today tells us where the consumer mind-set has been or a little about where it is, not where it is going.

With these caveats in mind, let's follow the steps and see where they take us.

Step 1. Popular business media tell us about what the culture finds interesting at a specific point in time.

Popular business books and other media allow us to see what the business community finds appealing and acceptable information. Eight books that definitely allow us to understand the business mind-set in recent years are as follows:

   (i) *The Tipping Point,* Malcolm Gladwell (2000)
  (ii) *The Experience Economy,* Pine and Gilmore (1999)
 (iii) *Good to Great,* Jim Collins (2001)
  (iv) *Who Moved My Cheese?,* Spencer Johnson (1998)
   (v) *The FIVE Dysfunctions of a TEAM,* Patrick Lencioni (2002)
  (vi) *Blink,* Malcolm Gladwell (2005)
 (vii) *Freakonomics,* Steven Levitt and Stephen Dubner (2005)
(viii) *Outliers,* Malcolm Gladwell (2008)

What all these books hold in common is that they have had very long runs at being popular purchases. A broad range of people mention that they have read the book and that there is something noteworthy in the book. Three of the four books provide analysis of patterns of events that the writers found lead to change or success. Stories within the books are well told, and at points almost take on aspects of the mythical. They tell of success or quest that leads to a better place. The fourth book (*Who Moved My Cheese?*) is a more simplified story, childlike in telling, but leading the reader to understand that people can get themselves to a better place—*if* they want. So these books tell us that the leitmotif for popular business thinking is the *story of the quest,* documented or not, that leads us to understand the human work situation better. This is where business mind-set is today.

What part of the consumer mind-set does Step 1 provide? Through the popular business media we are exposed to the thinking of a wide range of people who have taken the time or the money to try to describe those situations that have or might influence business. Today

Step 1, the personal story, has become a popular tool for large and small consulting firms. Step 1 works far more quickly in business than articles in refereed journals. In actuality the story provides us with a perspective of where academic research is going but has not completely gone there yet. In some cases, such as the "From Experience" portion of the *Journal of Product Innovation Management,* or the "Consultants Corner" in the *Journal of Sensory Studies,* the academic literature recognizes the value of stories as a complement to the harder science that the journal fosters and reports.

Step 2. The trendiness of trend-spotting.

A process to track what are considered trends, along with a systematic way to map and study the phenomenon, has become de rigueur today, standard operating procedure for any business that wants to stay in the "game." Moskowitz (Beckley, 2002) has suggested that a consumer mind-set is itself a trend before it reaches the end point manifested as a visible trend. With that thinking, he suggests that we act and make choices in ways that are very specific to us and our needs, based on our mind-set. As people work through situations, either cultural or behavioral, these mind-sets create the phenomenon of either a fad (short-term trend) or a true longer-term trend. Thus, we see the compelling nature of Facebook®, Twitter®, and, YouTube® or whatever trendy social media site is now hot.

There are many books, Web sites, newsletters, blogs, tweets, and conferences that discuss trends and/or the future. Indeed trends are no longer about the future; they are about where today is going. Most people engaged today in trend-spotting in companies generally do not have titles of "futurist" because the future is today's reality, albeit a reality that will occur tomorrow morning or perhaps in 30 minutes. People engaged in trends have titles that reflect the normal world, whereas they have hearts and minds that think about new and different worlds and where products and people might go.

Michael Tchong, a trend analyst, has developed a structured approach to trends, called trendscaping, and examines not only the typical listing of trends (trend categories) but also delves deeper with "metatrends"—phenomena that ripple through society connected and interconnected to a collection of subtrends (Tchong, 2003). Tchong is only one of many; a search through Google® and through Amazon® will reveal an amazing array of individuals, companies, books, and databases that deal with trends (e.g., a Web site like www.trendwatching.com).

Where did this fascination with trends come from? Most agree that the first trend-spotters were actually science-fiction writers such as Issac Asimov, Ray Bradbury, Frank Herbert, and Arthur C. Clarke. Engineers in the 1980s were told to read science fiction literature when they wanted to "see" really creative thinking about the future. Both Alvin Toffler (*Future Shock*) and John Naisbitt (*Megatrends*) crafted a more business-like approach. Faith Popcorn, with *The Popcorn Report* and her Brainreserve, followed this path. Since then, it has given full growth to what we have today. The Internet and multiple points of access to media and people (blogs, webcams, etc.) have created an opportunity where the common person with an interest in the world can become a trend-spotter or *cool hunter*.

Understanding trends, developing an interpretation for them, and understanding what parts of a consumer mind-set they represent is required to piece together yet one more part of the consumer mind-set. In the 1950s, *Time* magazine was able to capture much of the metaphorical mind-set of America. Perhaps the trend-spotters and cool hunters of today comprise a blend of science fiction and *Life* magazine.

Thinking and reading about trends is significant for the food and beverage world as well, not just for those in fast-moving products such as consumer electronics. The Institute of Food Technologists' most popular issue of *Food Technology* is generally the publication in which Elizabeth Sloan publishes her trend predictions. Her monthly column is well worth the read. Her trends capture what is happening today, in different parts of the world, in restaurants, in small companies, and reads more like travelogue in the food world than a prediction of things to come.

What part of the consumer mind-set does Step 2 provide? Trend-spotting and cool hunting are the formal approaches to looking at the public and private worlds in which people live with the objective to establish meaning and understanding. It is absolutely critical to look as broadly as one can at what is happening in one's local region, country, and world in order to understand the mind-set of the consumer.

Step 3. Demographics and the US Census; psychographic information.

Demographics is destiny, or so the popular phrase goes. The use of demographics to understand all aspects of a population has become standard practice. Demographics is defined as the size, structure, and distribution of population. Individuals who are engaged in the field believe that the study of demographics can explain about two-thirds of *everything* (Blackwell, Miniard, & Engel, 2001).

Demographics help describe consumers. To the astute researcher, demographics may suggest, although not establish, whether there are "top-level/top-of-mind" patterns lurking in the data. For example, demographic data have been used to assist in the prediction of changes that may be expected to occur in choice and demand as the structure of the consuming population evolves in the path dictated by the demographics (e.g., aging, moving to the suburbs and exurbs). Typical demographics required of any person conducting survey research include age, gender, location, length of time in an area, type of employment, income level, number of individuals in a household, ethnicity, and number of children living in the home.

Today, however, most researchers believe that demographics alone do not tell a complete picture regarding critical factors related to consumer mind-set and choice behavior (Moskowitz, Porretta, & Silcher, 2005). Even though most evolved researchers and businesspeople have begun to recognize that standard demographics can mislead as often as they inform, acquiring demographic data and using it to organize study results has become a ritual that most survey research typically goes through in the reporting phase. The power of demographics is simply overwhelming, the inertia they exert even more so.

There are some aspects of demographics that are critical, perhaps because they are intimately involved in behavior. Geodemographics have been found extremely useful in many areas of retailing when franchises and multiple unit operations move into their expansion modes. Geodemographics describe demographics that are aggregated to specific geographies and neighborhoods. A number of organizations compete in what is estimated to be a 100 million dollar subset of the marketing information landscape. All use the most recent US Census data as their foundation and promote their businesses through enhanced analytical products and services. A wide range of Web sites that are associated with both government and industry or organizations (like AMA, ESOMAR, DDB, and Interpublic) are very useful in finding suppliers, tools, and consultants who work in all of these areas of demographics.

The dialog continues to be lively and heated regarding the value, use, and misuse of demographic information and its ability to demonstrate trends or to generate predictions. An example of this is an article written in *Marketing News* for October 13, 2003, in which T. Semon suggests that an essay written in *American Demographics* (now defunct) titled "The wealth effect—who has the money and where you can find them" *was not about wealth at all but rather about income*. Semon concludes his column by providing a recommendation to modify the approach by obtaining two measures: how much wealth a person has, and how interested the person would be in spending that wealth on a given product (Semon, 2003). Interestingly, *Forbes* magazine, a year later, had started building a database of affluent consumers to try to understand just that particular question Semon had proposed.

Following the same idea, but this time using the "mind" as the source of demographics, many companies now obtain "psychographics," the psychological equivalent of demographics. Obtaining personal and emotional values in much the same way as fundamental demographics has become increasingly popular. In the past, this work was classified as psychographic data; increasingly we see it as part of a larger domain of knowledge called *behavioral modeling*. Today, a variety of approaches are taken for this behavioral modeling in which large samples of consumers are surveyed, values are measured using some sort of a scale, and then a number of proprietary segmentation schemes are applied to the data, varying from the simple to the elaborate segmentation analysis. These segmented populations are then used to understand usage patterns with products and have been organized to develop fairly high-level strategic market analysis (Neff, 2004).

Such metrics that examine the beliefs and values of consumers enable companies to better understand specific reasons why consumers purchase one brand over another and track the development of consumers' preferences in relation to changes in beliefs and values (Reibstein *et al.*, 2006).

Diary panels and other forms of usage studies (so-called A & U or attitude and usage studies) allow broader understanding of how people use products in their homes and to provide a systematic way to view the data. NFO (now called Mysurvey.com and Lightspeed Research, a part of WPP) was one of the original developers of diary panels. These tools have been adapted to include both mail and Internet-based reporting. Demographic information has also been used as a tool in examining consumers' increasing use of the Internet and its products (Ranaweera, Bansal, & McDougal, 2008). Network analysis appears to be the trend emerging at the end of the first 10 years of the new millennium directed at this evolved body of information. The bottom line is that counting is something we humans have done for a very long time. Seeking patterns in numbers is also a long tradition of humans. The use and abuse of demographics will continue. Businesspeople and researchers like the concreteness and the specificity of numbers.

What part of the consumer mind-set does Step 3 provide? Numbers are an extremely useful construct for forming thinking. What is important is that much of the work in Step 3, our census and segmentation data, is by its very nature a rearview mirror and may not necessarily explain today's behavior. Numerical data, by its very mass, need to be boiled down to generalities. Such generalities, although easy to present in meetings and making nice sound bites for the popular press, might in reality miss the actual individual reasons for action. Within those individual reasons might lie the hope for future ideas and products. In the scientific parlance of experimental psychologists, the normative, the nomothetic, the group might not hold the key to the individual, the ipsative, the idiographic. We must proceed further if we are to arrive at our goal of understanding the consumer.

Step 4. Syndicated research, diary studies, pantry panels.

Syndicated research in consumer-packaged goods has generally been identified as the big databases of scanner information that is obtained from two sources, SymphonyIRI Group (originally Information Resources, Inc.) or Nielsen (originally AC Nielsen). Most of the early syndicated data were collected in order to help understand movement of products at a point of sale. These syndicated data show what has been purchased. A number of analytical approaches for understanding these data at deeper consumer levels are conducted. In the past, the price (expensive) and need (mandated) to have this information for one's business planning were so important that they overwhelmed other forms of spending in business research, often driving out the custom research in favor of the contracts demanded as part of the data offering.

Syndicated research began in earnest in the 1970s and was driven by a new technology—the bar code. Most companies still purchase data from their categories. Several years ago a large retailer (Wal-Mart), no longer allowed their scanner data to be shared. This caused a large gap in the information and created desire on the part of manufactures to generate a more complete picture of the market. Furthermore, syndicated information services are entering a period of flux as a result of advances in information technology. A technology that appeared promising a few years ago—RFID (radio frequency identification)—appears to be one of those potential advances that has not worked out, at least yet.

Other forms of syndicated research were being created at the same time as the scanner work in order to obtain more information regarding usage patterns of products. Popular forms have been diary panels. NPD is perhaps one of the best-known panels, with presentations given yearly regarding eating patterns in America. Many areas in the food industry have been added to this type of panel work and conducted in specific areas (snack foods, beverages and spirits, restaurants, etc.). The value of these data lies in the ability to show the performance of different "areas of consumption" across years, thus revealing new areas that emerge or have begun a decline. Whereas these types of research are widely deemed to be useful, the most frequent negative comment from users is that the information is somewhat "rearview mirror" and does not explain why people do what they do. Other regular research such as the comprehensive studies offered by Roper Starch Worldwide (now NOP World, a division of United Business Media plc) have been "upgraded" with reports focusing on specific categories as compared to broader areas.

The ease of consumer knowledge acquisition and analysis through the Internet is fostering new types of studies, whose possibilities and long-term implications are still being realized. Many firms are becoming increasingly interested in technology and techniques by which they can examine "unstructured data" about consumers found online, such as the content of blogs and advanced search inquiries by consumers. These new "voices of the consumer" promise to convey the voices of many online users (Chou et al., 2007; Peters, 2010), and thus provide a new avenue for insights.

Although expensive, the benefits of syndicated work are legion, specifically structure and consistency of the data over time. The structure forces a foundation to underlie the knowledge, a key benefit for the marketing department, which has to fight battles every day, and which does not have the wherewithal to create this foundational knowledge. In previous times, when the syndicated reports presented a complete picture of sales, those data provide the single best independent metric that marketing, sales, and, by inference, product development had done a good job with developing a new product.

Syndicated research provides so much rich and granular data, that the development group often finds it hard to know what to do with all of it. If you are using SymphonyIRI or Nielsen to tell you your product is successful or weak in the marketplace, then the professional knows what to do with the numbers. But to really understand in one's specific business where to go with the results, what specifically to do next, demands much deeper thinking and the time to think. Time, unfortunately, whether to think or just to waste, is no longer a widely available commodity in today's business environment. Consequently, the time needed to digest the information in syndicated studies simply has disappeared. The studies must be reduced to sound bites, presented and quickly forgotten. Much of the value then dissipates.

What part of the consumer mind-set does Step 4 provide? Syndicated data and reports have comprised a popular format in which to create metrics about large numbers of products in a category. The next step is to understand the individual mind-set from syndicated tools. Beyond that—it's not clear at this writing.

Step 5. Secondary research services.

The most noteworthy secondary research service at the start of the new millennium comes from a privately held British company called Mintel International Group (www.mintel.com). Mintel's Global New Products Database® (GNPD) is a favorite with corporate R&D groups. Mintel has put in place a sufficiently flexible approach to pricing of their databases to make it feasible for purchase by a wide range of organizations (corporations, universities, and other institutions). It is possible that Mintel saw the gap in syndicated research and addressed it with their Mintel reports, which can be broad (Eating Habits—United States) or more narrow (Beer—Pan-European Overview). Mintel has partnered with a number of food organizations to provide multiple levels of their publications, from both a semicustom to general profession perspective.

In the past, the type of reports written by a group like Mintel were done on a custom basis by individual assistant brand managers or consulting firms. This more commoditized approach to reporting is changing the value and structure of market summaries. Reports like those from Mintel give the businessperson a lot of knowledge about the events in which people participate and what people do within the context of those events. These reports "report"; they do not seek to understand the "why" beyond the report. Such understanding about the "why" is left to the reader or researcher. In a sense, these reports are further "rearview" mirrors, but with a little more of the human touch.

What part of the consumer mind-set does Step 5 provide? These reports provide a way to get simple analyses and interpretations about quantitative information. Step 5 information can help to foster new business thinking, or in some occasions may be used as outsourced analysis, replacing staff members who previously performed this type of basic analysis.

Step 6. Custom research.

Custom research comprises the second wave of bringing in the mind-set of the consumer. Individual companies conduct custom research projects when they want to understand more deeply issues and topics arising out of Steps 1 through 5.

Custom research can take on many forms. Research may include tracking studies, attitude and usage studies, rounds of qualitative research, and brand equity, market structure, or need state studies. Once a nature of the problem has been identified, and the product category specified, custom research from both brand and product perspectives is almost always conducted. Such research involves concept screening and refinement, product optimization, product selection, standards development, and shelf life measurement for the limits of consumer acceptance.

Today's Internet technology is changing the nature of custom research. The use of the Internet for much of survey research has grown significantly since 1996. Better or faster computing, better software and hardware, better ways of finding consumers and much higher penetration of electronics and computers into the average household have made this form of data collection more popular. To see just how much things have changed, the prestigious EXPLOR award, established by the Nielsen Center and administered by AMA to identify the cutting edge applications for the use of the Internet, was awarded to Hershey Foods Corporation in 2001 for its use of online testing (AMA, www.marketingpower.com). In 2008, the award was given to MyProductAdvisor.com and General Motors for creating a new method to gauge consumer preference data. So the process evolves as the tools and technology present opportunities.

The mind-set of the consumer comes in during direct exposure to stimuli that the researcher presents to the respondent. The respondent's ratings or reactions are processed to create information about how the respondent reacts. The data can then be compared to expectations, or to data of a similar type from previous experiments. The data can also be used to create a "picture" of the mind of the respondent. Such data lend themselves to systematization, and in turn to benchmarking. Benchmarking of this information is customary on the market research side of the business but is seen much less frequently on the sensory analysis side.

What part of the consumer mind-set does Step 6 provide? Custom research generally provides the commissioning company with information it wants to know. Compelling insights beyond the mind-set of the executives involved in commissioning the research may be limited, however. The limitation comes from the limited focus of the project on the one hand, and the nature of the research design (not all designs and approaches are created equal). Most research is designed in order to answer one key question only—answers beyond the immediate focus are "nice to know" but rarely appreciated except in hindsight. That hindsight often occurs, but years later.

Step 7. Semi-syndicated research.

These are studies that run on a regular basis and have some specific purpose in mind, yet allow clients to modify or adjust the research somewhat to customize their needs. A long-running study of this nature is called HealthFocus® (www.healthfocus.com; a division of Irwin Broh & Associates). For several years HealthFocus® has conducted a periodic study on food and health issues. Sponsor companies that fund the study are given the chance to ask a limited, additional set of questions within the syndicated portion of the study, with the data from those questions reserved for the sponsor alone. This business model, a combination of syndicated data and private data, has proven to be a successful model. HealthFocus® research is similar to traditional tracking studies but with a subject focus. Standard question and answer formats are used for this annual semisyndicated research.

We can contrast this business model for health data to one that has grown up over the time between 2000 and 2010. Whereas Healthfocus was one of the first in the health arena, Natural Health Institute (NHI) is another example of a group that has grown in influence and prominence due to their specialization, consumer trends, and affiliation with certain health organizations. Both NHI and companies like Ipsos create the opportunity for customization in their omnibus work. Customization emerges from the series of questions that are provided by a client company and reported to the client company along with the omnibus information. In that way, the purchaser can get generalized insights with some customization.

Another pioneering, and now increasingly respected approach to the semisyndicated model is a body of research to understand the "algebra of the customer's mind." The approach, now called Mind Genomics or the mental algebra of everyday life, was begun a decade ago by two of the authors (Moskowitz & Beckley). The actual research approach was developed by author Moskowitz, through his Internet firm, i-Novation, using the sophisticated research approach of conjoint analysis and driven by specially written software called IdeaMap®.net. The Mind Genomics approach, developed in partnership with The Understanding & Insight Group, created a series of semi-syndicated research called It!™ Foundation studies. Moskowitz, Porretta, and Silcher (2005), and Moskowitz, German and Saguy (2005), describe the approach in detail and provide some representative data from different studies. The It! studies are run on an annual or biannual basis depending on the category. What is unique about this research is that through the use of conjoint analysis and mind-set segmentation, the results from this body of work appear to *anticipate* consumer desires as compared to looking into the rearview mirror, as do other studies.

The aforementioned It!™ studies are foundational. It!™ creates a fundamental, new-to-the-world body of knowledge about the consumer mind that that complements the other research data currently available. These It!™ Foundation studies work on three levels:

(i) *Latent interest in a topic or food category*: First the different studies in the It! database for a measure latent interest in topics (e.g., different foods, different beverages, fast food situations). This latent interest in a topic is meaningful when we realize that a single It!™ database might comprise 30–50 studies, of the same type, albeit with different products and different but similar elements across the products. The It! database begins by allowing a respondent to select a topic area of interest, with all available studies being presented as choices on a "wall" on an Internet page. The respondent chooses the study that is most interesting. For example, in the Crave It!™ study, a respondent could choose a study topic ranging from cola through steak and cheesecake to chips and even salad and fruit. The simple act of selecting one study topic (i.e., one specific conjoint interview) over another study topic reveals a top-of-mind choice.

(ii) *The granular algebra of the mind for a topic or food category*: When the consumer then works through the conjoint study, which is then followed by a more typical survey, the data analyst, marketer, or product developer immediately discovers the choices and thus the trade-offs that the consumer respondent makes when evaluating the different test vignettes. The compound nature of these vignettes presents the respondents with different messages, a sort of "blooming, buzzing, confusion." One happy consequence is that consumers cannot assign politically correct or "right answers" during the conjoint study. Consumer participants are forced to act in a sense and respond mode. What one begins to measure through this research is what is working in their unaware (unconscious) mind and driving them to accept or reject certain cues.

(iii) *The integrated design of the different studies within one It! database encourages a new, potentially powerful decision-making property to come into play*: The structure of a single foundation study is set up so that all of the conjoint designs across the different topics are parallel. Furthermore, the classification part of the study, coming after the conjoint portion with vignettes, is virtually identical from one study to another. With the data set, one is able to go with a foundation study across products, or even across a series of these foundation studies, to actually understand how behavior translates from category to category. This benefit (patent pending) allows for the creation of true cross-category innovation by being able, quantitatively, to compare behaviors across shopping categories like sports cars, shoes, and power drills in Buy It![TM] (Ashman et al., 2003) to issues of anxiety in Deal with It![TM] (Ashman, Teich, & Moskowitz, 2004) and more.

As these It![TM] Foundation studies began to be run year after year, what became apparent is that they represent the first piece of research that links to current syndicated research, yet at the same time embraces trends and behaviors. These It![TM] studies begin to *predict* what will happen. The It![TM] tool to understand and structure the consumer mind-set, when combined with many of the traditional tools listed integrates both rearview learning with forward-looking knowledge, is evolving into an interesting new tool for companies that seriously wish to understand and predict future behavior.

A sense of the type of data from the 36 studies conducted in Crave It![TM] 2001 appears in Table 5.3. What is interesting to observe is the value for the McDonald's brand and what it did to wanting a craveable hamburger in 2001. Within the next 6 months, and for the first time in their business, McDonald's reported significant decreases in sales at the store level (decline of 1.6%) (Forster, 2002). During the next few years McDonald's took corrective action. Since that time, running of the Crave It![TM] study has shown gradually increasing values that track specifically with the better consumer acceptance of brand McDonald's and its offers (Beckley et al., 2004).

The increasing pervasiveness of the Internet and the acceptance of Internet-based research data, enabled the creation of other series of studies such as the RDE approach (Rules Directed Experimentation) (Moskowitz, 2007). When this approach is linked to key factors that influence the consumer mind-set, the subsequent typing tool allows clearly identifying the traits of individuals who will find certain offers or products to be entirely compelling due to the link with their predisposed mind-sets. (Moskowitz & Onufrey, 2008; Moskowitz et al., 2009, 2010a, 2010b; Moskowitz & Gofman, 2010).

The It![TM] Foundation work presents a whole new way of imagining the mind of the consumer and how to look at the mind in a go-forward way, rather than relying on rearview mirrors. As such, these studies differ from, and complement the studies just presented such as HealthFocus® and other semicustomized surveys like those created by Ipsos and PERT Research.

At the time of this writing, it is becoming trendy to see whether neural imaging will reveal even more about consumers' mind-sets and how consumers make choices. This is an area that has been studied quite a lot over the last 20 years. Technology has made the research opportunities somewhat less costly than in the past. For the moment, the media interest (Wells, 2003) seems greater than the ability to understand the consumer mind-set, although if the past is any predictor of the future, there will be much more interest on neural imaging and so-called neuromarketing in the years to come (Lee, Broderick, & Chamberlain, 2007).

What part of the consumer mind-set does Step 7 provide? The tools in Step 7 are newer and are being used by a wide range of individuals within companies. The opportunities to explore well outside of one's standard categories force a new way to think, which may or may not happen in the next few years—even though calls for "new ways of thinking" are rampant in today's business buzz. The strategy of imposing a structure on a survey design,

Table 5.3 Impact scores for individual phrases for the total hamburger category.

Element		Impact
	Additive constant	15
E09	Lots of crispy bacon and cheese cover the grilled juicy hamburger on a lightly toasted bun	15
E01	Fresh grilled hamburger	9
E06	A char-grilled hamburger with a taste you can't duplicate	9
E07	A grilled aroma that surrounds a thick burger on a toasted bun	7
E34	Fresh from the grill, made especially for you	7
E17	So juicy…you practically have to wipe your mouth twice after each bite	6
E02	Burger smothered in onions and cheese	5
E03	Juicy burger with the crunch of lettuce and tomato	4
E05	Layers of burger, sauce, relishes, pickles, and lettuce on a moist sourdough sesame seed bun	4
E14	You can imagine the taste as you walk in the door	4
E11	With great-tasting french fries	3
E27	A joy for your senses…seeing, smelling, touching, tasting	3
E35	Simply the best burger in the whole wide world	3
E24	An outrageous experience	3
E08	Moist bites of bun, burger, and onion	3
E04	Gooey grilled burger with rich sauce and fresh lettuce and tomato	2
E12	With a cola and fries	2
E19	Fills that empty spot in you…just when you want it	2
E30	From McDonald's	1
E20	Makes you feel happy when you eat it	1
E26	Now you can escape the routine	1
E21	Quick and fun	0
E13	With all the relishes you want on a relish/trimmings table: jalapenos, pickle relish, whatever…	−1
E23	Ecstasy in your hand	−1
E29	From Burger King	−1
E10	When it's cold outside, and the burger is warm and inviting	−1
E28	From Wendy's	−1
E16	You can just savor it when you think about it during work and school hours	−2
E36	With the safety and care and sanitation that makes you trust it and love it all the more	−2
E22	It feeds *the hunger*	−2
E25	When you are sad, a hamburger makes you glad	−3
E18	Extra sides of lettuce and tomato…all the time…with salt too	−4
E32	From Jack in the box	−5
E31	From Carl's Jr.	−7
E15	With horseradish sauce	−7
E33	From White Castle	−7

Note: Each number shows the conditional probability that the concept for the fast food hamburger will be rated craveable (rating 7–9 on the 9-point scale), if the element is present in the concept. The additive constant shows the expected percent feeling a hamburger will be craveable if no elements are present, and can be used as a baseline value.
Source: Used with permission from It!™ Ventures, LLC.

Table 5.4   Points to consider when incorporating the consumer mind-set in a company.

(i)   Adopt a business mind-set that provides the capacity to *discover* new and powerful insights into the human condition by learning what isn't working as opposed to assumptions of what is working.
(ii) *Understand the difference* between statistically reliable and "share of heart."
(iii) Try to *eliminate "truth toxins"* environments that hinder one's ability to make wise business decisions.
(iv) Be willing to *examine the true implications* of the decisions you are making regarding consumer mind-sets.
(v)   Structure the consumer mind-set in an environment that is action-oriented, teaching-oriented, nonhierarchical, humble, interdisciplinary, truth-seeking, forward thinking, and democratic.
(vi) Look for ways to energize the groups that need to think about consumer mind-set by creating impactful visual images, concrete expressions of the consumer voice, emotional connections, and methods that allow the consumer mind-set to be understood with the authenticity and originality that drives it.
(vii) Cultivate relationships with others (outside of your company) that provide you mirrors. The mirror allows you to understand the choices you make and the reframing you do, when exploring the consumer mind-set.
(viii) Develop a database system that allows for rapid retrieval of both quantitative and qualitative information with a context.
(ix) Be honest about how you handle the consumer mind-set. If you adapt what is told to you or how you view the information within your business, deal with this honestly. Do not be duplicitous.

Source: Adapted from work in Pigott, 2002, and Ambach and Hess, 2000.

whereas at the same time in some places in the study, epitomizes the trend in the public world for mass customization. The structure of semisyndication and semicustomization creates cost-effective use of strategy and research funds. Step 7 represents one of the growth arena within business, consumer research, and sensory research.

CLOSING REMARKS

The seven steps provided here are a point of view of "how to do it." They do not constitute the only point of view, however, just those of the writers. What is clear is that we need to incorporate the consumer mind-set into the corporate culture and have that mind-set play a definitive role in decision-making. There are differences of opinion regarding what should be included in a framework and how extensive the framework should be. The sections in this chapter have attempted to highlight issues involved in trying to hear the consumers, understand their thinking, and put *those* consumer mind-sets on equal footing with the opinions of company decision makers. This chapter has attempted to summarize what needs to occur to allow the consumer mind-set to truly be a part of the product design and development cycle. We can end the chapter with nine points to consider when incorporating the consumer mind-set into the company culture (Table 5.4).

REFERENCES

Anonymous. 2001. Are you leading or lagging? *IQ Magazine* November/December:72–73.
Ambach, G. & M. Hess. 2000. Measuring long-term effects in marketing. *Marketing Research* Summer:23–27.

American Marketing Association. 2001. A Bold Commitment Pays Off. A Case Study on an Award Winner for the EXPLOR Award. 2001. www.marketingpower.com/.

Ashman, H., S. Rabino, D. Minkus-McKenna, & H.R. Moskowitz. 2003. The shopper's mind: What communications are needed to create a "destination shopping" experience? Proceedings of the 57th ESOMAR Conference "Retailing/Category Management—Linking Consumer Insights to In-Store Implementation." Dublin, Ireland.

Ashman, H., I. Teich, & H.R. Moskowitz. 2004. Migrating consumer research to public policy: Beyond attitudinal surveys to conjoint measurement for sensitive and personal issues. ESOMAR Conference: Public Sector Research, May, Berlin, Germany.

Beckley, J.H. 2002. Mindsets: Uncovering consumer wants and needs . . . quickly. Chef to Shelf Conference, 20–21 May, Chicago, IL.

Beckley, J.H., H. Ashman, A. Maier, & H.R. Moskowitz. 2004. What features drive rated burger craveability at the concept level? *Journal of Sensory Studies* 19(1):27–48.

Beckley, J.H. 2010. Consumer Research on a Shoestring Budget. Presented at the 41th Annual Intermountain IFT Meeting, Sun Valley, ID. March 17. Available on Intermountain's Web site.

Bianco, A. 2004. The vanishing mass market. *Business Week* July 12:61–68.

Blackwell, R.D., P.W. Miniard, & J.F. Engel. 2001. *Consumer Behavior*, 9th edn. Mason, OH: Thomson South-Western.

Bowman, J. 2009. Crisp Insights. *Research World*. July/August:37.

Byron, E. 2010. Wash Away Bad Hair Days. Rinse. Repeat. *New York Times*. June, 30.

Chou, C., K. Gummaraju, M. Prabhakaran, & V. Siva. Semantics Driven Consumer Insights. www.infosys.com.

Collins, J. 2001. *Good to Great*. New York: HarperCollins.

Drucker, P.F. 2001. *The Essential Drucker*. New York: HarperCollins.

Eisenfeld, B. 2009. Research on-the-cheap. *Quirk's Marketing Research Review*. February :48–51.

Evans, P. & T.S. Wurster. 2000. *Blown to Bits*. Boston, MA: Boston Consulting Group.

Forster, J. 2002. Thinking outside the burger box. *Business Week Online*, September 16.

Frankwick, G.L., B.A. Walker, & J.C. Ward. 1994. Belief structures in conflict: Mapping a strategic marketing decision. *Journal of Business Research* 31:183–195.

Freeman, L. 2001. Small, smaller, smallest: New analytical tools can slice markets too thin. *Marketing News* September 24:1.

Gladwell, M. 2000. *The Tipping Point*. New York: Back Bay Books/Little, Brown.

Gladwell, M. 2005. *Blink*. New York: Back Bay Books/Little, Brown.

Gladwell, M. 2008. *Outliers*. New York: Back Bay Books/Little, Brown.

Gerzema, J. & E. Lebar. 2009. The trouble with brands. *Strategy + Business*. Issue 55, Summer 2009. pp. 49–57.

Gupta, S., D.R. Lehman, & J. Ames Stuart. 2004. Valuing customers. *Journal of Marketing Research* 41(1):7–14.

Hampp, A. & R. Parekh. 2010. Why Microsoft killed Kin after just six weeks. Advertising Age. June 30.

HealthFocus International. www.healthfocus.com/reports.htm.

Helm, B. 2010. Procter & Gamble in bind over Mom's web attack on Pampers. www.bloomberg.com. May 14.

Johnson, L. 2010. Augmented Reality for Marketers: Mapping the Future of Consumer Interactions. American Research Foundation, Web 2.0 Expo, September 28.

Johnson, S. 1998. *Who Moved My Cheese?* New York: G.P. Putnam's Sons.

Kuczynski, A. 2009. Pulp Friction. *New York Times*. Spring 2009.

Lee, N., A. Broderick, & L. Chamberlain. 2007. What is "neuromarketing?" A discussion and agenda for future research. *International Journal of Psychology* 63(2):199–204.

Lencioni, P. 2002. *The Five Dysfunctions of a Team*. New York: Jossey-Bass.

Lerer, L. 2010. Under attack, pollsters debate their methods. *Bloomberg Businessweek*. July 26-Aug. 1:30–31.

Levitt, S. & S. Dubner. 2005. *Freakonomics*. New York: Harper Perennial.

Mintel International Group. www.mintel.com.

Moskowitz, H.R., J.B. German, & I.S. Saguy. 2005. Unveiling health attitudes and creating good-for-you foods: The genomics metaphor, consumer innovative web-based technologies. *Critical Reviews in Food Science and Nutrition* 45(3):165–191.

Moskowitz, H.R. & A. Gofman. 2007. *Selling Blue Elephants. How to make great products that people want before they even know they want them*. Upper Saddle River, NJ: Wharton School Publishing.

Moskowitz, H.R. & A. Gofman 2010. Improving customers targeting with short intervention testing. *International Journal of Innovation Management* 14:435–448.

Moskowitz, H.R., A. Gofman, M. Bevolo, & T. Mets. 2010a. Decoding consumer perceptions of premium products with rule-developing experimentation. *Journal of Consumer Marketing* 27(5):425–436.

Moskowitz, H.R., A. Gofman, L. Ettinger Lieberman, I. Ray, & S. Onufrey. 2009. Sequencing the genome of the customer mind by experimental design and short-intervention testing. Presented at the Society for Marketing Advances Conference: Embracing Challenges and Change: A Global Perspective, November 4–7, New Orleans, LA.

Moskowitz, H.R., A. Gofman, L. Ettinger Lieberman, I. Ray, & S. Onufrey. 2010b. Sequencing the genome of the customer mind by experimental design and short-intervention testing. *Journal of Academic and Business Ethics* 3. Available online at http://www.aabri.com/manuscripts/10574.pdf and in print.

Moskowitz, H.R. & S. Onufrey. 2008. Rethinking segmentation. *ABA Banking Journal,* October (cover story).

Moskowitz, H.R., S. Porretta, & M. Silcher. 2005. Creating an Integrated Database from Concept Research. In: *Concept Research in Food Product Design and Development*. Ames, IA: Blackwell Publishing Ltd.

Naisbitt, J. 1988. *Megatrends*. New York: Grand Central Publishing.

Neff, Jack. 2004. J&J considers next step for oral, skin-care business. *Advertising Age* August 30:3.

PERT. www.pert.com.

Peters, J. 2010. At Yahoo, Using Searches to Steer News Coverage. *New York Times,* July 5.

Pigott, M. 2002. Looking beyond traditional models. *Marketing Research* Fall:8–11.

Pine, B.J., II & J.H. Gilmore. 1999. *The Experience Economy.* Boston, MA: Harvard Business School Press.

Popcorn, F. 1992. *The Popcorn Report*. New York: Harper.

Ranaweera, C., H. Bansal, & G. McDougall. 2008. Web satisfaction and purchase intentions: Impact of personality characteristics during initial web site visit. *Managing Service Quality* 18(4):329–348.

Reibstein, D., N. Bendle, P. Farris, & P. Pfeifer. 2006. Marketing Metrics: Understanding market share and related metrics from *Marketing Metrics: 50+ Metrics Every Executive Should Know*. Philadelphia, PA: Wharton School Publishing.

Semon, T.T. 2003. Wealth not income indicates spend capability. *Marketing News* October 13:15.

Smith, C.S. 2004. A French employee's work celebrates the sloth ethic. *New York Times*, August 14.

Stengel, J. 2004. Stengel's call to arms. *Advertising Age* February 16:1.

Tchong, M. 2003. *Trendscape 2004*. San Francisco: Trendscape.

Toffler, A. 1970. *Future Shock*. New York: Random House.

Wells, M. 2003. In Search of the Buy Button. www.Forbes.com. (September 1).

Wittink, D. 2004. Editorial. *Journal of Marketing Research* 41(1):2.

Unilever. 2006. US launch of Sunsilk hairspray line of hair products.

Zahay, D., A. Griffin, & E. Fredericks. 2003. Sources, uses and forms of data in the new product development process. Paper presented at PDMA Conference, 4–5 October, Boston, MA.

6 Developing relevant concepts

The first step in product development is to identify what the consumer wants. To this end, a whole discipline of concept development has arisen in marketing research, which is now migrating to the corporate sensory group on the one hand, and to the outside ingredient supplier on the other.

WHAT IS A CONCEPT AND WHY IS IT IMPORTANT IN EARLY STAGE DEVELOPMENT?

A concept is simply a description of the product or service. A concept is nothing more. What is important to know about the concept is that it comes in at least two flavors—product concepts versus positioning concepts. Concepts range from the very simplest description to the very complex, well-articulated description. Anyone in the corporation can make a concept, but quite often the creation of concepts is left to specialists, who may or may not be able to do a good job.

Examples of product concepts and a related positioning concept appear in Figures 6.1, 6.2, and 6.3. Note that the concept comprises words and pictures. Fundamentally, the concept describes the product and/or presents reasons why the product should be purchased. Rather than looking "bare-bones" like the concepts in Figures 6.1, 6.2, and 6.3, sometimes the concepts may look more like complete paragraphs, of the type one might read in a newspaper. At the end of the day, however, the concept is really the idea about the product embodied in text. The rest, the actual physical form of the concept, varies from company to company, often at the whim of the product developer, the marketer, or the advertising agency.

The concept sets the expectations about the product. For most product developers the concept constitutes the descriptive blueprint about what the product should look like, feel like, and taste like. Most concepts do not go into such depth that they constrain the product developer to follow a certain path, but rather suggest the sensory attributes of the product. The product developer uses his or her insight and experience to create a physical prototype (called *protocept*) that delivers what the concept ostensibly promises. In many cases there

Sensory and Consumer Research in Food Product Design and Development, Second Edition.
Howard R. Moskowitz, Jacqueline H. Beckley, and Anna V.A. Resurreccion.
© 2012 Blackwell Publishing Ltd. Published 2012 by Blackwell Publishing Ltd.

A product concept (text + graphics)

All natural ingredients
Rich fruit pieces
And a toy in each box

Figure 6.1   A product concept with text and graphics.

A product concept (text only)

All natural ingredients

Rich fruit pieces

And a toy in each box

Figure 6.2   A product concept with text only.

is a substantial leeway in the developer's options, because no one, consumer or developer alike, knows what to expect.

When the concept goes beyond the description of the product to certain types of health claims, constraints may emerge because of legal and technical limitations. Some health promises limit the product developer in terms of formulations because they promise a specific number of calories, a certain type of ingredient, or a certain set of constraining health benefits that can only be delivered by specific ingredients, respectively.

Whether the concept is product oriented or positioning oriented, the concept comprises a fundamental step in the development process. Without concepts, there is no systematic way

A positioning concept (text only)

A great breakfast

For kids and adults alike

And good for an afternoon snack too

Figure 6.3   A positioning concept with text only.

to guide the developer, and no way to plan an array of products consistent with the corporate business objectives. Concepts guide, constrain, and represent the business objectives.

HOW SENSORY ANALYSTS GOT INVOLVED IN CONCEPT RESEARCH

For a very long time, sensory analysis concentrated only on products. The sensory researcher abandoned concept research to the marketer and marketing researchers. In some respects, this abandonment of concept research to other specialists in the corporation can be easily understood. Having come from scientific specialties, sensory analysts would not have been expected to understand the nuances of communication, nor would they even have been considered when it came to evaluating the goodness or badness of a concept. The sensory analyst would be expected to work with product development to create the *protocept*, or prototype corresponding to the concept.

As is always the case, corporate boundaries are permeable when it comes to products and concepts. It is almost impossible to exclude interested corporate members in product development initiatives, especially since the products are the lifeblood of the corporation. Therefore, it should come as no surprise that interest in concept research emerged in the sensory world as the professional became involved in issues beyond the pure product. When the sensory professional was limited to expert panels and quality control the notion of concept was irrelevant since concepts had little to do with the current product and with product description. As soon as the sensory specialist got involved with new product development, was stimulated by marketing, and came into contact with consumers, interest in the concept piece of new product development began to skyrocket. Sensory specialists were not, however, invited particularly often to ideation sessions until much later.

With the growing integration of different departments dealing with early stage development, the sensory analyst has been given a "seat at the table." This "seat," an overused phrase to describe corporate importance, is well deserved. Sensory analysts are responsible for early stage development guidance. How can they be expected to guide development when they are not privy to the nature of the product as conceived of by the marketer? Surprisingly, for decades sensory analysts were content to work in the dark, woefully ignorant of the marketing objectives behind the product, perhaps because their intellectual history from the expert panels blinded them to the importance of factors other than scientific purity.

DOING SIMPLE CONCEPT RESEARCH BY TESTING "BENEFITS"

Concept research comes in many types. The simplest version of concept research tests single ideas to identify which ones are acceptable, unique, and perhaps fit a certain end use. Whether the approach is called *benefits testing*, *promise testing*, *idea screening*, or any of a dozen other names, the objective is quite simple. The researcher sifts through a variety of alternative ideas in order to discover those that offer promise for the next step, such as full concept testing or even prototype development. The ideas may deal with features, with health statements, packaging ideas, and so forth.

Table 6.1 Results of benefit testing: mean ratings on 9-point scales for interest, uniqueness, fit to brand.

		Interest	Fit to brand	Unique
A	It can do everything a solid margarine can—but better	6.4	7.1	6.3
B	It can do everything a solid margarine can—but it's quicker, handier, and cleaner	6.0	6.5	4.5
C	The first product that gives a delicious taste, nice brown color and wonderful gravy at high temperatures without burning	5.8	4.3	6.6
D	The packaging has a transparent stripe and measuring marks at the side of the pack for quick and controlled dosing	4.9	4.6	7.1
E	For busy people	4.1	4.2	3.5
F	The packaging allows easy opening and reclosing	5.5	6.7	4.5

A sense of the benefits testing data can be seen in Table 6.1, which shows some different benefits, features, and so on, and the ratings of interest, uniqueness, and fit to the brand. There is no particular fixed set of questions that a researcher uses for promise testing, except perhaps for interest or some other measure of acceptance. Acceptance is the key evaluative criterion. On the other hand, most researchers also use attributes such as uniqueness because in business one of the goals is to create a new yet different product. If the promises or phrases that emerge most strongly are not unique, then the exercise will not be productive. Most companies want something that differentiates them. The particular execution of the promise test may be in person-to-person interviews, mail-out interviews, Internet-based interviews, and so forth. The key to the entire process is to have a large number of statements and allow the respondent to rate each on a variety of relevant attributes.

TRADITIONAL CONCEPT TESTING

Concepts constitute fully formed ideas, presented in either a simple fashion, or in some cases presented in a fully executed fashion with high-quality artwork in color. Concept testing, in contrast to promise testing, is often a hotly disputed territory in corporations. Traditionally, marketing researchers were the ones involved in concept testing because they were the gateway to marketing. The rationale was that after the promises had been screened and explored, the project should be turned over to those professionals who could measure concept performance.

Many market researchers went beyond simple concept scores, looking for added information and insight. Concept scores alone, showing performance, do not tell the researcher "why" or "and then what." Diagnostics tell the researcher what factors are driving concept acceptance/rejection, and play a similar role to product diagnostics. Market share simulations tell the marketer about how many cases will be sold, or how much of the market the new product is expected to capture. Market share analysis invokes other types of information, such as estimates of awareness, distribution, the effects of pricing and advertising, and normative baselines against which to compare concept performance.

We can get a sense of the type of information obtainable in concept tests from looking at the data in Figure 6.4.

Figure 6.4 Results for a single concept, including open ends.

The study dealt with a single concept. The respondents were selected to match the consumer profile of expected users. The respondents read the concept, evaluated it on a set of diagnostic attributes, and then provided open-ended responses (likes and dislikes). The data is typically presented in terms of responses by total panel and key subgroups. Looking at subgroups prevents the researcher from selecting a concept that polarizes the respondents so that some groups like the concept, whereas other groups dislike the concept. The subgroups can usually be created from the classification questionnaire that accompanies the concept.

This particularly basic structure to concept testing can be varied many times, such as presenting the respondent with different concepts, varying the concepts, and so forth. In the end, the objective is to determine whether the concept is good enough to proceed with. It is worth remarking that the concept test is done with a relatively large number of respondents (often more than 100, in some cases more than 300). The large base size is a legacy from sociological research. In concept studies there is no systematic variation of elements, and no way to search for patterns among stimuli in order to gain deeper understanding. Most concept tests of the sort shown here simply look for the winning concept. Diagnostics are useful to identify problems in the concept, but do not provide concrete direction.

SYSTEMATIC DEVELOPMENT OF CONCEPTS—CONJOINT ANALYSIS

In the early 1970s, researchers in marketing began to realize that the experimental design of concepts might lead to better products and services. Experimental design is a well-known research strategy in the physical sciences, and has been popular for more than 60 years (Box, Hunter, & Hunter, 1978). Chemists and physicists use experimental design to identify the conditions under which the yield of a process is maximized, subject to imposed constraints. The application of experimental design to more subjective aspects, such as concepts and products, simply needed individuals brave enough to apply proven methods in one discipline to problems in the other.

Conjoint measurement, one instantiation of experimental design, was developed in the 1960s by mathematical psychologists interested in the topic of fundamental measurement of subjective issues (Luce & Tukey, 1964). The real use of conjoint measurement began in the 1970s with the pioneering adaptations by Paul Green and Jerry Wind at the Wharton School of Business, University of Pennsylvania (Green & Srivnivasan, 1978). Both individuals were and continue to be researchers in marketing, who have made enormous advances in the field. Their contribution to concept testing was recognizing that the experimental design of concepts could easily identify the drivers of concept acceptance or communication. Rather than having the respondents attempt to communicate these drivers, Green and Wind suggested that one could easily identify such drivers simply by presenting the respondent with a set of concepts whose components vary in some design structure. When the respondents like some concepts more than others, or feel that some concepts or more unique, communicate some benefits more strongly, and so forth, then, in turn, this difference across concepts can be traced to the different elements. Since, however, the elements or concept components vary in a known way, statistical methods such as regression analysis are able to identify the part-worth contribution of each element.

The marketing research community quickly recognized the elegant simplicity of conjoint measurement. It would take many years for conjoint measurement to penetrate the sensory analysis field, primarily because conjoint measurement was connected with concepts, rather than with products. Linear thinking would reject concepts as "marketing-related" rather than "product-related." Sensory analysts were becoming familiar with experimental design and analysis through formal courses such as those taught at DuPont, such as DOE (design of experiments), or offered by consultants. The late arrival of conjoint measurement into sensory analysis circles was not an issue involving designed experiments per se. Rather, like so many other trends in science, it took the maturation of the sensory researcher, and the growth out from pure product work to understanding the consumer mind, to bring conjoint measurement into the sensory world. In the first decade of this twenty-first century, conjoint measurement has begun its acceptance in the R&D-oriented sensory world, not so much for advertising concepts or for concepts at all, but rather as a way to understand the mind of the consumer in a structured, quantitative fashion, and in turn recognizing this guidance for product development purposes.

Conjoint measurement comes in two types, motivated by different intellectual heritages. At one end of the spectrum are those conjoint studies wherein the respondent evaluates pairs of statements in terms of preference. The paired data are processed to generate a set of utilities that embody the choices. This paired comparison approach echoes the paired comparison methods in product testing where the respondent is instructed to choose which of the two stimuli is liked more, or is more intense. The paired comparison tradition comes from the original world view of Fechner (1865), the founder of psychophysics, who struggled with the problem of measuring perceptions and concluded (albeit perhaps incorrectly) that the human subject was simply unable to act as a measuring instrument. In Fechner's opinion and in subsequent approaches, one way around this inherent limitation was to have the respondent make paired comparisons of two stimuli. The degree of confusion or overlaps was inversely proportion to the closeness of the two stimuli on some underlying scale. Fechner's way

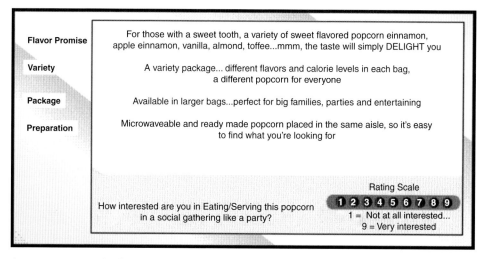

Figure 6.5 Example of a concept created by experimental design, showing the underlying architecture.

of thinking about its subsequent translation to Thurstonian scaling or scaling by "error" (Thurstone, 1927) led to descendent methods of paired comparisons and pairwise trade-off analyses. These two topics will not be dealt with further in this book.

A far more productive approach in concept measurement, but also in measurement of physical stimuli, requires the respondent to rate complete or partial profiles (Wittink & Cattin, 1989) These profiles, test concepts, or vignettes as practitioners refer to them, comprise systematically varied combinations. To the respondent the combination appears to be a concept, or perhaps a set of phrases strung together, such as Figure 6.5. The respondent typically "fills in" the connectives and has no trouble responding to this string of elements. To the researcher, however, each particular combination results from combining the elements according to the architecture dictated by the experimental design.

Once the researcher creates the elements, then the combinations according to a design plan, and then collects the ratings, it is a straightforward statistical task to estimate the part-worth utilities of the elements using ordinary least squares (OLS) regressions. OLS is by this time considered to be off-the-shelf software, widely available in most statistical packages. Table 6.2 shows an example of combinations, the representation of the elements in binary format, the statistical analysis of the data by regression analysis, and then the final utilities, which are the coefficients of the individual elements in the regression model.

There are a variety of issues involved in conjoint measurement, despite the methodological simplicity afforded by the regression analysis. These include the element creation, experimental design, methods to estimate the utilities of untested elements for larger studies, segmentation, and concept optimization, respectively. And there may be more issues, which emerge as conjoint measurement becomes increasingly popular.

Besides the methodological considerations are those involving the field execution, such as the number of concepts a respondent can test, the form of the concepts, the number of rating attributes, the issue of boredom and fatigue, and the detection of malingering. We will deal with each of these in a short fashion. Other, more comprehensive treatments provide the reader with the necessary background information (Moskowitz, Poretta, & Silcher, 2004).

Table 6.2 Combinations, statistical models, final utilities.

Part 1: Concepts structure
Concept color text

1	Blue	New
2	Blue	Great
3	Blue	Low
4	Green	New
5	Green	Great
6	Green	Low
7	Red	New
8	Red	Great
9	Red	Low
10	Blue	None
11	Green	None
12	Red	None
13	None	New
14	None	Great
15	None	Low

Part 2: Concept elements expanded as binary (yes/no) as well as rating of interest (Int)

Concept	Cblue	Cgreen	Cred	Tnew	Tgreat	Tlow	Int
1	1	0	0	1	0	0	7
2	1	0	0	0	1	0	6
3	1	0	0	0	0	1	4
4	0	1	0	1	0	0	4
5	0	1	0	0	1	0	2
6	0	1	0	0	0	1	5
7	0	0	1	1	0	0	9
8	0	0	1	0	1	0	4
9	0	0	1	0	0	1	6
10	1	0	0	0	0	0	8
11	0	1	0	0	0	0	2
12	0	0	1	0	0	0	5
13	0	0	0	1	0	0	7
14	0	0	0	0	1	0	5
15	0	0	0	0	0	1	4

Multiple R: 0.80 Squared multiple R: 0.64

Adjusted squared multiple R: 0.38

Effect	Coefficient	Standard Error	t	P (2 Tail)
CONSTANT	5.11	1.40	3.66	0.01
CBLUE	0.97	1.24	0.79	0.45
CGREEN	−2.03	1.24	−1.64	0.14
CRED	0.72	1.24	0.58	0.58
TNEW	1.72	1.24	1.39	0.20
TGREAT	−0.78	1.24	−0.63	0.55
TLOW	−0.28	1.24	−0.22	0.83

	Analysis of variance		
Source	Sum of squares	df	Mean square
Regression	36.34	6	6.06	$F = 2.42$
Residual	20.06	8	2.51	$P = 0.12$

C, color; Int, interest; T, tail.

METHODOLOGICAL ISSUES WITH THE STUDY ITSELF

(i) *Creating elements*: How are elements created? This question is quite easy to answer—one can just imagine the different features of the product, and list them. In actuality, however, most researchers act stymied when they select elements. It appears to be a formidable job, with a great deal of responsibility, and somewhat predisposed to failure, or so it might seem. To jumpstart the process and assure success one needs only to plow through the competitive frame of communications in the product category as well as allied or similar categories, identify the different elements, try these out in a preliminary conjoint study, return with the winning elements, expand them, and then run a second study. A little practice will soon reveal how straightforward is the task of creating elements.

(ii) *Experimental design*: Many researchers today are not particularly familiar with experimental designs for concepts, because the researcher either comes from a liberals arts background, or has been educated in science but thinks of experimental designs as being appropriate for the more complex, physical phenomena. Indeed, experimental designs are the meat and potatoes of some sciences. In sensory analysis, the notion of experimental design has only recently become widespread, and the idea of experimental designs applied to language and communication still represents *terra incognita* for the research community.

(iii) *Beyond physical variables to language*: Experimental designs come in a variety of incarnations. Most experimental designs require that the independent variable take on a set of selected values, which lie on a continuum. Language does not have a continuum, however, so the translation of experimental design to concepts immediately reaches an impasse. However, if one reconsiders experimental designs as a binary problem, so that each variable of several levels really represents several yes/no options, then things become quite simple. The experimental design becomes a series of yes/no options, or binary options. Every variable has a set of options, only one option of which can enter the combination. Quite often the experimental design can be defined so that one of the options is the null option—when that element appears in the experimental design, then no option in that variable appears (see Table 6.2).

Tables 6.3 and 6.4 show two different experimental designs that are commonly used. The experimental design in Table 6.3 comprises a five-level Plackett–Burman screening design (Plackett & Burman, 1946). Each category comprises five levels: 1, 2, 3, 4, 5. We can translate the design in a set of 25 combinations, 5 variables, with each variable appearing in 5 of the 25 combinations. Option 1 for each category is defined as the null category so that no element from the category appears. Each option either appears or does not appear. This stratagem of having some concepts that have no elements from a particular category enables the regression model to estimate the absolute magnitudes of the utilities using what is conventionally known as "dummy variable regression modeling." Dummy variable modeling treats the independent variables as either "0" if the element is missing from the concept, or "1" if the element is present in the concept.

Other main effects designs can be used as well, such as that shown in Table 6.4, comprising 36 elements combined in 60 combinations. We will use the design in Table 6.4 to explicate a conjoint analysis study. It is important to note, however, that Tables 6.3 and 6.4 need not limit us. There are many standard and customizable designs that could be used as well to deal with conjoint problems.

Table 6.3 Plackett–Burman screening design used to combine concept elements efficiently into relatively few concepts.

Concept	C1	C2	C3	C4	C5
1	4	1	3	1	1
2	0	4	1	3	1
3	3	0	4	1	3
4	3	3	0	4	1
5	2	3	3	0	4
6	3	2	3	3	0
7	4	3	2	3	3
8	1	4	3	2	3
9	2	1	4	3	2
10	2	2	1	4	3
11	0	2	2	1	4
12	2	0	2	2	1
13	4	2	0	2	2
14	3	4	2	0	2
15	0	3	4	2	0
16	0	0	3	4	2
17	1	0	0	3	4
18	0	1	0	0	3
19	4	0	1	0	0
20	2	4	0	1	0
21	1	2	4	0	1
22	1	1	2	4	0
23	3	1	1	2	4
24	1	3	1	1	2
25	4	4	4	4	4

Table 6.4 Example of experimental design with 36 elements in 60 combinations (first 20 of 60 combinations shown).

Concept	Category A	Category B	Category C	Category D	% top 3 box
Concept 1	0	1	0	9	34
Concept 2	9	1	1	1	27
Concept 3	0	4	4	7	45
Concept 4	0	5	0	8	17
Concept 5	8	8	8	0	33
Concept 6	5	2	5	0	29
Concept 7	5	8	0	2	32
Concept 8	7	7	7	0	41
Concept 9	4	4	7	0	19
Concept 10	5	5	0	5	27
Concept 11	4	0	6	7	54
Concept 12	6	0	4	0	39
Concept 13	0	7	4	0	47
Concept 14	1	9	0	3	18
Concept 15	7	6	0	7	28
Concept 16	2	2	5	0	39
Concept 17	6	7	6	4	58
Concept 18	2	5	8	2	46
Concept 19	8	2	8	8	32
Concept 20	0	6	7	4	18

In Table 6.4 each of four categories (A, B, C, D) comprises nine elements. An element may or may not appear in a specific concept. Each concept comprises 2–4 elements, with each category contributing at most one element to the concept. Occasionally, and according to the design the category may contribute no elements to the concept. The dependent variable "% top 3 box" shows the proportion of respondents from the set of 262 individuals who rated each particular combination 7–9 on a 9-point scale. This information (of course, for all 60 concepts) suffices to allow dummy variable, OLS regression to be performed on the responses from one single individual.

(iv) *OLS regression*: The goal of experimental design is to ensure that elements appear independently of each other, so that the researcher can estimate what each element contributes. The experimental design ensures that the elements are independent by the nature of the array. The independence permits the researcher to use methods such as regression analysis and logit analysis to estimate what every element contributes (Agresti, 1990). We will concentrate here on the conventional method, called OLS, because OLS results are easy to obtain and easy to interpret. Most statistical packages have embedded within them programs for OLS.

(v) *Setting up the data and thinking through the dependent variable—models for level of interest versus models for percentage of people interested*: The study deals with cola beverages, popular drink. Each of 262 respondents evaluated the 60 combinations or test concepts, rating each combination on a scale of interest. The scale is anchored, with 1 = not at all interested and 9 = extremely interested. There are other scales, other anchor points, but those remain for a different discussion, and are not germane here. Each respondent rates the 60 different combinations in a unique order. The order randomization is set up in order to minimize any chance of bias due to order. When one concept always follows a specific other concept, there is the perennial chance that ratings of one concept will bias the ratings assigned to the other. We don't know the nature of the bias, nor even whether or not the bias exists, but nonetheless the randomization removes this bias by canceling it out. After running the study the researcher is left with a database that looks like the data in Table 6.4.

We can perform two simple transformations on the data. One transformation simply multiplies the ratings by a convenient number, such as 11, to make the scale bigger. This multiplicative transformation does not affect the ratings, but does allow the researcher to work with larger numbers. Another transformation, much more profound, changes the nature of the data. Rather than working with a 9-point scale, the transformation changes the scale to a binary scale, to reflect concept rejection or concept acceptance. Marketing researchers are more accustomed to this type of scaling. One conventional, but not necessary cut-point, is 7. Ratings of 7–9 are transformed to "100." Ratings of 1–6 are transformed to "0." The transformation creates a binary scale in place of a category scale, and reflects concept rejection/acceptance. The transformation comes from the intellectual history of the researcher; sociologists and marketers are more accustomed to proportions of people showing a behavior than to the degree of interest of a single individual or the average level of interest across a population of individuals.

(vi) *Interpreting the results for the total panel—persuasion model*: Once the researcher runs the regression on the total panel data the results become instantly clear. The additive constant for the persuasion model shows the baseline interest in concept acceptance when no element is present in the concept. The additive constant is clearly an estimated parameter, since every concept comprised 2–4 elements by design. The coefficients or utilities represent the increment level of concept acceptance when the element is introduced into the concept. The coefficient can be large or small, positive or negative. Large coefficients mean that when the element is introduced into the concept it increases the acceptance by a great deal. Small coefficients mean that when the element is introduced into the concept it increases the acceptance, but by only a small degree. The sum of the additive constant and the coefficient of the particular concept elements represent the best estimate of the level of acceptance that the concept will enjoy. It is obvious that poorly performing elements can reduce the level of acceptance. Table 6.5 (column C) shows the results for cola beverages, using this type of analysis.

The results in Table 6.5 (column C; persuasion) are fairly straightforward to interpret. We already know that there were 262 respondents who participated in the study. We further know that the concept elements were arranged to allow for dummy variable regression. The concept elements are statistically independent of each other, making the regression straightforward and statistically valid.

The output of the regression is an equation of the form:

$$\text{Rating} = k_0 + k_1(\text{Element 1}) + k_2(\text{Element 2}) + \cdots + k_{36}(\text{Element 36})$$

We can interpret the results very simply for the *persuasion model* in a simple way, as follows:

(i) The additive constant is the expected rating on the 9-point (recall, multiplied by 11), corresponding to a concept with no elements. Of course all of the concepts comprised 2–4 elements in this particular experimental design. Thus, the additive constant is an estimated parameter. It is often interpreted as a baseline value. The constant is 5.9 on the 9-point scale. A constant of this level means that even without elements the idea of cola beverages is very attractive.

(ii) The elements vary. They tend to be quite small, however, for the positive elements that contribute to acceptance. Thus, the highest performing elements are the "traditional" type of elements ("A classic cola . . . just the way you like it"; "from Coca-Cola®"). These elements add just about a half point or slightly less on the 9-point scale.

(iii) Some elements do poorly, such as the lesser-known brands, or the mention of "Diet cola with a slice of lemon . . . the world's most perfect drink!" The poor performance might be due to the expected poor taste of diet cola, not necessarily linked to any brand.

(vii) *Interpreting the data—interest model*: As noted several times in this book, market researchers have a different intellectual history from sensory analysts. Market researchers can trace some of their history to sociology, with interest in the behavior of groups of individuals, rather than interest in the behavior of a single individual. The interest in laws governing groups is often referred to as "*nomothetic*," from the Greek word *nomos*, or law. The interest in individual responses is referred to as idiographic, from *idios* or individual. The consumer researcher typically reports the data in terms of "top 2 box" purchase intent, which refers to the number of respondents who give an answer. For conjoint measurement, this analysis may be done by one of two methods. When the respondents evaluate the same set of concepts, then instead of recording the level of interest on a scale, the researcher records the percent of respondents who, having seen the concept, rate the concept as interesting (e.g., 4–5 on a 5-point scale; 7–9 on a 9-point scale). When each of the respondents evaluates different concepts, as they do in

Table 6.5 Results from experimentally designed studies with cola concepts showing the winning and losing elements.

A	B	C	D
		Persuasion	**Interest**
	Base size	262	262
	Additive constant	5.9	47
Element01	A classic cola . . . just the way you like it	0.5	7
Element32	From Coca-Cola®	0.4	8
Element02	Carbonated, sparkling Cola . . . just the right amount of taste and bubbles	0.3	3
Element26	Pure satisfaction	0.3	3
Element34	Icy cold	0.2	4
Element27	It quenches *the thirst*	0.2	4
Element18	So refreshing . . . you have to drink some more	0.2	1
Element07	A thick slushy of cola and ice	0.2	4
Element24	Looks great, smells great, tastes delicious	0.1	2
Element10	Drinking cola is so inviting	0.1	1
Element22	Relaxes you after a busy day	0.1	2
Element17	You can imagine the taste even before you drink it	0.1	1
Element35	Resealable single serve container . . . to take with you on the go	0.1	0
Element14	So refreshing you want to savor how it makes you feel	0.1	2
Element20	When you think about it, you have to have it . . . and after you have it, you can't stop drinking it	0.1	3
Element13	Premium quality	0.1	1
Element23	A great way to celebrate special occasions	0.1	0
Element21	Simply the best	0.0	1
Element25	A wonderful experience . . . shared with family and friends	0.0	−1
Element33	Multiserve containers . . . so you always have enough!	0.0	−2
Element19	Quick and fun . . . ready to drink	0.0	−1
Element15	100% natural	0.0	0
Element36	With the safety, care and quality that makes you trust it all the more	0.0	−1
Element08	Cola . . . the perfect mixer for everything you drink	−0.1	−4
Element03	A perfect beverage . . . with breakfast, lunch, a break, or dinner	−0.1	−4
Element31	From Pepsi®	−0.2	0
Element09	An ice cream float—cola, ice cream . . . chilled and tasty	−0.3	−2
Element16	Flavored colas . . . cherry, vanilla, lemon . . . whatever you're looking for	−0.5	−7
Element11	Fortified with important vitamins and minerals for your body	−0.5	−9
Element30	From Dr. Pepper®	−0.5	−5
Element05	Cola . . . all the taste but only one calorie	−0.6	−8
Element04	Cola . . . the dark brown color, faint smell of vanilla, and bubbles tell you, you have real cola	−0.7	−12
Element12	With twice the jolt from caffeine . . . gives you just the added energy you need	−0.7	−9
Element28	From Royal Cola®	−0.9	−14
Element29	From Tab®	−1.7	−23
Element06	Diet cola with a slice of lemon . . . the world's most perfect drink!	−2.0	−24

Source: Data courtesy of It!™ Ventures, llc.

Note: The first data column (C) of the table shows the utility values from the persuasion model. The second data column of the table shows the utility values from the interest model, where ratings of 1–6 were transformed to 0; ratings 7–9 were transformed to 100.

IdeaMap®.net, then the foregoing strategy cannot work, because the respondents do not evaluate the same set of concepts. An alternative, and equally powerful strategy, records a 100 for the rating of the concept when the respondent accepts the concept (7–9) or records a 0 when the respondent rejects the concept (1–6). It is important to keep in mind that this binary cut of the 9-point scale is arbitrary and is done according to convention set up ahead of the study to denote a concept acceptor versus a concept rejector.

Looking at Table 6.5, but this time concentrating on the interest model (column D), we would conclude the same thing as we did previously, but the numbers would mean different things. This time the additive constant, 47, means that a cola concept without elements would score 47 on the top 3 box. This 47 means that without the help of elements, approximately 47% of the respondents would rate the "theoretical" cola concept 7–9, on the 9-point scale. Brand name "Coca-Cola®" generates an additive constant of 8, meaning that an additional 8% of the respondents, or a total of 55% (47 + 8 = 55) would rate the concept between 7 and 9 if the concept were to have the name of Coca-Cola®, and nothing else.

We could continue this process of creating a new concept about cola, either by selecting winning elements, or even by looking at concept elements that fit the strategy, such as flavor or tradition, respectively. The researcher can estimate the acceptance of the concept by adding together the additive constant and the individual utility values.

For scientific purposes, it is important to limit the number of concept elements to what a respondent can comprehend in a short reading. Occasionally, the sum of the additive constant and the individual element utilities will be very high, especially when the researcher "*cherry picks*" the elements so that only the high-performing elements enter the concept. At the extremes, where all of the elements are high, the estimation of the level of acceptance may not be accurate, simply because by cherry picking the best elements the researcher has gone outside of the limits obtained in the actual study. At those outside limits, the estimation of concept performance will be by the nature of the data simply less accurate.

It is important to recognize that this type of information would not have been obtainable from conventional concept tests because conventional evaluations test one version of a complete concept. They cannot identify the part-worth contribution of the element. Furthermore, the information could not be obtained from evaluation of single concept elements in isolation (promise testing) because it is difficult for the respondent to evaluate single elements such as price without a context. Conjoint analysis generates its own context, and forces the respondent to evaluate a combination of elements, each of which competes with every other in the concept. This competition, and the respondent's inability to be totally analytical at a conscious level force the respondent to react intuitively to the elements, thus giving a more honest answer.

SEGMENTATION

People differ from each other. The classic research strategy to address these interindividual differences looks at the responses from different groups of respondents, with defined differences, such as age, gender, market, product usage, and so forth. Over the last several decades, however, consumer researchers have discovered that these ever-present individual differences cannot be accounted for simply by the differences in such geodemographic or brand usage patterns. Some deeper pattern exists. The notion of psychographic segments has emerged from the marketer's frustration with conventional ways that fail to account for the differences. Psychographics assumes that in some yet unknown fashion the consumers differ because of their mind-sets or values. Marketers don't necessarily want to know how these differences come to be, as might a scientist, but rather seek statistical methods that allow them to break apart this apparently homogeneous population into different groups, showing different behaviors (Wells, 1975).

The conventional way to segment respondents is by standard geodemographics (e.g., gender, age, income, market), or by psychographic profiles (e.g., self-stated profiles that are assumed to tap into deeper aspects of the mind). These conventional ways to divide respondents arrive at similar results to the total panel. There are always the odd variations from one subgroup to another (e.g., ages 61–75 are different than the others), but by and large winning elements in one group are winning elements in the other groups. We can see this similarity clearly in Table 6.6, which presents the data from our cola study, this time looking at different subgroups of consumers. The subgroups were generated based on how the respondents profiled themselves.

Another way to divide respondents looks at how they describe themselves. For this exercise we divide the respondents into three groups (Traditional, Add Excitement, Total Experience). These three classes of individuals were developed from an analysis of response patterns to different foods in the Crave It!™ Studies (Beckley & Moskowitz, 2002). The Crave It!™ studies looked at 30 different foods, and 36 elements for each food. On the basis of the utility values from more than 4,000 respondents, these three segments emerged. The respondents in that study may not have been aware of the segmentation, nor had any idea whether or not they belonged to a segment. However, as an organizing principle the segmentation made a great deal of sense.

When we instruct the respondent to classify himself or herself into one of the three segments, based simply on a statement about the segment, we find that most people in the cola study classified themselves as "Traditional," slightly fewer as "Add Excitement," and far fewer as "Total Experience." Yet, when we look at the winning elements from this self-defined psychographic segmentation (Table 6.7), we find their utility values to be quite similar. Thus, this type of psychographic segmentation does not divide the respondents into the different groups as we might have hoped.

The segmentation becomes more productive when the researcher divides respondents by the pattern of concept acceptance. There are a variety of different methods by which to segment respondents. In one popular method, requiring only the actual utilities themselves from the persuasion model, the researcher uses an index of similarities of patterns across the different respondents, based on the element utilities. The approach, embodied in off-the-shelf statistical programs (e.g., Systat, 2004), follows these simple six steps:

(i) *Create the model relating the presence/absence of concept elements to the rating.* This is the aforementioned persuasion model. The persuasion model is the better model to use in segmentation because it shows the *intensity* of a respondent's feeling toward a concept. The Persuasion model is "efficient," because it uses all the data, rather than recoding the data to a binary format as the interest model does.

(ii) *Create the persuasion model for each respondent.* For our cola data this means creating 262 regression models, each modeling relating the presence/absence of the same concept elements to the acceptance rating. Each respondent generates 36 regression coefficients, or utilities, one coefficient for each of the 36 concept elements for cola. The systematic approach to concept design ensured that each respondent would test 60 concepts, sufficient to estimate the utility of each of the 36 elements.

Table 6.6 Utilities for cola based on self-described geodemographic groups.

	Total	Frequency		Gender		Age				
		Daily	< Daily	M	F	21-30	31-40	41-50	51-60	61-75
Base size	262	185	51	70	192	32	62	87	61	20
Additive constant	47	47	51	46	48	48	51	48	50	27
E32 From Coca-Cola™	8	7	**10**	**12**	6	**16**	2	9	8	6
E01 A classic cola … just the way you like it	7	8	7	8	7	6	7	5	**12**	6
E34 Icy cold	4	6	0	6	3	0	1	5	5	**16**
E07 A thick slushy of cola and ice	4	5	2	0	5	4	5	3	5	−1
E27 It quenches *the thirst*	4	4	3	2	4	3	1	4	7	2
E02 Carbonated, sparkling cola … just the right amount of taste and bubbles	3	4	1	4	3	−3	4	4	5	2
E20 When you think about it, you have to have it … and after you have it, you can't stop drinking it	3	3	2	2	3	5	1	2	1	**16**
E26 Pure satisfaction	3	2	6	0	4	5	2	4	−2	8
E22 Relaxes you after a busy day	2	1	6	2	2	1	1	2	5	6
E14 So refreshing you want to savor how it makes you feel	2	2	3	4	2	−6	2	3	4	7
E24 Looks great, smells great, tastes delicious	2	3	−1	2	2	3	1	0	2	6
E17 You can imagine the taste even before you drink it	1	1	1	−1	2	0	2	0	4	−4
E13 Premium quality	1	1	1	−1	2	−3	1	4	1	−3
E18 So refreshing … you have to drink some more	1	1	1	2	1	−2	−3	2	5	5
E21 Simply the best	1	1	0	−1	2	3	−1	−1	1	8
E10 Drinking Cola is so inviting	1	0	2	0	1	−6	−1	3	2	0
E35 Resealable single serve container … to take with you on the go	0	0	1	2	0	2	4	−3	−2	7
E31 From Pepsi™	0	0	0	3	−1	−7	9	−3	−4	**10**
E15 100% natural	0	0	−1	6	−3	−2	2	−2	−2	6

E23	A great way to celebrate special occasions	0	0	0	-2	0	-4	-4	3	-2	7
E36	With the safety, care and quality that makes you trust it all the more	-1	0	0	3	-2	-5	2	-2	-2	7
E25	A wonderful experience...shared with family and friends	-1	-1	0	-7	2	3	-4	3	-6	3
E19	Quick and fun...ready to drink	-1	-2	-1	-1	-1	2	-7	-1	2	0
E33	Multiserve containers...so you always have enough!	-2	-1	-4	-4	-1	4	-2	-3	-1	-3
E09	An ice cream float—cola, ice cream...chilled and tasty	-2	-3	0	-1	-3	-1	-4	-4	0	0
E08	Cola...the perfect mixer for everything you drink	-4	-4	-4	-4	-4	-9	-5	-4	-1	0
E03	A perfect beverage...with breakfast, lunch, a break, or dinner	-4	-2	-9	2	-6	**-11**	-1	-2	-1	**-16**
E30	From Dr. Pepper™	-5	-5	-4	1	-6	**12**	-5	**-10**	-6	-1
E16	Flavored colas...cherry, vanilla, lemon...whatever you're looking for	-7	**-10**	1	-3	-8	**-12**	-6	-7	-9	4
E05	Cola...all the taste but only one calorie	-8	-7	**-10**	**-11**	-7	-9	-7	**-12**	-9	**13**
E11	Fortified with important vitamins and minerals for your body	-9	**-11**	-3	-8	-9	-9	**-10**	**-10**	-8	-3
E12	With twice the jolt from caffeine...gives you just the added energy you need	-9	-8	**-11**	-6	**-10**	-1	-6	**-11**	**-16**	2
E04	Cola...the dark brown color, faint smell of vanilla, and bubbles tell you, you have real cola	**-12**	**-13**	-8	-9	**-13**	**-19**	**-13**	**-11**	**-13**	**-11**
E28	From RC Cola™	**-14**	**-15**	**-14**	1	**-20**	**-16**	-9	**-19**	**-16**	0
E29	From Tab™	**-23**	**-24**	**-22**	**-18**	**-25**	**-19**	**-22**	**-23**	**-29**	**-23**
E06	Diet cola with a slice of lemon...the world's most perfect drink!	**-24**	**-25**	**-20**	**-20**	**-25**	**-31**	**-32**	**-20**	**-26**	7

Source: Data courtesy of Irl Ventures.
Note: The boldface cells correspond to elements whose utility values are greater than or equal to +10, or less than or equal to −10. The boldfaced type highlights the pattern of differences among the groups on key attributes.

Table 6.7 Utilities for cola based on three self-described psychographic segments, and three mind-set segments.

		Total	Traditional... straight-forward... why mess a good thing?	Plain and simple can be so boring... add some excitement with variety, accessories... lots of stuff	It is the total experience that matters most... the anticipation, the sharing... a good time with family and friends	S1	S2	S3
	Base size	262	73	61	128	98	137	27
	Constant	47	48	52	45	44	52	38
	Self-described Traditional							
E32	From Coca-Cola™	8	7	9	8	19	2	−5
E01	A classic cola...just the way you like it	7	6	10	7	5	9	9
	Self-described Add excitement							
E01	A classic cola...just the way you like it	7	6	10	7	5	9	9
E32	From Coca-Cola™	8	7	9	8	19	2	−5
	Total experience							
E32	From Coca-Cola™	8	7	9	8	19	2	−5
E01	A classic cola...just the way you like it	7	6	10	7	5	9	9
E34	Icy cold	4	1	1	7	2	6	3
E07	A thick slushy of cola and ice	4	3	−3	7	10	−1	7
E27	It quenches *the thirst*	4	−2	5	6	2	6	1
	Mind-set Segment 1— traditionalists							
E32	From Coca-Cola™	8	7	9	8	19	2	−5
E07	A thick slushy of cola and ice	4	3	−3	7	10	−1	7
E09	An ice cream float—cola, ice cream...chilled and tasty	−2	1	−7	−2	10	−9	−13
	Mind-set Segment 2-excitement seekers							
E31	From Pepsi™	0	−2	−2	3	−18	10	19
E01	A classic cola...just the way you like it	7	6	10	7	5	9	9
E34	Icy cold	4	1	1	7	2	6	3
E27	It quenches *the thirst*	4	−2	5	6	2	6	1
	Mind-set Segment 3—the experientials							
E06	Diet cola with a slice of lemon...the world's most perfect drink!	−24	−20	−19	−28	−16	−38	22
E31	From Pepsi™	0	−2	−2	3	−18	10	19
E01	A classic cola...just the way you like it	7	6	10	7	5	9	9
E14	So refreshing you want to savor how it makes you feel	2	−1	0	5	2	1	9
E07	A thick slushy of cola and ice	4	3	−3	7	10	−1	7
E13	Premium quality	1	0	0	2	−1	2	6

Source: Data courtesy of It!™ Ventures.

(iii) *Create a measure of distance or dissimilarity between each pair of respondents by using the statistic (1-R), where R = the Pearson correlation coefficient.* The value of R goes from a high of $+1$ (when two sets of 36 utilities are perfectly linearly related to each other), down through 0 (when they are not at all related), and further down to -1 (when they are inversely related). Thus, a value of R of 1 means that two respondents show a perfect linear relation. Their distance is $1 - 1 = 0$, as it should be, because they are functionally identical to each other. In contrast, two respondents showing a correlation of -1 are as dissimilar as they could be. The elements that one respondent likes the other respondent dislikes. Their pairwise distance is 2, or $1 + 1 = 2$, as it should be. Once the pairwise distances are computed, clustering programs partition the respondents into a limited number of sets, in order to minimize some specified criterion. The researcher specifies the number of subgroups or segments, but beyond that the statistical program does the grouping of respondents into the individual clusters, based on the statistical criteria that are to be minimized.

(iv) *The segmentation that emerges from this type of analysis does not label each of the segments.* That task of interpreting is left to the researcher What the clustering/segmentation program does, however, is divide the respondents by the patterns of their responses, an approach that lies in between pure "attitudes" of "what I say I am" and pure behavior of "what do I do when I go shopping." The reaction to concepts can be likened to reactions to small advertisements about a product.

(v) When all is said and done, three new segments emerge from this analysis. We see these three segments in Table 6.7 (last three columns). The elements that perform well within a segment tend to display a clearer theme, although the theme is not a single-minded one.

(vi) *The segments are relatively homogeneous in what appeals to them.* Thus, there are no countervailing forces, wherein two groups of respondents with different mind-sets cancel each other out. We see some of this canceling effect in Table 6.7, even for the brand "Coca-Cola®." This brand has a utility value of $+8$ for the total population. The experiential segment (Seg 3) dislikes it, with a utility of -5. The only reason that the Coca-Cola® brand scores so high is that the experientials constitute only a small (10%) proportion of the population. Were they to comprise say 50% of the population, then we would see far poorer performance of the element with the "Coca-Cola®" brand name.

WHAT CONCEPT RESULTS ARE USED FOR

By itself, concept research is simply an exercise in identifying responses to stimuli. These results find application in a variety of end-uses:

(i) *Knowledge building*: Responses to concept elements show how the consumer responds to the aspect of products. Quite often the researcher does not know how the consumer will react to product features, to brands, and so forth. In a highly competitive business such as carbonated beverages, it is impossible for management to predict how consumers respond to all the different aspects of the product when they are tested in concept format. Going into a concept study, no one really knows the value of the brand (i.e., the utility of the brand name), or the relative importance of messages such as health versus taste. Conjoint measurement provides that type of information.

(ii) *Product and packaging guidance*: The elements can deal with concrete features of a product. These include options for new products, which comprise combinations of features. In some product areas there are dozens, or even hundreds of alternative ideas that can be used for new products. In other product areas, such as cola beverages, the range is constrained. The product developer can put together new combinations of features at the concept level to identify winning new ideas. A sense of different product concepts emerging from this exercise can be obtained from Table 6.8, which shows three concepts: one from the total panel, and one concept each from the respondents ages 21–30 and 61–75, respectively. Table 6.8 also shows the expected contribution of utility values from each subgroup to the total utility of the concept. It is from this type of data that the developer can simulate responses to alternative product ideas.

(iii) *Advertising guidance*: Several of the elements in the exercise deal with messages about why the respondent should buy the product, rather than what the product contains. These messages can be used to guide development because the developer can interpret the meaning of the elements. The messages can also be used to guide the advertising agency, which always looks for winning ideas that will break through the clutter of competitors. These messages may be either product statements or positioning statements.

Table 6.8   Three newly synthesized concepts for cola (without brand name), based on the results from the conjoint analysis.

		Total sample	Ages 21–30	Ages 61–75
	Base size	262	32	20
	Constant	47	48	27
	Total sample			
E01	A classic cola...just the way you like it	7	6	6
E27	It quenches *the thirst*	4	3	2
E02	Carbonated, sparkling Cola...just the right amount of taste and bubbles	3	−3	2
E01	A classic cola...just the way you like it	7	6	6
	Sum of utilities	69	60	42
	Age 21–30			
E01	A classic cola...just the way you like it	7	6	6
E26	Pure satisfaction	3	5	8
E20	When you think about it, you have to have it...and after you have it, you can't stop drinking it	3	5	16
E33	Multiserve containers...so you always have enough!	−2	4	−3
	Sum of utilities	59	67	53
	Age 61–75			
E34	Icy cold	4	0	16
E20	When you think about it, you have to have it...and after you have it, you can't stop drinking it	3	5	16
E05	Cola...all the taste but only one calorie	−8	−9	13
E14	So refreshing you want to savor how it makes you feel	2	−6	7
	Sum of utilities	49	37	79

MESSAGE DECONSTRUCTION: SYSTEMATIC CONCEPT RESEARCH AS A STRATEGIC TOOL

The foregoing treatment of concepts considered them as descriptions of new products. Concept research is much more, however. Concepts need not be new ideas. Rather, concepts may comprise elements of the competitive framework, currently in the market. The researcher need only identify competitors, see what these competitors feature in their advertising, deconstruct the advertising into their components (concept elements), recombine, measure the response, and then identify the contribution of the component elements to consumer interest. The benefit to the researcher and the food industry is the concept exercise to identify what is currently working and what is not. A side benefit is the identification of new segments in the population, based on the response to the concepts, and the assessment of the degree to which current competitor offerings appeal to these different segments.

To better understand concept testing as a strategic tool, let us move from colas to weanling foods. When a baby begins to eat solid food, and is weaned from the breast, the mother wants to learn about products, benefits, and so forth. A number of web sites feature such weaning foods. By looking at these messages the researcher quickly identifies which messages "work" among the target population (mothers who are or who have recently weaned their babies). Table 6.9 shows the results of this exercise. Thus, in Table 6.9 we see that there are three segments, and that no competitor has a "lock" on a particular segment. This state of affairs comes about because most manufacturers want to appeal to as many target consumers in the population as they possibly can, even if the appeal is only modest to the different segments. Were the data to come out differently, with one manufacturer appealing strongly to a segment, we might conclude that the manufacturer had identified the segment, and targeted its product development toward that segment.

PACKAGE DESIGNS AS CONCEPTS

Most researchers think of concept testing as the assessment of text and graphics, with the graphics element being a picture. Recently, however, researchers have begun to look at graphics in the same structured way that they look at concepts. That is, to the researcher schooled in conjoint measurement, graphics design comprises combinations of graphics features in much the same way that concepts comprise combinations of text features. When approached in this fashion, package design research can be considered simply as another type of concept research.

For the food researcher the systematic approach can pay dividends. Whereas previously a great deal of package research concentrated on the evaluation of single test stimuli, the designed experiment allows the researcher to identify what features of the package drive acceptance and communication of specific impressions, such as "heathfulness." When the researcher also measures the response time to the packages, then this response time or latency can become a dependent variable, and the researcher discovers the amount of time each feature of the package requires to be processed.

A sense of this type of research approach can be obtained by looking at a simple study of packages for frankfurters. Figure 6.6 shows the components of the design. The different components can be systematically varied and then recombined by experimental design into a number of different packages. To the respondent the packages look as if they comprise

Table 6.9   Deconstruction of web sites dealing with weaning.

		Total	Seg 1	Seg 2	Seg 3
	Base size	200	69	58	73
	Constant	69	65	72	70
	Segment 1—control and process				
D1	Wait 3–5 days to add each new food to baby's diet…so you can detect any allergic reaction	6	**13**	6	−1
D4	Introduce oat or barley cereal once baby has adjusted to rice cereal	2	8	−3	−1
D6	Continue feeding cereal for the first 2 years, for a healthy start in life	−2	8	−1	−14
A8	A variety of organic baby food designed specifically to keep babies healthy	−1	8	−12	−1
D5	Begin solids with one feeding a day…if baby doesn't like it, try again in a few days	2	7	−2	−1
D2	Making the transition from breast or bottle to solid foods is a big event for both you and baby	4	6	1	5
D7	Single-ingredient baby cereals like rice make a perfect introduction…fortified with the additional iron your growing baby may need	2	6	1	−1
	Segment 2—authority recommended practice				
B5	Baby food with DHA and ARA…nutrients found naturally in breast milk help support babies' mental and visual development	4	−2	9	5
B2	Lifelong eating habits and tastes are learned early	2	−4	7	3
C7	The American Academy of Pediatrics recommends breastfeeding exclusively for the first 6 months…gradually introducing solids while continuing to breastfeed for 6 more months	0	2	7	−7
D8	Start with one or two tablespoons until baby gets used to swallowing solids	0	−3	6	−3
D1	Wait 3–5 days to add each new food to baby's diet…so you can detect any allergic reaction	6	**13**	6	−1
	Segment 3—member of a group—"mummies"				
B8	Become a label reader…check the ingredient list, storage, and serving instructions of all baby food	1	−3	−1	7
D2	Making the transition from breast or bottle to solid foods is a big event for both you and baby	4	6	1	5
C1	Nutritionists recommend five servings of fruits and vegetables a day	2	1	1	5
A6	Pear Berry Medley and yummy Banana Pineapple Dessert made with hand-peeled bananas…guaranteed baby will love them	−1	−2	−8	5

Note: The elements come from various web sites. The elements from the different sites were mixed and matched as if they were ideas for a new product.

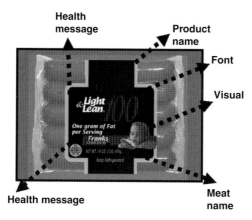

Figure 6.6 Package design for meat showing components.

elements that belong together. To the computer doing the combination the features are overlays, superimposed on each other according to the design (i.e., the blueprint of the package), and then presented to the respondent. The output is the same, namely the part-worth contribution of the different packages. Indeed, once the equivalence is created between package features and concept elements, virtually every analysis is similar. Table 6.10 shows the results of this analysis for the frankfurter data. Rather than belaboring the actual results, it suffices to say that:

(i) Brandmarks make a difference in acceptance, but less so as a driver of expected healthfulness and expected good taste.
(ii) Product name is also very important as a driver of acceptance, but less so as a driver of healthfulness and expected good taste. Nonetheless, product name can drive healthfulness when the fat level is very low.
(iii) The "flash" message (e.g., hardwood smoked) makes little difference to the ratings.
(iv) The health messages do about equally well as drivers of acceptance and of healthfulness. The health messages do not drive expected good taste, however.
(v) The pictures do not perform strongly, except for the small logo of the company and a picture of a kid eating a hot dog.
(vi) The visual elements in a single category show gradations, from strong performers to weak performers.

IdeaMap™ AND THE SCALABILITY OF CONJOINT ANALYSIS TO MANY MORE ELEMENTS

The reader may have noticed that a unifying theme of this chapter is the value of designed experiments, in which the researcher systematically varies the concept elements in order to estimate the part-worth contribution of each element. The conventional experimental design and the subsequent analysis by OLS regression require substantially more combinations (test concepts) than there are concept elements, in order to validly estimate the utility values. Thus, for 30 elements, there may be 45–90 combinations required. For 60 elements there may be 75–120 combinations or even more.

Table 6.10   Results from systematic analysis of package design features.

Graphic element	Interest buying	Expected tastiness	Expected healthfulness
Additive constant	0	44	50
Main label			
Main label with shoulders	9	5	4
Main label, no shoulders	0	0	0
Brandmark			
Larger brandmark	12	6	4
Smaller brandmark	10	5	4
Current size brandmark	0	0	0
Product name			
Light & Lean 100	19	9	10
Light & Lean 99	16	7	8
Current Light & Lean 97	15	8	8
Light & Lean 95	13	6	8
97 WHITE	11	3	1
Large Light & Lean, small 97	10	5	6
Light & Lean only	10	5	4
Only 97	7	3	1
Flash			
Great for kids	6	1	3
Hardwood smoked	5	3	1
Hot dogs	5	1	2
Franks	4	2	3
Wieners	1	−1	0
Health message			
97% Fat free	12	4	12
1 g of fat per serving	12	4	12
1 g of fat/one great taste	11	6	11
Extra lean	10	4	7
Fat free	10	2	13
Picture			
Small logo	7	2	3
Kid eating hot dog	6	5	3
Picture of a heated hot dog	5	5	4
Current runners	4	3	4

What happens when there are 200, or 300 elements, rather than 30 or so? The typical conjoint design requires 250 or 350 combinations at a minimum. Each person can only test a very small fraction of the experimentally designed combinations. Beyond 50 or so concept elements it becomes almost impossible to create an individual level model, for each particular respondent. One could, of course, have each respondent evaluate just a portion of the combinations, and for each combination use as the rating the average of respondents who saw the combination, or the proportion of respondents who rated that combination 7–9 (namely, use the proportion of concept acceptors). This strategy of allocating the respondents across the many combinations will work, but it will not generate a model for each respondent.

Individual level models are important for issues such as segmentation. It would be good to have methods in conjoint analysis by which to create these individual level models, even with very large numbers of concept elements. Furthermore, working with more than 36 elements, such as several hundred, could free the researcher to evaluate many aspects of products and

services. Furthermore, once the researcher can attack several hundred concept ele
database becomes significantly richer.

Almost 20 years ago, Moskowitz and Martin (1993) suggested a method by
could work with several hundred concept elements, but at the same time each respondent
would test a limited number of combinations. The method, called IdeaMap™ estimates an
individual level model for a particular respondent based on the combinations that the respondent evaluated, but then estimates the utility values for the untested elements. Table 6.11
presents the algorithm as the authors described it. It is important to keep the IdeaMap™ approach in mind for studies where the goal is to investigate several hundred concept elements,
but create an individual model, and work with only a relatively small base size (100–200
respondents, rather than thousands). IdeaMap™ allows the researcher to deal with hundreds
of new ideas, or to dissect a large number of messages in a category to identify what elements

Table 6.11   The IdeaMap™ method.

Step	Action	Rationale
1	Create a collection of elements	These are the "raw materials for the concepts." There may be 10–400 of these elements
2	Edit the elements	Make sure that these elements comprise "stand alone," declarative statements, or small pictures
3	Classify the elements into categories	Categories or "buckets, silos" comprise like-minded elements. Categories will be used for bookkeeping purposes. A category might be benefit, ingredient, price, etc.
4	Identify restrictions	Restrictions comprise pairs of elements from different categories that cannot appear together, either for logical reasons (e.g., they contradict each other)
5	Dimensionalize the elements on the set of semantic scales. Use about 4–8 scales	Locate the elements as points in a semantic space, on semantic scales. A semantic scale might be: 1 = More for children to 9 = More for adults
6	Create small concepts for a given respondent, by selecting a limited number of categories and elements	Each respondent evaluates a subset of categories and elements. The set of elements are chosen to conform to the pairwise restrictions (Step 4). Each respondent evaluates a unique set of elements
7	Run the evaluation with each respondent, using that respondent's individual set of concepts	Present the concepts in a randomized order for the respondent, and acquire the ratings. In any particular respondent's mind these are the only concept elements for the study
8	Create the individual model for that respondent using dummy variable regression	This is appropriate because each respondent evaluated a set of concepts that in themselves conformed to a valid experimental design
9	Estimated untested elements	The algorithm has already been explained (Moskowitz & Martin, 1993). Briefly, at an individual level replace elements that were untested by the average of the nearest eight elements in the semantic space, but do not replace the coefficients or utilities of the tested elements. Following this algorithm 20–30 times, always reestimating utilities of untested elements but not changing utilities of test elements, will generate a convergent solution. The result is a utility model for each respondent for tested and untested (i.e., estimated or imputed) elements
10	Analyze the results	Aggregate the individual data according to subgroups of interest

work and what do not. The IdeaMap™ method is particularly valuable in its ability to create segments (Moskowitz & Krieger, 2001).

"DOING IT": UP-FRONT DESIGN AND IN-FIELD CONSIDERATIONS IN CONCEPT RESEARCH

Since most concept research has emerged from business rather than from science, there are few hard and fast rules that govern the design of these projects. Thus, the up-front considerations tend to be based on executional considerations rather than on strict scientific principles. Despite the lack of good science underpinnings, the accumulated expertise of researchers doing practical concept evaluation has produced these eight caveats:

(i) *Do not make the concepts too long.* Shorter concepts are easy to read. When the concept is overly long the respondent typically does not read the full concept, but rather skims the concept to get the key points. Superficial reading rather than detailed reading is the hallmark of the consumer respondent, who does not follow directions. Few respondents pay attention to the concept in its entirety, despite instructions to do so. There are the occasional compliant respondents, but these do not represent what a person does outside the test situation, when confronted with this type of information in say a magazine, or on a FSI (free standing insert).

(ii) *Do not make the individual elements too long.* Strive to make the concept elements crisp and short, because otherwise the respondent may not comprehend the element. Most respondents will not complain. Rather, they will comply with the task, but not pay attention.

(iii) *Make the concept evaluation task interesting to the respondent.* Paying the respondent to participate often works. Although marketing research has been founded on low-cost interviews, it is better to motivate the respondent by a cash payment, or a chance to win a sweepstakes. Remunerating the respondent for time spent in the task does not lead automatically to uprated concepts. Just because a respondent is paid does not mean that the respondent will say that he or she likes the concept. Indeed, the respondent may actually feel obliged to be more honest.

(iv) *Do not ask the respondent to rate the concept on a set of repetitive attributes.* Marketing research has a history of asking the respondent to rate stimuli on a laundry list of attributes. This approach comes from the belief that with one or two stimuli the answer lies in the rating scale. In actuality, the better approach is to present the respondent with more stimuli, systematically varied if possible, and instruct the respondent to use only one or two rating scales. Beyond the first few rating attributes the respondent becomes bored. The boredom may not show up immediately, but it does emerge if the respondent is forced to rate a relatively large number of concepts on many attributes.

(v) *Choose a sufficient number of respondents, but not an overly large number.* In studies dealing with the base size in concept testing it appears that with 50 respondents the data begins to stabilize. That is, beyond 50 respondents, and in many cases even beyond 30 respondents, the results are similar. Therefore, the additional respondents do not add any more stability to the results. Figure 6.7 shows how an increasing number of respondents leads to stability.

(vi) *Do not make the perfect the enemy of the good.* Often the researcher engineers the study in order to remove biases. The researcher may require the respondent to wait

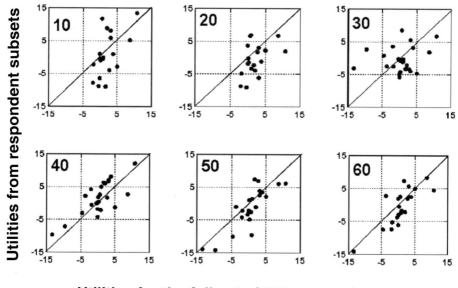

Utilities for the full set of 200 respondents

Figure 6.7   How increasing the number of respondents increases the stability. Stability usually emerges after the data from 50 respondents has been aggregated. (Data from Moskowitz *et al.*, 2000.)

10 minutes between concepts, or put other onerous constraints on the respondent. The researcher may attempt to fit different adjectives to each scale, so that every scale is different. In the end this attempt to make e study perfect produces a relatively boring, occasionally offensive study. In concept work, as in product work, consumers give their first impressions, and do not pay the detailed attention to their ratings as one might think, listening to researchers who try to anticipate everything. It just does not work like that.

(vii) *Make the scales easy to use, and if possible, keep the same scales for all attributes to lighten the respondent burden.* To the degree that the study is easy to do, the researcher will obtain better data. Even when the researcher makes every effort to tell the respondent to use each scale individually according to its scale points, people get tired of changing focus. Respondents adopt a strategy in order to simplify their task. This often means adopting one way of thinking about the scale, and applying it to all scales. In view of the respondent tendency to do this, it is probably best to use a common scale. One has to consider, however, the opposite tendency. Rather than confusing the respondent, the researcher who uses a single scale format may end up with correlated data from one scale to another. The resolution of this issue—different scales versus the same scale—must be considered in light of the desire for good data, and the strategy of respondents to simplify the task and minimize their efforts.

(viii) *Consumers are not marketers: They do not overtly differentiate between product concepts and positioning concepts unless forced to do so.* It is marketers and developers, not consumers, who differentiate between types of concepts. To a consumer, a concept is a concept. The typical consumer reaction is like/dislike, or accept/reject. All other ratings are intellectual exercises that the respondent can perform, but which do not come naturally.

ANALYTIC CONSIDERATIONS IN CONCEPT RESEARCH

Concept research is as much an art as a science. Just as there are no hard or fast rules for concept design or fielding, so there are no fixed rules for analysis. There are caveats and generally accepted practices:

(i) *When dealing with complete concepts, try to make the concepts comparable in format.* Complete concepts can range from simple line drawings to fully executed copy that closely resembles print advertising. Often the execution of the concept is as important as the basic idea. Unfortunately, there is no simple correction factor that transforms the score achieved by a simple, unadorned text concept to the expected score for the fully executed concept. Thus, when comparing concepts it is important that they be presented in similar formats. Otherwise the scores may confound the message of the concept with the execution of the concept.

(ii) *The persuasion model typically provides the same direction as the interest model.* Recall the persuasion model uses the relation between the presence/absence of elements and the actual rating of the concept. The interest model uses the relation between the presence/absence of the elements and the binary rating of interested versus not interested. Although one might feel that the transformation radically changes the data, in actuality both types of dependent variables (interest, persuasion) generate the same direction. Winning elements remain winners, losing elements remain losers.

(iii) *Spotting bad data.* One of the key issues in concept as well as product research is the need to spot and perhaps to then eliminate bad data. Unlike objective measurements, which can go wrong because the wrong measure is being taken or a poorly calibrated instrument is being used, subjective measurements come with a third problem—honesty of results. Many respondents candidly report that they do not take the study seriously. This in itself is not a problem. However, often the respondents say that they get bored, and fill up the questionnaire with random ratings. This is a problem. With conjoint measurement, however, one can identify bad data. The trick is to use a measure of consistency. In conjoint measurement the same element is repeated several times, and the variables are independent of each other in a statistical sense. One can fit the data by an equation. The goodness of fit of the equation to the data can be quantified by a well-accepted statistic such as the multiple R^2. This statistic, in turn, can be used to weed out bad data from good data. As Figure 6.8 shows, about 80% of the data shows a fairly high R^2 value (>0.64), across 20 different food products. Moskowitz *et al.* (2001) showed, however, that with this type of approach, leaving in some of the poorer performing individuals who are less consistent, changes the results only marginally, as long as there is a preponderance of respondents who pay attention to the task.

(iv) *A lot of the insights come from the segments, not from the conventional subgroups.* Although researchers spend a great deal of time thinking about the response of different, conventional subgroups to the concept, in actuality the real "action" comes from the segments. The high-scoring elements come from the segments. In the total population often these segments counteract each other. What one segment likes the other dislikes. As a consequence, no element does really as well as it should. When the mutual suppression is removed by segmentation the really strong performances emerge. The removal of suppression becomes clearly obvious in most data sets, after segmentation is done. Furthermore, the segments tend, in general, to make a great deal of intuitive sense.

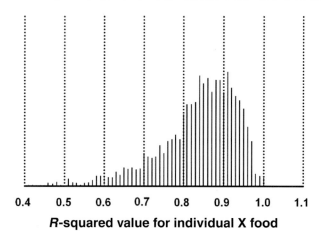

R-squared value for individual X food

Figure 6.8   The measure of consistency, R^2. The measure of consistency shows how well the persuasion model for a respondent actually fits the data. About 80% of the data shows a fairly high R^2 value (>0.64) across 20 different food products.

(v) *The additive constant give a sense of how important is the basic idea.* High-additive constants mean that the basic idea is acceptable. Low constants mean that the basic idea is not particularly acceptable. Often consumers cannot articulate this measure of relative importance. Table 6.12 lists the additive constant for different foods and beverages from a variety of studies, run under similar conditions (the It!™ Databases). The additive constants are clearly different.

Table 6.12   Additive constants from the 2001 Crave It!™ Studies.

Food	Constant
BBQ ribs	44
Steak	41
Cheesecake	41
Potato chips	40
Chicken	40
Ice cream	40
Chocolate	39
Coffee	39
Cinnamon roll	38
Cola	37
Peanut butter	37
Cheese	36
Olives	36
Nuts	35
French fries	34
Tacos	34
Hamburger	30
Pizza	30
Pretzel	29
Tortilla chips	27

Source: Data courtesy of It!™ Ventures.

(vi) When creating new concepts from the data, it is best to incorporate only a few concept elements rather than many, with these concept elements showing high utility values. The business aspect of concept research, and especially conjoint analysis, is both understanding and application. Application means using the data to create new and more powerful concepts. One way to do that strings together elements to achieve a high utility value. The utility value for the concept is the sum of the component utility values. To get a high utility value one can string together many elements in order to reach the goal, or string together just a few "breakthrough" elements. It is better to use just a few elements, each of which scores well, than to put together many elements, each of which scores modestly. In concept creation we look for the breakthrough ideas. Too many elements may not work well because the concept is too big. Typically concepts comprising 2–4 elements perform the best, especially when the individual elements are strong performers.

(vii) When creating new concepts, it is important to put together concept elements that make sense, rather than just selecting the winning elements and mindlessly stringing them together. Concepts have to make intuitive sense. Although it is tempting to create concepts "by the numbers," choosing those limited number of elements that do well, it is important to use judgment. There are no hard and fast rules about the concept. However, the researcher should suggest combinations that provide a brand name, a meaningful message with support, and if necessary a package and a price. The researcher should be on the lookout for combinations that could be confusing, or that in some way contradict each other.

BUSINESS CONSIDERATIONS

(i) *What type of elements to include?* Researchers, marketers, and product developers argue again and again about the nature of the elements to include. Marketers like to put in competitor elements, product developers like to put in "stretch elements," and advertising agencies like to put in emotion and executional elements. It is important at the outset of the study to clarify the reason for the study. When the study is designed to identify new product ideas, there is no reason to put in brand names and emotional statements. When the study is designed to identify the language that drives consumers to be interested in buying the product, there is every reason to put in brand names and emotional statements. Fortunately, conjoint measurement is sufficiently robust a tool to allow mixtures of different kinds of concept elements in the same concept. It is not that consumers need clarity. They do not. They react to what they see. Rather, it is the involved parties who need clarity in the goals of the study. This clarity will then ensure the proper elements that best address the question.

(ii) *The number of concept elements.* Researchers suffer the perennial issue of whether to make a concept study very large to encompass all possible elements, or make the concept study modest or even small to execute it rapidly. This business consideration can be rephrased as whether the researcher wants to accomplish everything in one large study, or work in an iterative mode. Many researchers feel that in a simple concept study they need test only one or a few concepts, but when it comes to a conjoint study they must work with as many elements as possible because they have to take advantage of the power of the conjoint method. With large-scale studies, there are often many vested interests and parties, each wanting his opinion listened to, respected, and followed. In the end many of these very large studies fall apart. Sometimes the researchers demand that the conjoint study deal with pairwise interactions. As the studies get larger these demands become increasingly onerous, impossible to fulfill in a rigorous manner, and occasionally some researchers throw up their hands and do focus groups. Other researchers feel that they can do several small conjoint studies in sequence, building the subsequent conjoint study from the results of the previous one. There is no correct strategy here, but rather a matter of preference.

(iii) *The number of respondents.* The number of respondents, or base size, is an important consideration. Conjoint studies typically stabilize at approximately 30–50 respondents, per key subgroup. However, by about 15 respondents one can begin to see the patterns, although the individual utility values are not stable, and may wobble around. By the time the base size of 50 is reached, for a specific subgroup, it is unlikely that any additional respondents will influence the utility values if they are chosen from the same subgroup of individuals. The foregoing applies to the scientific and the statistical considerations. In business, validity is also a matter of opinion. Popular opinion in research holds that management will believe data from many people more readily than they will from just a few people. For management to believe data it is often best to have a minimum of 100 respondents and a minimum of 50 in each key subgroup. Although the results might stabilize at 30, it makes more sense to assure management by using the large base size. The only time that management will readily accept a small base size instead of a large base size is when the cost differential is high. At that point management often beats a hasty retreat, and accepts the base size that it feels it can afford.

(iv) *Small-scale studies versus large-scale studies.* In science there is the perennial problem of how to address and answer a problem. Should the researcher opt for a series of little, iterative, low-risk studies, with each study comprising a limited number of stimuli, and therefore provide a relatively small amount of information? Or should the researcher opt for one large study, with many people, many elements, so that the selfsame study can answer many current and presumably future questions? This is a recurring issue that simply needs to be surfaced. It cannot be answered one way or the other on a general basis. When addressed in the fashion of "what will we get out, and what do we need," the choice in any specific situation should become clearer and easier to make.

(v) *Self-authoring studies for concept development and the iterative approach.* This author feels that the *iterative approach* is better than large-scale studies, because iterative research is cost-effective and manageable. A better strategy is to make the research steps small, and manageable, by using conjoint measurement, albeit in a more restrictive, less encompassing format. Self-authoring systems do well in this situation. The self-authoring system allows the researcher to set up small-scale studies, launch them on the computer, acquire the respondents (often by Internet), and then have the data automatically analyzed.

REFERENCES

Agresti, A. 1990. *Categorical data analysis.* New York: John Wiley & Sons.
Beckley, J. & H.R. Moskowitz. 2002. *Databasing the consumer mind: The Crave It!, Drink It!, Buy It! & Healthy You! Databases.* Anaheim, CA: Institute Of Food Technologists.
Box, G.E.P., J. Hunter, & S. Hunter. 1978. *Statistics For Experimenters.* New York: John Wiley & Sons.

Fechner, G.T. 1860. *Elemente der Psychophysik.* Leipzig, Germany: Breitkopf and Hartel.

Green, P.E. & V. Srinivasan. 1978. A general approach to product design optimization via conjoint analysis. *Journal Of Marketing* 45:17–37.

Luce, R.D. & J.W. Tukey. 1964. Simultaneous conjoint measurement: A new type of fundamental measurement. *Journal Of Mathematical Psychology* 1:1–27.

Moskowitz, H.R. & D.G. Martin. 1993. How computer aided design and presentation of concepts speeds up the product development process. Proceedings Of The 46th ESOMAR Conference, 1993, Copenhagen, Denmark, 405–419.

Moskowitz, H.R., A. Gofman, B. Itty, R. Katz, M. Manchaiah, & Z. Ma. 2001. Rapid, inexpensive, actionable concept generation and optimization – the use and promise of self-authoring conjoint analysis for the foodservice industry. *Food Service Technology* 1:149–168.

Moskowitz, H.R., A. Gofman, P. Tungaturthy, M. Manchaiah, & D. Cohen. 2000. Research, politics and the web can mix: considerations, experiences, trials, tribulations in adapting conjoint measurement to optimizing a political platform as if it were a consumer product. *Proceedings of Net Effects[3]* 109–130.

Moskowitz, H.R. & B. Krieger. 2001. Financial products: Rapid, iterative and segmented development by conjoint measurement and self-authoring iterative procedures. *Journal of Financial Services Marketing* 4:343–355.

Moskowitz, H.R., S. Poretta, & M. Silcher. 2004. *Food Concepts.* Ames, IA: Iowa State University Press.

Plackett, R.L. & J.D. Burman. 1946. The design of optimum multifactorial experiments, *Biometrika* 33:305–325.

Systat. 2004. *Systat, the system for statistics. User Manual for Version 11.* Evanston, IL: Systat, Division of SPSS.

Thurstone, L.L. 1927. A law of comparative judgment. *Psychological Review* 34:273–286.

Wells, W.D. 1975. Psychographics: A critical review. *Journal of Marketing Research* 12:196–213.

Wittink, D.R. & P. Cattin. 1989. Commercial use of conjoint analysis. *Journal of Marketing* 53:91–96.

7 High-level product assessments

The original product research in sensory analysis dealt with tests of difference (same/different) for quality control, and tests of acceptance (degree of like/dislike). These tests were typically commissioned to winning products from a set of alternatives, or to maintain product quality in the light of ingredient substitution on the one hand and product stressors (e.g., storage time and temperature) on the other.

The world of the sensory professional has changed, and product testing with it. Today's high-level product assessments range from analyzing the entire product category in order to uncover sensory segments and liking drivers (category appraisal), and onto product optimization with systematically varied prototypes to identify the best possible product subject to constraints. In the larger sense, product testing has evolved from being simply a laboratory test to a full-blown study of the consumer and how that consumer reacts to products.

TRADITIONAL PRODUCT TESTING—ORIGINS, RATIONALES, INTERPRETATION

Today's sophisticated methods for product evaluation were not always in place in the corporation. It is instructive to read the scientific literature regarding product testing of say more than 70 years ago. The reader might well be surprised. By today's standards the literature around 1940 and before can be best described as scattered, stunted, and simplistic (see Amerine, Pangborn, & Roessler, 1965). There really was no literature to speak of, although one might find discussions of difference tests in different subject-related journals. There was no coherent field of sensory analysis. When one looked at journals devoted to meat studies, brewing, perfumery/flavor, milk, and so forth, from time to time one might encounter an article that dealt with one or another aspect of discrimination testing. Looking back from our vantage point, we fail to realize that the field of sensory analysis is fairly young, although it has undergone grand growth. What we read in the literature that we call "historical" is not sensory analysis, per se, but rather the description of some practical tests to answer practical, simple problems perception raised in other contexts, but dealing with what we now call sensory issues.

Sensory and Consumer Research in Food Product Design and Development, Second Edition.
Howard R. Moskowitz, Jacqueline H. Beckley, and Anna V.A. Resurreccion.
© 2012 Blackwell Publishing Ltd. Published 2012 by Blackwell Publishing Ltd.

Product testing received its impetus from three clear sources, all reaching their public apex in the 1920s to 1940s. One source was the US government, and the desire to measure food acceptance. This source led to methods to measure food acceptance. Meiselman and Schutz (2003), as noted in Chapter 1, describe these efforts. The other source is the pioneering work by the Arthur D. Little Inc. consulting company in Cambridge, MA. Cairncross and his protégé, Sjostrom, spent a considerable amount of time developing a business to describe the sensory characteristics of foods. This "Flavor Profile Method," described by Caul in 1957, but discussed a almost decade earlier by Cairncross and his colleagues (Cairncross & Sjostrom, 1950), created a business and in turn a profession out of sensory analysis. There were certainly methods to describe the sensory characteristics of products such as beer (Clapperton, Dagliesh, & Meilgaard, 1975), but no concerted effort to turn this descriptive work into a scientific field. The final source was the continuing effort to improve methods of discrimination, and thus be able to show differences between products. The difference testing comes from the inspiration of psychologists such as Fechner (1860) and Thurstone (1927), and represents the contribution of experimental psychology to the nascent field of sensory analysis.

Most of the traditional research in product evaluation used simple descriptive and inferential statistics. The objective was to find the average across respondents, and where necessary, to determine whether the products were different from each other, or different from some gold standard value. Thus difference testing involved T tests and analyses of variance. There was little in the way of multidimensional scaling, brand mapping, or any of the other modern tools. There were correlation analyses to show that some specific sensory attribute was related to a physical level of an ingredient/treatment, or an objective instrumental measure.

WHAT DOES THE CORPORATION REALLY WANT TO GET OUT OF PRODUCT TESTING?

Corporations are not universities, and they are not in the business of funding basic research in sensory analysis unless they can see a payout from that investment. Corporations are in the business of creating products and services that they can sell on the market to produce a profit. Unlike basic science in such areas as biology/genetics, materials science, and so on, improved methods in sensory analysis have to be harnessed to business decisions. In and of themselves, sensory analysis methods for product testing are only exercises in research methodology, without any visible benefit to the company.

Armed with this somewhat harsh but realistic point of view, not necessarily shared by other practitioners, we can see the evolution of product testing methods within the framework of the profit-oriented company. Discrimination testing held sway for six decades or more because it allowed the company to maintain quality. In a period when there was greater demand than supply, the company had to ensure that the products it was selling were up to standard in terms of sensory quality and acceptance. Hence the popularity of discrimination tests was high, because discrimination tests provide a simple yes/no statement about quality versus a standard. Even in today's complex world, one need only look at the correspondence on the European Sensory Network for the past few years to discover how important discrimination testing remains. Many of the questions deal with ways to measure differences between products.

Along with discrimination testing, descriptive analysis has held sway, for pretty much the same reason. Whereas discrimination testing provides a single measure of same/different, descriptive profiling creates a so-called *signature* of the product. A great deal of descriptive analysis training emphasizes expertise with respect to a current gold standard. Indeed, for the most part descriptive analyses have been used to determine in what ways and to what degree a product differs from the gold standard.

In a more competitive environment, however, discrimination testing and descriptive analyses fall short of creating new products. What does the developer do with these test results to create the new product? How can these test methods be incorporated into the new product process? Some rethinking of test procedures and statistical analysis has to be done. Profiling produces only part of the information that a corporation needs. If the corporation's sole interest were to maintain the *status quo*, then profiling and discrimination testing would be quite adequate. Corporations need something different, and beyond these tests of same/different and profiling.

TYPES OF ATTRIBUTES AND WHAT REALLY MATTERS ONCE THE ATTRIBUTE DATA ARE OBTAINED

One of the key jobs of a company is to produce products that consumers will accept. This job entails identifying the features of the product that drive acceptance. Beyond profiling is the task of identifying how sensory attributes drive acceptance and image ratings, and even further, how ingredients drive sensory, acceptance, and image ratings.

Some definitions are in order here:

(i) *Sensory attributes*: Amount of a characteristic, such as amount of graininess, degree of sweetness. Sensory attributes do not involve an evaluation of "good/bad." For example, a product that is very sweet can be liked or disliked. Sensory attributes simply measure the amount of a characteristic. A sensory attribute can be quite simple, such as sweetness, or more complex, such as "real apple flavor." A sensory attribute can be intuitively obvious or may require explanation and ostensive definition (i.e., a reference that represents the sensory attribute). Sensory attributes abound by the hundreds; a variety of lists have appeared, varying from the simple to the comprehensive, and from lists that attempt to deal with supposed sensory primaries to lists that are created to deal with the array of sensory experiences (Crocker & Henderson, 1927; Brandt, Skinner, & Coleman, 1963; Harper, Bate-Smith, & Land, 1968; Szczesniak, Loew, & Skinner, 1975; Civille & Lyon, 1995).

(ii) *Sensory directional attribute*: A sensory attribute for which the respondent rates the degree to which the product overdelivers or underdelivers relative to a mental ideal. Thus, for the case of sweetness, a beverage may be too sweet, just right, or not sweet enough. The respondent can rate the degree of "too sweet," or simply say that a product is "too sweet." The sensory directional attribute is not a sensory attribute. Instead, it is a sensory attribute that has been made far more complex by the insertion of the ideal point, so it is not a simple description of the product, per se (Moskowitz, 1999).

(iii) *Image attributes*: These attributes describe the product, but they cannot be called sensory attributes. The image attributes are not evaluative either. They involve more complex cognitive responses. Attributes such as "natural" fall into this category. It is occasionally difficult to determine whether an attribute is a simple sensory attribute or an image attribute. In matter of fact, such distinctions do not add anything and need not be made for most applied issues.

Table 7.1   Some hedonic scales and their verbal descriptors.

Scale points	Descriptors
2	Dislike, unfamiliar
3	Acceptable, dislike, (not tried)
3	Like a lot, dislike, do not know
3	Well liked, indifferent, disliked (seldom if ever used)
5	Like very, like moderately, neutral, dislike moderately, dislike very
5	Very good, good, moderate, tolerate, dislike (never tried)
5	Very good, good, moderate, dislike, tolerate
9	Like extremely, like very much, like moderately, like slightly, neither like nor dislike, dislike slightly, dislike moderately, dislike very much, dislike extremely
9	FACT Scale (Schutz, 1964): Eat every opportunity, eat very often, frequently eat, eat now and then, eat if available, don't like—eat on occasion, hardly ever eat, eat if no other choice, eat if forced

Source: After Meiselman, 1978.

(iv) *Evaluative attributes*: The respondent judges the product in terms of good versus bad. Evaluative or acceptance scales need not incorporate the term liking. Acceptance can be attributes such as purchase intent or FACT scale. Table 7.1 shows some examples of these evaluative attributes. For companies, evaluative scales are the key measures that quantify product performance. Most corporate decisions using consumer research involve evaluative attributes, and performance versus competition or versus a benchmark level (Meiselman, 1978).

The sensory analysis literature, including texts and scientific articles, is replete with discussions about how to profile products on attributes, and examples of analysis (Meilgaard, Civille, & Carr, 1987). However, when all is said and done, attributes are just indicators of a subjective response to the product. When the researcher looks at the magnitude of difference between two products on a profile of attributes, the researcher will have created a report card. This report card may demonstrate that the products differ from each other on all, many, some, or just a few attributes.

The important thing to keep in mind is what one does with these attributes, beyond the act of profiling. It is the relation between attributes that matters, and the way this interattribute relation is modified by external conditions that lead to good insights. This point needs to be stressed again and again, because people have a tendency to do "busy work" rather than thinking. Sensory profiling can reach unexpected heights when providing direction, but just as easily and unfortunately far more frequently may generate into mindless, mundane busywork.

VENUES FOR SENSORY TESTS—THE "STERILE BOOTH" VERSUS THE CENTRAL LOCATION VERSUS THE HOME

When product testing first began, tests were conducted in whatever venue was available. The focus of these very early studies, conducted in the 1930s and 1940s, appears to be on the technical issues. Whether the test was done in the laboratory, at a state fair, in a school room, among soldiers in a barracks, there was no sense that for valid results one absolutely had to test the samples in a professional testing venue, with white booths, controlled lights, sources of water, and sanitary facilities for expectoration. There is a delightful innocence about these early reports that makes reading the papers a wonderful trip into the past. The lack of self-consciousness of the investigators is apparent from these papers, as they struggled to deal with the substantive problems they were addressing, rather than with the exquisiteness of facilities and field execution. Perhaps some of this naiveté can be traced to the fact that these early investigators were not particularly aware that they were founding a new field.

Naiveté, sad to say, cannot last. Even scientists become self-conscious of what they do, and play to the crowd. Over the years, and with the systematic professionalization of sensory analysis as both a practice and a science, there arose a cult centered around the testing facility. Some of this can be traced to the explosion of interest in science in the late 1950s and 1960s, when advances in instrumentation were all the rage, and when the sensory analyst had to move beyond simple "taste tests" to demonstrate their scientific integrity. There were no journals as yet devoted to "taste testing," but companies began to recognize the value of such testing to assure product quality. In the absence of a full-fledged science, it appeared judicious and politic to create a scientific image based on execution. That image was expressed by the adoption and soon proliferation of exquisitely designed, sterile testing facilities—the so-called taste test booths. One did not have to demonstrate a high degree of scientific aptitude to achieve recognition in the field. Rather, one had to purchase, install and then publicly proclaim that one had the latest "taste test facilities," custom built, for one's sensory analysis. By the end of the 1960s, research directors across the United States, and even worldwide, were competing with each other about who had the better facility.

The truth of the matter is that the facility is not particularly important. Certainly for various types of tests it is vital to control ambient odors, lights, and make mouth-washing facilities available. Yet, for the most part respondents find it disconcerting to spend hours in a sterile white booth. In the main these exquisitely outfitted facilities are used for in-house panelists, and for small consumer panels, and often simply for show. Sensory specialists work with these booths; market researchers do not, and often do not understand what is to be gained by such control in such expensive surroundings.

There are two other venues in which to conduct product tests; central locations at malls or in an office, and at home. These two deserve explication and differentiation. Both venues are now used by sensory researchers working in product guidance at the R&D level, and by marketing researchers who do tests for marketing, often to measure acceptance prior to launch. The American Society for Testing and Materials (ASTM) has issued guidelines for sensory testing facilities (ASTM, 1973):

(i) *Central location facilities*: These facilities are typically modest rooms, rented in a mall, or created in a corporate office in a building away from the client R&D development or manufacturing facility. Sometimes the corporation owns the test facility, which is located close to the R&D facility but sufficiently far away to mask any clear relationship. Occasionally the facility is quite up to date, with large and well lit rooms, and a large kitchen. More often, however, the facilities are located in a strip mall or in a large enclosed mall, where the respondents can be recruited from the local mall traffic. Despite the lack of style, these facilities are the norm, and do quite well for most studies that do not demand stringent facilities for product preparation. The central location facility offers the chance for the field service that conducts the study to supervise the evaluation, and for the researcher to observe the field service as the interviewers conduct the interviews. Tests run in the central facility, the so-called CLTs (central location tests), are popular among both sensory researchers and market researchers, because of the ability to monitor the study, and the ability to test multiple products efficiently.

(ii) *Home use*: Sometimes it is important for the respondent to prepare the food, and serve it in the fashion to which the respondent has become accustomed. In this case, the central location facility is not appropriate because it does not typically allow the consumer to prepare the food. Some facilities allow for respondent preparation, but these facilities are unusual. The respondent can prepare and evaluate the sample(s) at home, fill out a questionnaire, and either phone in the answers, mail in the answers, or even log on to a computer and email the ratings. Home use tests, or HUTs as they are abbreviated, are very popular, especially among market researchers who want to measure responses to the product under "natural" conditions.

RESPONDENTS – EXPERTS OR CONSUMERS

Sensory analysis traces one of its origins to the expert, whose deep knowledge of a particular product category guided new product development and quality control. Old habits and ways of thinking are hard to displace, nor must one necessarily try to displace them. Expert panels, emerging from this history of experts, represent one of these old ways of thinking, yet a way of thinking that enjoys a great deal of merit.

Over the years various researchers involved in product work have involved themselves heavily in the training of panels. Beginning with the Flavor Profile™, continuing with the QDA™ method, and finally with offshoots such as Spectrum™, these researchers have spent countless dollars and hours creating panelists who could register the sensory notes of a product, and quantify the intensity. It is necessary to do so in many cases because there are no other ways to register the sensory characteristics of products. A lot of the "taste" is flavor. Flavor has many nuances, but does not have a set of "objective standards" against which it can be referred. Whatever nomenclature is developed, it is vital to have people who can point to reference stimuli as exemplifying that attribute. Thus, arose the need for the trained panel.

The problem is not so much the trained panel in descriptive analysis and in the emotions surrounding the rationale for and the ultimate contribution of the panel. What is problematic is the cadre of individuals who aver that the consumer cannot adequately describe his perceptions, and therefore the only task a consumer can do validly is rate degree of liking. According to this very vocal, opinionated cadre, it is the expert panelist and *only* the expert panelist who can do an adequate job profiling the attributes.

This debate about the trained versus untrained respondents continues, albeit in a much more muted fashion. Psychophysicists who trace their intellectual history from experimental psychology have shown that the untrained respondent can assign numbers to match the intensity of stimuli. To a great degree the untrained respondent can behave as a reasonably good measuring instrument. This is the world of "scaling," where the observer acts as a measuring instrument. When the ratings of 5–10 respondents are averaged, the data are even better, and the correlation with external varied intensity is even higher, because the noise is averaged out. So what about the argument of consumer versus expert? It is not that the consumer is incapable of assessing stimuli as much as the consumer simply does not have the language, and so does not know what to attend to. By virtue of training, the expert panelist learns to attend to nuances in the stimulus, knows what these mean, and therefore is more focused. In other cases, where the attributes are simple, the expert and the consumer do equally well (Moskowitz, 1996).

At the practical level, when using consumers to test products, it is always a good idea to instruct them about the testing procedure and the meaning of attributes. This instruction does not automatically make a consumer into an expert. Rather, the instruction ensures that the respondent knows what is expected. One would be horrified to discover on an after-the-fact basis that the respondent did not know how to evaluate products, and that the respondents evaluated the product in the wrong way at the wrong time. Even something as simple as the evaluation of cereal can be incorrectly executed if the respondent does not know when to evaluate the product "in the bowl," after milk has been added. It is also beneficial to explain the meaning of the attribute to the respondent. Often respondents do not know the meaning of some of the attributes, even if those attributes seem clear and relevant to the researcher. The explanation offered to the respondent need not be long, but it should be done, simply to ensure that the respondent who is confused has a chance to ask a question. If there is no chance to ask, most respondents will simply not ask.

Finally, there is the ever-present question about how many respondents to work with in a study. Market researchers like to work with a base size of 100. Sensory analysts work with fewer researchers. A lot of the scientific work deals with the data from 10, 15, or 20 respondents. Which is right? Are market researchers spending too much money on respondents? Do scientists report results that are unstable because they work with too few individuals?

Studies on base size suggest that by the time the base size reaches 30 or more individuals the results are stable (Moskowitz, 1997). The demonstration of this has been done by sampling the data from the original study, pulling base sizes of specific numbers of respondents, and comparing the mean of the ratings of the sample to the overall mean. By the time the base size comprises data from 30 respondents (i.e., 30 ratings per product) the correlation between the original, full set of data from all respondents and the sample data becomes very high, when the correlation is computed on the mean ratings of a set of different products. This base size of 30 appeared to hold for both sensory and liking attributes. The base size of 100 respondents espoused by market researchers in their recommendations for studies may well be the result of the need to make the data appear robust to the ultimate client—the marketer. Base sizes of 100 or more certainly appear more robust than base sizes of 30, and the results obtained with the larger base size would therefore stand a higher chance of being believed. In contrast, academics do not have the money to spend on their studies, and thus opt for the lowest number that generates believable results, but doesn't cost too much. In practice, base sizes of 10, 20, or 30, respondents respectively from psychophysical studies have migrated into work by sensory analysts, although for the most part sensory analysts like to work with more than 30 respondents because they can use the conventional methods of inferential statistics.

SEARCHING FOR PATTERNS—WHAT DRIVES LIKING?

As one of the most important attributes for consumers, overall liking or attribute liking is quite often at the center of corporate research. A recurrent question is "what drives liking?" That is, what physical variables or the subjective variables appear to co-vary with liking.

A direct way, but not necessarily the best way, to find drivers of liking asks the respondent to rate the relative importance of different attributes. *This is the attitudinal approach.* The rating may happen in the context of product evaluations, or separate from product evaluations (e.g., just as a questionnaire that stands on its own). The problem with this direct method is its divorce from actual behavior. People do not really know what is important. They may think they know, and a hallmark of research is the way that respondents will confidently answer questions put to them, even if they do not have the faintest idea what is the correct answer. Ask a person what is important and the person might blurt out something such as "good taste." It is hard to go beyond this, into specific aspects of taste, or even the fact that the so-called taste is really good aroma, appearance, or mouthfeel, or a specific taste/flavor/odor quality.

A more productive approach looks at the pattern of ratings, and is known as R-R or response-response analysis. *R-R analysis is a behavioral approach.* The R-R approach uses the respondent behavior, and makes inferences. We cannot easily get into the subconscious of the consumer, but the reality is that we don't have to. We can present the consumer with a set of similar products, obtain the profiles of these products on a set of attributes, either from experts or consumers, and then measure consumer acceptance. This "raw material," ratings of liking and ratings of sensory attributes on *different products will reveal what drives liking.*

Let us look at some data from a study of margarine. The respondents evaluated 40 different samples, assigning ratings to the separate products. Each of the 120 respondents in the test evaluated 20 margarines, over two days, for a total of 60 ratings per margarine. The respondents rated all attributes on anchored 0–100-point scales. Some of these attributes were sensory characteristics, others were liking attributes, and still others were image attributes, respectively. Table 7.2 shows the type of data for the study.

The foregoing data allow the researcher the opportunity to perform least two analyses that will identify what drives liking:

(i) *Plot overall liking versus attribute liking, and compute a simple, linear correlation.* Figure 7.1 shows the scatter plot for liking versus several sensory liking attributes. Keep in mind that this type of scatter plot is best used when the researcher works with several products, and when the data points correspond to the different products. The figure shows that liking of attributes highly co-varies with overall liking for margarine, as it does for most products. *This co-variation suggests that attribute liking does not really generate new knowledge. Rather, attribute liking values should be used to identify problems with specific products, and not to drive new product development.* This caveat applies to those margarine products in Figure 7.1 that lie at the low value of the abscissa, corresponding to low liking values for sensory attributes.

(ii) *Plot the quadratic relation between liking and sensory attribute level.* Basic research over the past 40 years suggests that as a sensory attribute increases, liking first increases, peaks and drops down. This quadratic function has been demonstrated for simple model systems such as sugar or sodium chloride dissolved in water (Moskowitz, 1981). The relation is not new, having been anticipated almost 80 years ago by Engel (1928) and even earlier by Wilhelm Wundt, the founder of experimental psychology (Boring, 1929; Beebe-Center, 1932). The quadratic relation also applies to food as well, and therefore provides a good organizing principal. Figure 7.2 shows the quadratic curve. The independent variable is the sensory attribute and the dependent variable is overall liking. From this type of analysis the researcher identifies the rate at which liking

Table 7.2 Example of data from the margarine study.

	Summary statistics across all 40 products				Products 101–105				
	Min	Max	Range	StDev	101	102	103	104	105
Unwrap softness	48	76	28	7.0	48	48	48	53	56
Unwrap stickiness	41	71	30	7.1	43	41	50	46	46
Like appearance	43	77	34	9.8	45	49	49	55	44
Like color	42	75	33	8.3	48	52	48	58	52
Like spread	35	82	47	12.0	64	67	35	75	72
Like appearance on bread	42	79	37	8.1	58	47	42	64	57
Like melt	28	69	41	8.7	56	50	28	48	51
Like aroma	47	71	24	4.7	53	52	47	57	53
Like taste	45	71	26	6.8	45	47	48	49	52
Like aftertaste	39	67	28	6.5	39	45	43	42	47
Like texture	40	70	30	6.8	40	45	48	48	50
Easy spread	33	88	55	15.7	87	88	33	87	86
Fast melting	14	66	52	10.9	63	51	14	38	46
Soggy bread	8	25	17	3.7	25	20	10	19	18
Aroma strength	35	60	25	5.3	52	52	35	56	50
Aroma buttery	32	63	31	5.6	47	49	32	53	49
Flavor strength	39	65	26	5.2	53	52	39	50	54
Flavor buttery	39	60	21	5.8	42	47	39	39	45
Flavor salty	17	43	26	5.8	25	27	17	27	22
Flavor sweet	29	45	16	4.1	30	42	29	33	32
Flavor cheesy	9	26	17	2.9	14	20	9	26	18
Flavor bitter	11	27	16	3.8	25	24	21	22	27
Flavor sour	9	26	17	3.9	26	19	17	25	26
Flavor aftertaste	40	63	23	4.8	46	53	40	56	55
Texture hard	11	54	43	11.7	11	17	54	25	23
Texture buttery	34	63	29	7.5	34	37	46	39	41
Texture creamy	45	75	30	7.0	69	69	45	70	66
Image calories	36	63	27	6.9	48	50	47	40	36
Image fat	36	61	25	5.7	47	49	50	44	36
Image similar to butter	34	62	28	7.4	34	39	44	39	35
Overall	46	71	25	6.5	46	47	49	52	52
Factor scores									
F1	−3.11	−1.73	−1.98	−2.31	−0.40	0.65	−3.11	1.29	0.43
F2	2.19	1.75	2.00	2.13	−1.71	−1.10	−1.73	−1.60	−1.29
F3	5.30	3.48	3.98	4.44	1.86	0.19	0.61	1.55	2.00
F4	1.00	1.00	1.00	1.00	1.73	0.96	−1.77	−0.19	0.01

StDev, standard deviation.

changes with respect to the sensory attribute level, and the sensory level at which liking reaches its maximum level. Whether the relation is a nice quadratic function or some other type of function, it is still the case that the optimum level often lies within the sensory range covered by the different products, rather than lying at the highest level or lowest level.

(iii) *Create a more comprehensive model, showing several sensory attributes as drivers of liking.* We know that liking is driven by the interplay of several sensory attributes. We might choose two sensory attributes, and plot them together on a sensory-liking surface, where the X and Y coordinates are the sensory attributes, and the Z coordinate

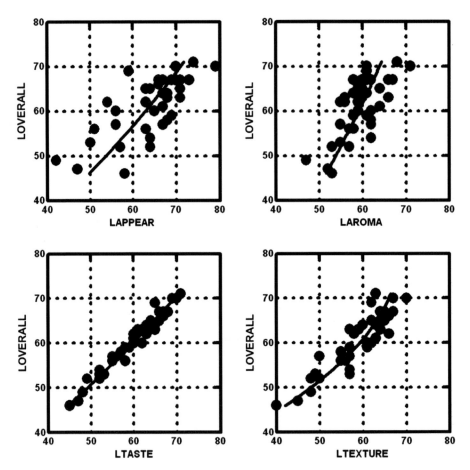

Figure 7.1 Scatter plot showing how overall liking co-varies with attribute liking. Each darkened circle corresponds to one of the test margarines. LAPPEAR, liking of appearance; LAROMA, liking of aroma; LTASTE, liking of taste; LTEXTURE, liking of texture; LOVERALL, overall liking.

is liking. Figure 7.3 shows an example of this. One could do this exercise for many different sensory attributes versus liking. The problem becomes to make sense of the data, and not just to spend one's time plotting curves. The data must tell the corporate researcher a story, from which one gains wisdom about the product.

CREATING A "PRODUCT MODEL" USING THE "CATEGORY APPRAISAL" APPROACH—UNRELATED PRODUCTS

A "Product Model" comprises a set of equations relating independent variables to ratings. The independent variables are either ingredients or factor scores; ingredients when the products are systematically varied, and factor scores when the products come from the competitive frame where the linkages across products are based on the sensory attributes.

The product model represents an action-oriented approach to product testing that integrates different sources of data about a product, and provides the manufacturer with concrete

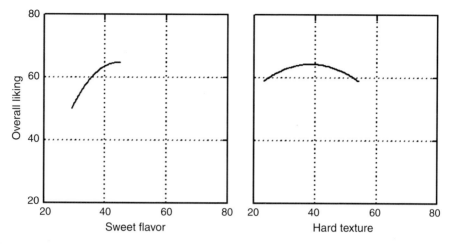

Figure 7.2 Quadratic smoothed functions for sensory versus liking margarine.

guidance. Fundamentally, the product model attempts to integrate all of the different types of attributes into a set of equations, with the property that the equations can be easily manipulated to achieve two objectives:

(i) Identify the profile of products that maximize some objective (e.g., overall liking), subject to constraints.
(ii) Match a specific profile as input.

Product modeling began with the field of response surface analysis, more than 40 years ago, when statisticians began to move beyond difference testing to modeling (Gordon, 1965). Product modeling came to the fore in research some years after discrimination testing and profiling reached their zenith. Neither discrimination testing nor profiling "prescribe" next steps; product modeling attempts to do so, either by pointing out target products, or more productively by recommending optimal levels of process and formula variables.

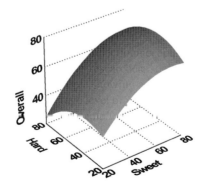

Figure 7.3 The sensory-liking surface for margarine.

Product modeling is a child of the computer age. Statistical computation became easy in the middle and late 1960s with the introduction of statistical programs onto mainframe computers (e.g., the BMD system; BMD, 1973). It should come as no surprise that the practical outcomes of product modeling, direction and specification, would emerge rapidly as statistical computing matured.

Product modeling follows a set of steps described in Table 7.3. The approach uses either factor scores or ingredients. When there are no ingredients to serve as independent variables,

Table 7.3 Steps to create a product model.

Step	Activity
1	**Data preparation:** Array the data set in a rectangular format, with rows corresponding to products and columns corresponding to formulations, consumer sensory ratings, consumer liking and image ratings, expert panel ratings, and instrumental ratings. Some data sets may not have formulations or instrumental ratings.
2	**Create a basis set:** Let the basis set correspond to ingredients, or if the products are not connected by design, then create the basis set by performing a principal components analysis on the sensory attributes. Use the factor scores as the basis set, rather than product ingredients. Follow Steps 3a and 3b to create this basis set.
3a	**Principal components analysis:** When using sensory attributes, perform a principal components analysis on the sensory attributes identified in Step 2, save the factor scores (after rotation to a simple solution, e.g., through quartimax rotation; Systat, 1997). For the margarine data, four significant factors emerged, accounting for 80% of the variability. More factors could have been extracted, but at the cost of having fewer degrees of freedom in the regression modeling.
3b	**When there are ingredients:** Use the ingredient levels.
4	**Finalize the matrix:** Adjoin the factor scores to the data matrix.
5	**Model:** Create a quadratic model for each attribute in Step 1, relating that attribute to the linear, square, and relevant cross-terms of the independent variables (the basis set). The goodness of fit of these models can be assessed by the "adjusted" multiple R^2 statistic, which shows the percent of variability accounted for by the equation. The statistic corrects for the number of predictor variables in the equation. Each attribute (sensory, liking, image) generates its own statistic for goodness of fit. It is important to reiterate that the factor scores, serving as the independent variables, were developed using sensory attributes alone from the margarine data. These factor scores were then used as independent variables to predict liking and image attributes.
6a	**Optimize:** Discover the highest achievable level of an attribute by determining the specific combination of the variables in the basis set that, in concert, generate the optimum, subject to imposed constraints. The imposed constraints take on two conditions. Condition #1, called the convex hull, requires that no sensory attribute of the optimum lie higher than the highest level of that attribute achieved in the study, nor lower than the lowest level of that attribute achieved in that study. Condition #2 allows that a sensory attribute or the factor scores can be further constrained by the researcher to generate even tighter limits than the limits imposed by the convex hull.
6b	**Reverse engineer a profile:** Using programming techniques (e.g., multiobjective goal programming), set a target profile for a set of attributes. This is the "goal" profile. Identify the combination of ingredients or factor scores that yield that "goal," using the equations. Make sure that the constraints (convex hull, additional imposed constraints) are again obeyed.
7	**Estimate full profile:** Identify the levels of the independent variables (ingredients, factor scores) that generate the solution for either the optimization problem or the reverse engineering problem. Then use the model (Step 5) to estimate the full set of attributes corresponding to the solution.

the researcher has to bootstrap the independent variables. In either case, the computational approach creates the equations interrelating each of the dependent variables to the independent variables, their squares and cross terms.

Let us follow the direction in Table 7.3 to create the product model for margarine. For this illustration we assume that the test products do not share any common levels of ingredient. We will deal with systematic variations and the traditional RSM (response surface method) afterward. We will use the margarine data found previously in Table 7.2.

When creating a product model it is important to sample many products and ensure that these products differ from each other on a sensory basis, at least to consumers. Quite often experts are used for this task as well. Whether experts or consumers are used, the researcher obtains a sensory profile of the products, and clusters the products so that products with similar sensory profiles appear in the same cluster. Then, in order to reduce the number of products to a manageable few, the researcher selects one product from each cluster. This strategy ensures that the products differ from each other, but does not ensure that the correct products have been selected. That knowledge of "correctness" comes at the end of the study, although the experienced researcher can probably intuit it ahead of time.

The second step creates a *basis set*, or set of factor scores, using principal components factor analysis. This second step is necessary, because the researcher did not systematically vary the formulations. If the researcher had varied the formulations, then by definition the formulations would be independent (through design), and the second step would not be necessary. The factor scores are surrogate ingredients. They are independent of each other through the statistical reduction to factors. They are parsimonious—for margarine there are only four of the factors. They can be used as independent variables in an equation (see Step 3). *One need not name the four factors in order to use them.* However, they do co-vary with some sensory attributes in a meaningful way:

Factor 1 (26% of the variance): Aroma, creamy, easy spread, not hard
Factor 2 (24% of the variance): Buttery, salty, butter texture, flavor intensity
Factor 3 (15% of the variance): Bitter, sour, not sweet
Factor 4 (15% of the variance): Melt rate, soggy on cracker

The third step creates an equation relating the factors, their squares, and their cross-terms to each attribute. Each of the attributes, sensory, liking, image, expert panel, instrumental measure, even market performance, generates its own equation. With the availability of regression packages on the PC, creating the equation is quite straightforward. The researcher should keep in mind the limitations, such as the number of factors to use in the equation, the use of quadratic terms and interaction terms, and so on. By and large, however, most researchers now know how to use the regression procedures in PC-based statistical packages, so the modeling is fairly straightforward. Creating equations with the powerful packages becomes nothing more than point and click, with the selection of dependent variables and independent variables. Not that the equations are done completely automatically, however; they are not. It is just that today's computation systems make regression analysis quite automatic. Table 7.4 shows the equation relating the factor scores (F1–F4) as independent variables to the rating of overall liking as the dependent variable.

The fourth step optimizes one equation, subject to constraints applied to both the independent variables and on the dependent variables, respectively. For example, the independent variables (i.e., the factor scores) must lie within the range actually achieved across all of

Table 7.4   Equation for liking versus factor scores (F1–F4).

Dependent variable = overall liking
Squared multiple R = 0.79
Adjusted squared multiple R = 0.72
Standard error of estimate = 3.45

Regression equation

Constant	62.40
F1	1.22
F2	4.54
F3	−2.39
F4	0.62
F1*F1	−0.77
F2*F2	−0.43
F3*F3	0.36
F4*F4	−0.73
F1*F2	1.54
F2*F4	1.28

the margarine products. This explicit constraint is easy to obey, because the researcher need only check the upper and lower limits of the 40 margarine products. Since each product has its own factor scores, there must be a single highest and a single lowest factor score across the full set of products. The sensory attributes also must be constrained, but for a different reason. The constraints, known as implicit constraints, are also set up so that they encompass the highest and lowest sensory levels achieved in the study. By setting these constraints, the researcher ensures that the sensory profile to be achieved through optimization lies within the achievable range, or the "convex hull." Table 7.5 shows an example of these limits. Once the limits are set on the independent variables and on the sensory attributes, the optimization algorithm systematically searches through the different combinations of factor scores until it discovers a combination of factor scores that achieves the highest possible level of the attribute being optimized (e.g., overall liking), while at the same time ensuring that the sensory attributes lie within the constrained range.

Table 7.5 shows the results of this empirical search:

(i)   The lower and upper limits of the factor scores, and the values of the factor scores corresponding to the optimal product.

(ii) The rating to be maximized (overall liking).

(iii) The attributes, their expected level corresponding to the factor scores, and whether they act as constraints. The sensory attributes act as constraints.

(iv) Note that the sensory profile is not the profile of any single margarine product. Rather, it is a composite, corresponding to the factor scores that define the optimal product. One can search the data set for each sensory attribute in order to identify the one or several products that have the requisite sensory level corresponding to the optimal. The product developer then uses this information to construct "targets" or "landmarks" to guide development.

(v)   This approach *does not* produce a formulation. It cannot. The formulation can only come from studies that relate formulation/processing variables to ratings (see the succeeding text).

Table 7.5   Attribute profile of the "optimal margarine" synthesized by optimizing liking within the range achieved by the factor scores, and by the sensory attributes.

Factor	Low	High	Optimal
F1	−3.00	2.00	0.09
F2	−1.70	1.70	0.96
F3	−2.00	2.00	−1.65
F4	−2.00	2.00	1.26

Predicted values

Like overall	= 72	Optimization Equation	
Soft	= 76	Constrained 48 to 76	P38
Sticky	= 63	Constrained 41 to 71	P31
Like appearance	= 74		
Like color	= 69		
Like way it spreads	= 79		
Like appearance on bread	= 72		
Like the way it melts	= 64		
Like the aroma	= 64		
Like the taste	= 71		
Like aftertaste	= 63		
Like texture	= 68		
Easy to spread	= 80	Constrained 33 to 88	P28
Fast melting	= 53	Constrained 14 to 66	P38
Makes bread soggy	= 18	Constrained 8 to 25	P04
Aroma strength	= 52	Constrained 35 to 60	P17
Buttery aroma	= 53	Constrained 32 to 63	P17
Flavor strength	= 59	Constrained 39 to 65	P23
Buttery	= 59	Constrained 39 to 60	P26
Salty	= 37	Constrained 17 to 43	P25
Sweet	= 43	Constrained 29 to 45	P25
Cheesy	= 14	Constrained 9 to 26	P17
Bitter	= 13	Constrained 11 to 27	P11
Sour	= 14	Constrained 9 to 26	P11
Aftertaste	= 53	Constrained 40 to 63	P11
Hard	= 31	Constrained 11 to 54	P15
Buttery texture	= 62	Constrained 34 to 63	P04
Creamy	= 70	Constrained 45 to 75	P16
Versus margarine	= 56		
Calories	= 50		
Fat	= 50		
Similar to butter	= 60		

Note: Each optimal sensory level also has a specific product to serve as a landmark (e.g., P24, or product 24).

The foregoing approach for optimization can be readily used with any combination of factor scores in order to estimate the likely sensory profile. Let's look at an example. Figure 7.4 shows the distribution of products across the different factor scores. Each product is a circle in the two (or higher) dimensional map. When the marketer wants to create a new product, one strategy might identify "holes" in the map, corresponding to regions where there are no products. Once the factor scores corresponding to the holes are identified, it is straightforward to estimate the sensory profile of the product corresponding to the hole, and in turn to find target sensory landmarks.

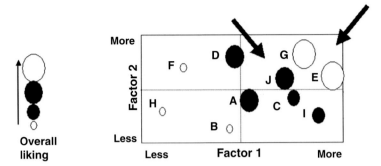

Figure 7.4 Location of in-market products in a two-dimensional factor map, the acceptance of these products, and the location of promising areas for new products ("holes in the map").

RESPONSE SURFACE MODELING—WORKING WITH KNOWN INGREDIENTS AND PROCESS CONDITIONS

When the researcher varies the formulations, builds a model, and either optimizes or reverse engineers the product (see the succeeding text), the output comprises a set of numbers corresponding to the optimum, as it was for category appraisal described in Section "Creating a 'product model' using the 'category appraisal' approach—unrelated products." The fortunate thing here is that the results are stated directly in terms of actionable formula variables that the developer can use. It should come as no surprise, therefore, that the original use of product modeling was the optimization of formula variables. The logic of the approach is precisely the same, with one exception. That exception is abandoning the factor analysis to generate the "basis set" or independent variables. Whereas in category appraisal the researcher has to create this basis set, in conventional response surface modeling the researcher need not. The basis set is created at the time of the study set-up, by virtue of using experimental design.

We can see the immediate applicability of response surface models by looking once again at margarine data. This time, however, we look at systematically varied formulations of margarine, rather than in-store products and single prototypes. The formulations comprise different levels of fat, coloring, and flavoring. The researcher lays out these combinations in a design, so that the variables are independent of each other. Depending upon the nature of the relation between the independent and dependent variables, and the number of independent variables, the design may comprise two levels of each independent variable, three levels, or even more. The classical treatment of the problem appears in Box, Hunter, and Hunter (1978). The recent proliferation of PC-based analyses allows researchers to use computer programs that design the necessary combinations (e.g., Systat, 2004). Once the design is set up, the researcher can run the study, and do the regression analysis just as was done with the category appraisal. The design ensures that the regression model will run.

The typical problems encountered when one does response surface research is the reaction to the effort. In a simple phrase, there's a lot of work involved, and whenever there is work there are the inevitable reactions to what is being demanded. It is easy for researchers to buy products "off-the-shelf" for category appraisals, or even to make "best-guess" prototypes. Developers can create these prototypes because that is their job. When it comes to creating systematically varied combinations, however, one gets a variety of responses from developers, researchers, and marketers. A few of these are:

(i) *It is too much effort to make the prototypes.* The best way to address this issue is to show the benefits of getting products "right the first time." The benefits include lower marketing costs, increased chance of market success, less time iterating to new products, etc.

(ii) *We will choose the best products and limit our evaluation to these promising prototypes.* This type of response is to be expected from anyone with experience. It assumes, not necessarily correctly, that the developer "knows" the consumer's mind. Often that is the case. Often, however, consumer preferences differ dramatically from what might have been expected. It is up to the researcher to identify what wins, not the developer, although the developer and the marketer can automatically dictate, by fiat, what they will test. They aren't necessarily right, even when they have the power to do so.

(iii) *We have other methods that will get us there, so we don't really have to make all these prototypes. We can use analysis instead of doing the development.* This response is becoming increasingly popular. For example, rather than doing experimental design that takes time and effort, someone will suggest that JAR (just-about-right) scales can show where the product needs fixing, and that with these data the developer "should know" what to do. It's not the case, however. JAR scales don't deliver actual formulations. Nonetheless, giving someone a less costly, less effortful alternative often derails the RSM exercise, usually in the interests of budget and expediency.

Researchers are familiar with experimental designs (Box, Hunter, & Hunter, 1978), which lay out the combinations. The choice of experimental design depends upon a number of factors, and whether or not the researcher wants to create equations using nonlinear terms (e.g., quadratic terms, pairwise cross-terms (interactions). Table 7.6 shows schematics for just two of the many hundreds of experimental designs from which a researcher can choose:

(i) *The Plackett–Burman design allows the researcher to investigate two levels of each of many different variables, in an efficient manner.* With 15 variables one needs only 16 "runs" or combinations. However, most prudent researchers use this type of design with fewer variables (e.g., 8–10 variables in 16 runs). The more prudent approach enables the researcher to fit a model with more degrees of freedom, and thus reduces the chance of fitting error in the data.

(ii) *The three-level, Box–Behnken design (a variation of the central composite design) is a very popular design for optimizing.* The design allows the researcher to investigate surfaces that have curvature. One consequence is that the model allows the researcher to optimize attributes such as acceptance, which reach their highest point in the middle range of ingredients. The three-level design also allows for pairwise interactions to be modeled by the equation.

(iii) *There are two variations of the design.* The full replicate requires 25 products. The full replicate allows estimation of all pairwise interactions, as well as quadratic terms. The half replicate eliminates eight of the combinations (shown by the darkened combinations). By giving up estimation of half of the pairwise interactions, the design saves eight combinations, reducing the effort of the product developer.

When the researcher wants to optimize the formulation, the next step is to select a design, and identify the variables under the developer's control. In the margarine project the developer

Table 7.6 Two schematics for experimental designs, allowing the researcher to relate independent variables to responses.

Two-level Plackett–Burman screening design for up to 15 variables

Prod	A	B	C	D	E	F	G	H	I	J	K	L	M	N	O
1	1	1	1	1	0	1	0	1	1	0	0	1	0	0	0
2	1	1	1	0	1	0	1	1	0	0	1	0	0	0	1
3	1	1	0	1	0	1	1	0	0	1	0	0	0	1	1
4	1	0	1	0	1	1	0	0	1	0	0	0	1	1	1
5	0	1	0	1	1	0	0	1	0	0	0	1	1	1	1
6	1	0	1	1	0	0	1	0	0	0	1	1	1	1	0
7	0	1	1	0	0	1	0	0	0	1	1	1	1	0	1
8	1	1	0	0	1	0	0	0	1	1	1	1	0	1	0
9	1	0	0	1	0	0	0	1	1	1	1	0	1	0	1
10	0	0	1	0	0	0	1	1	1	1	0	1	0	1	1
11	0	1	0	0	0	1	1	1	1	0	1	0	1	1	0
12	1	0	0	0	1	1	1	1	0	1	0	1	1	0	0
13	0	0	0	1	1	1	1	0	1	0	1	1	0	0	1
14	0	0	1	1	1	1	0	1	0	1	1	0	0	1	0
15	0	1	1	1	1	0	1	0	1	1	0	0	1	0	0
16	0	0	0	0	0	0	0	0	0	0	0	0	0	0	0

Three-level, Box–Behnken design for four variables

Prod	Full replicate				Prod	Half replicate			
	A	B	C	D		A	B	C	D
1	1	1	1	1	1	1	1	1	1
2	1	1	1	−1		1	1	1	−1
3	1	1	−1	1		1	1	−1	1
4	1	1	−1	−1	2	1	1	−1	−1
5	1	−1	1	1		1	−1	1	1
6	1	−1	1	−1	3	1	−1	1	−1
7	1	−1	−1	1	4	1	−1	−1	1
8	1	−1	−1	−1		1	−1	−1	−1
9	−1	1	1	1		−1	1	1	1
10	−1	1	1	−1	5	−1	1	1	−1
11	−1	1	−1	1	6	−1	1	−1	1
12	−1	1	−1	−1		−1	1	−1	−1
13	−1	−1	1	1	7	−1	−1	1	1
14	−1	−1	1	−1		−1	−1	1	−1
15	−1	−1	−1	1		−1	−1	−1	1
16	−1	−1	−1	−1	8	−1	−1	−1	−1
17	1	0	0	0	9	1	0	0	0
18	−1	0	0	0	10	−1	0	0	0
19	0	1	0	0	11	0	1	0	0
20	0	−1	0	0	12	0	−1	0	0
21	0	0	1	0	13	0	0	1	0
22	0	0	−1	0	14	0	0	−1	0
23	0	0	0	1	15	0	0	0	1
24	0	0	0	−1	16	0	0	0	−1
25	0	0	0	0	17	0	0	0	0

identified four key variables; Flavoring/Coloring blend A, Flavoring/Coloring blend B, fat level, and process speed. These four variables generated 25 combinations, following the Box–Behnken design shown in Table 7.6.

The consumer evaluation of systematically varied prototypes follows the same chore-ographed steps as the consumer evaluation of products in a category appraisal. In addition to the conventional response data from consumers, however, the researcher often incorporates other data, such as ratings by expert panelists, measures from machines, and cost of goods based upon the ingredients. These new, objective, dependent variables take on more meaning because the formulations are under the researcher's control. Thus for each of the products that the developer creates one can create or obtain a set of objective measures that can be later used in the optimization. For example, the optimization might comprise the maximization of liking, subject to constraints on overall cost. The optimally acceptable product may simply cost too much for the developer, and thus an alternative product would be desirable.

The models relating the formula variables under the developer's control to the dependent variables (consumer, expert, instrument, cost of goods) are estimated using the same type of conventional, off-the-shelf software. The approach is ordinary least squares analysis. For our study of 25 systematically varied margarines, the researcher obtained all the necessary data. For each dependent variable the researcher then fits the individual regression model, of the form:

$$\text{Rating} = k_0 + k_1(\text{A}) + k_2(\text{A}^2) + k_3(\text{B}) + k_4(\text{B}^2) + k_5(\text{C}) + k_6(\text{C}^2) + k_7(\text{D}) + k_8(\text{D}^2)$$
$$+ k_9(\text{A} \times \text{B}) + k_{10}(\text{A} \times \text{C}) + k_{11}(\text{A} \times \text{D}) + k_{12}(\text{B} \times \text{C}) + k_{13}(\text{B} \times \text{D})$$
$$+ k_{14}(\text{C} \times \text{D})$$

A = Flavoring/Coloring A
B = Flavoring/Coloring B
C = fat level
D = nutritional level

Experimentally designed combinations for margarine generate optimal formulations that are expressed in ingredients, rather than in sensory levels. It is not necessary to identify landmark products to serve as targets. Optimal formulations make the developer's job easier. The optimization algorithm searches among the different combinations of ingredients in a logical and efficient fashion, until it discovers an ingredient combination that satisfies specific objectives (see Table 7.7). For example, the optimal formulation must lie within the ingredient range tested in the original study. This is the case for ingredient 2, Flavoring/Coloring "B." However, ingredient 1, Flavoring/Coloring "A" is at its allowable upper level. It is incorrect to project outside this range because there is simply no data to support the projection. The optimum might lie either near that top level, or far beyond that top level. The researcher should do another experimental to ascertain how high is "up," and at what level acceptance begins to drop down.

EXTENDING RESPONSE SURFACE MODELING TO OTHER BUSINESS ISSUES

Most product developers work within constraints, so that the formulations that they create are developed in order to satisfy some technological or business objective. Usually these objectives transcend the typical feasibility issue, and devolve to products that must satisfy a

Table 7.7 Optimal formulations for margarine for total panel, without constraints, and with sensory and cost of goods constraints imposed on the formulation.

A	B	C	D
Formula variable	**Total**	**Dark <30**	**Cost <220**
Flavoring/coloring A	3.00	1.00	1.00
Flavoring/coloring B	2.18	1.99	2.15
Fat level	2.07	2.74	2.24
Added nutritive ingredient	1.00	1.00	1.00
Expected ratings			
Cost of goods	258	240	220
Overall liking	61	57	59
Like appearance	41	49	49
Like flavor	35	38	40
Like texture	46	51	51
Image—natural	64	63	61
Image—expensive	46	78	46
Image—fat	23	39	33
Image—buttery	60	65	65
Image—fits the brand	34	38	41
Sensory—dark	47	30	38
Sensory—aroma strength	44	39	40
Sensory—flavor strength	31	28	29
Sensory—sweet	43	49	48
Sensory—creamy	45	48	48
Expert—attribute 3	26	27	34
Expert—attribute 4	32	36	34
Instrument—measure 1	19	37	37
Instrument—measure 2	25	32	31

given concept, and/or must cost less than a certain amount to produce. If the product fails to fit a concept, then the product and concept may be acceptable, but the concept will promise one thing and the product will deliver another. When the product is highly acceptable but too expensive, then the product does not make business sense. In either case, the traditional way to develop products that satisfy constraints is to make prototypes, determine acceptance, identify the degree to which they satisfy the constraints, and then move on. This back and forth method can take time, cost money, and may in the end be unproductive because the trial and error method may fail to identify a feasible formulation.

Experimental design can help the developer ensure that the product makes business sense. The design as described previously represents an array of products, each of which can be tested for "fit to the concept." There is every incentive for the developer to create a model for fit to the concept or brand, and a model for cost of goods, relating these to the independent variable. Then, for every product developed the researcher can immediately estimate the likelihood that the product formulation will fit the concept and/or the cost of the product. Even more reasonable is the strategy of imposing constraints on fit to concept and cost, just as there are constraints on ingredients and on sensory levels. Following that strategy, the developer can insure that no product emerges from the optimization that does not satisfy these imposed constraints. Furthermore, if after working with the model the researcher fails to find any combination that maximizes liking so it scores reasonably high, and has the requisite cost or fit to concept scores, then this is a clear indication that there is no product within the range of ingredients developed. It is better to learn that sad news early in the development than have it come as a surprise later on. Examples of constrained optimization for sensory attributes and cost are the following:

(i) *Imposed sensory constraint (column C).* Every sensory attribute has an equation as well. Thus the optimal product can be developed so that its sensory profile lies within a specific range. In this example perceived darkness of margarine has been lowered from the optimal level.

(ii) *Imposed cost constraint (column D).* Since the cost is also expressed as an equation across the ingredients, every tentative optimal level has an expected cost associated with it. By imposing a constraint on cost the researcher identifies the specific combination of ingredients that maximizes acceptance, subject to a limit on cost of goods.

Finally, it is important to consider optimization directed toward satisfying multiple targets. In previous years product optimization using RSM technology was principally limited to R&D facilities. The projects were relatively simple—for example, identify the optimum formulations for a product that had two or three variables under the developer's control. The targets to satisfy were consumers of the brand, tested by the research guidance panel. At the same time, however, marketing researchers were expanding the scope of product optimization and the RSM approach. These larger scale studies focused on more strategic issues, such as optimizing a product for several countries, for multiple sensory segments, for adults versus children. The issues involved when larger scale marketing and business considerations are introduced can be addressed fairly directly. The product model allows the researcher to impose explicit constraints on the ingredients, and explicit constraints on the dependent variables. By considering the key subgroups as constraints, one can identify the optimum product formulation such that it is highly acceptable to the total panel (at least highly acceptable given the range of liking ratings actually achieved), and at least minimally acceptable to key subgroups.

UNDERSTANDING THE PRODUCT IN DEPTH—SENSITIVITY ANALYSES

Many product researchers approach experimental design as an opportunity to learn about the effects of formulations on responses. Rather than seeking the optimum alone in a mechanistic manner, these researchers recognize that the change of response with the formulation teach a great deal about the dynamics of a product. The learning that emerges from such analyses reveals a lot about the product that can be useful for future work. Essentially, the learning provides in-depth information about the dynamics of the product, information in fact that is generally absent for most products, despite the proliferation of product research. The sensitivity analysis parallels the way a psychophysicist approaches the relation between stimulus and response. The psychophysicist wants to discover how the human being transforms the stimulus into a response, and does so by creating psychophysical functions. The independent variable is the stimulus level, and the dependent variable is the response. For conventional sensitivity analysis, inspired by psychophysical thinking, there is only one stimulus variable and one response variable. The other stimulus or independent variables are held constant at some predesignated level.

Table 7.8 Sensitivity analysis of ratings to changes in Flavoring/Coloring B.

Flavoring/Coloring A	3.00						
Fat level	2.07						
Added nutritive ingredient	1.00						
Flavoring/Coloring B	1.00	1.33	1.67	2.00	2.33	2.67	3.00
Cost of goods	223	233	243	253	263	273	283
Overall liking	48	54	58	60	60	58	54
Like appearance	48	43	41	40	41	44	49
Like flavor	43	38	35	35	36	40	45
Like texture	48	47	46	46	46	46	47
Image—natural	58	61	62	63	63	63	61
Image—expensive	48	47	47	46	47	47	48
Image—fat	33	26	22	22	25	32	42
Image—buttery	58	59	59	59	60	60	60
Image—fits the brand	32	29	28	31	38	47	60
Sensory—dark	53	46	44	45	50	59	71
Sensory—aroma strength	47	45	43	43	44	46	48
Sensory—flavor strength	39	35	32	31	31	34	37
Sensory—sweet	47	45	44	43	43	44	45
Sensory—creamy	51	47	45	44	45	47	51
Expert—attribute 3	31	28	27	26	27	28	30
Expert—attribute 4	34	36	36	34	29	22	13
Instrument—measure 1	18	18	18	18	19	19	20
Instrument—measure 2	19	21	23	25	25	26	25

Creating the sensitivity analysis is straightforward. For the experimental design used here, we know that the formula variables ranged between 1.0 and 3.0. Furthermore, we created a set of equations interrelating the formula variables and the ratings. Thus, for any combination of formula variables within the range tested, we can estimate the attribute profile, simply by solving the equation for that combination. Finally, we can begin at any combination of levels of the independent variables, hold three of the four variables constant at that starting combination, and vary the fourth variable in small increments across the entire range. When we do that, each new combination generates its own profile of ratings, as shown in Table 7.8. By looking at the change in each of the attributes in Table 7.8 we easily see how changes in the variable being studied, Flavoring/Coloring B, affects the response to the margarine. We have held the formula at the unconstrained optimal level (see Table 7.7, column B). For example:

(i) Liking increases, peaks and drops down. We might have expected that, since the optimal level of Flavoring/Coloring A lies at an intermediate level.
(ii) Some attributes increase, such as objective cost of goods, and the image of "fits the specific margarine brand."
(iii) Some attributes decrease, such as Attribute 4 from the expert panel.

Quite often statisticians go beyond the conventional sensitivity analysis to isointensity contours. Many of the common statistical programs have the capability of plotting these equal intensity contours. The contours represent those combinations of two variables that, in concert, produce a constant value of the dependent variable.

Equal intensity or isointensity contours trace their heritage from a different intellectual history. The contours come from engineering that shows relations between variables, rather

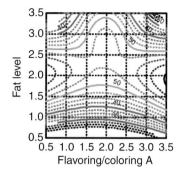

Figure 7.5 Equal intensity contour. The dependent variable is overall liking.

than from marketing. Marketing is interested in a numerical answer, engineering is interested in patterns and mechanisms.

Figure 7.5 shows an example of the equal intensity contour two of the ingredients, Flavoring/Coloring A and fat level, respectively. Each variable should be looked at when ranging from 1.0 to 3.0. The other two variables are held at their midpoints of 2.0 each. The contour plot shows the expected value of overall liking corresponding to these two variables at different levels. Contours are an attractive way to present the data, but become unwieldy when the researcher has to deal with more than three or four independent variables, or when the researcher has to deal with more than one dependent variable.

TURNING THE PROBLEM AROUND 180 DEGREES: REVERSE ENGINEERING

Now that we have dealt with the issue of optimization, let us consider a parallel problem. If optimization finds a point in the space that corresponds to a maximum level of a variable (e.g., highest liking), then what about finding a point in the space that corresponds to a particular profile of responses? Rather than maximize a variable, we minimize the distance between the profile that we want (the goal profile) and the expected profile of the product. When we specify the profile of the desired product we can search the space, looking for a set of ingredients that generate a profile lying very close to the profile we specify.

Three illustrative results (Table 7.9). Keep in mind that the modeling created equations relating the ingredients, their squares and pairwise interactions to each of the variables: cost, consumer liking, consumer sensory attributes, consumer image attributes, expert panel attributes, and instrumental measures. Using the metaphor of goal fitting, we set up a profile as to be the "goal profile," and then search for specific combination of ingredients that delivers that goal profile, or at least lies close enough to the goal profile. Depending upon the ingredient ranges, and imposed constraints (e.g., minimum allowable liking) the goal profile will be more or less approximated.

SURROGATE APPROACHES TO OPTIMIZATION – THE JAR SCALE

During the past fifty years, beginning in the 1960s, researchers have attempted to optimize products without necessarily doing the necessary "homework." Indeed, it was and remains

Table 7.9 Three results from reverse engineering a product formulation to fit a goal profile.

Ingredient	Simply fit the goal within ingredient range		Fit the goal, but make liking > 55		Fit a goal defined by instrumental measures	
	Ingredient set #1		Ingredient set #2		Ingredient set #3	
Flavoring/Coloring A	1.81		2.05		1.37	
Flavoring/Coloring B	1.80		1.94		1.24	
Fat level	1.00		2.48		1.00	
Nutritional addition	1.00		1.00		1.29	
	Goal profile #1	Expected profile #1	Goal profile #2	Expected profile #2	Goal profile #3	Expected profile #3
Cost of goods		180		248		169
Overall liking		41	>55	55		32
Like appearance		42		47		47
Like flavor		40		37		47
Like texture		50		46		52
Image—natural		68		67		65
Image—expensive		49		46		50
Image—fat		41		40		51
Image—buttery		66		68		65
Image—fits the Brand		45		37		55
Sensory—dark	20	28	20	42		29
Sensory—aroma strength	25	32	25	36		41
Sensory—flavor strength	30	22	30	27		32
Sensory—sweet	35	43	35	40		48
Sensory—creamy	35	36	35	44		44
Expert—attribute 3		28		32		29
Expert—attribute 4		39		37		27
Instrument—measure 1		31		25	30	29
Instrument—measure 2		25		7	30	30

tempting to cut corners because systematic experimental design requires time, resources, and an admission that perhaps experimentation is necessary. In business, if not in academia, time and resources are precious commodities. Anything that can reduce expenditures is welcome.

One popular way to identify direction for product change is known as the JAR or just-about-right scale. The JAR scale requires the respondent to rate the product in terms of overdelivery or underdelivery relative to an attribute. The respondent evaluates the one product, and does the rating, such as that shown in Table 7.10. The task is quite simple, and is often done in consumer research studies conducted by sensory analysts (product guidance) as well as by market researchers. The task generates tables showing either the percent of respondents who feel that the product over/under delivers, or an average magnitude of over/under delivery. The presumption is that if the respondents all feel that the product delivers perfectly, then the product should be optimized.

A sense of the type of data that might emerge from JAR scales comes from a study on frankfurters (Table 7.10). Each of 120 respondents evaluated 11 different frankfurters on a "blind" or "unbranded" basis. Each sample was rated on a set of attributes. One of the types of attributes was the sensory directional, where a rating of 0 denoted "far too little," a rating

Table 7.10 Average values for a 100-point JAR scale for 11 frankfurters.

	Sensory directional attribute				
Frankfurter	Dark	Smoky	Meaty	Salty	Greasy
101	6	−1	−8	2	−1
102	7	−4	−10	1	−3
103	−17	−8	−16	2	−1
104	−10	−11	−16	2	1
105	−21	−16	−14	5	−2
106	1	−8	−11	4	5
107	−31	−16	−13	−5	2
108	−4	−7	−10	−2	−4
109	−20	−13	−20	−7	−5
110	−29	−14	−17	−4	−1
111	−4	−8	−21	9	1

Note: The original JAR scale was anchored at 50 "just right." Positive numbers mean that the product overdelivers on an attribute. Negative numbers mean that the product underdelivers on an attribute.

of 50 denoted "just right" and a rating of 100 denoted "far too much." By subtracting 50 from the rating one can quickly see how the different frankfurters perform.

The results in Table 7.10 suggest:

(i) *From the directional information the developer can get a sense of where the products fail to do well, and the nature of the problem.* Unlike regular product tests where the sensory-liking curve must be constructed from several products, the directionality data suffices from one product. The developer only needs, presumably, a report from one single product to identify a problem. Indeed, it is the data from single, unconnected products, where directional data are most useful. Presented with one single product, the respondent identifies what is presumably wrong with the product.

(ii) *Many of the products have negative numbers, meaning that they underdeliver on attributes.* There are some positive numbers, however, such as +6 for frankfurter 101, meaning that it is too dark.

(iii) *There is a wide range of numbers.* A good rule of thumb adopted by the author is that a directional of −10 or +10 is worth paying attention to, because it signals a severe problem. On the other hand, a directional that is much smaller, such as a +5 or a −5 may signal that the product is not on target, but the deviation from what consumers want is not severe.

(iv) *Some attributes suggest severe departures from ideal.* An example of this in Table 7.10 is the attribute "meatiness." When this type of departure is seen again and again, in the same direction, one wonders about the nature of the attribute. No frankfurter in the study is sufficiently meaty. When in the course of a number of studies one sees that "in general" most products underdeliver on that attributes, one ought to suspect that the attribute "meatiness" is not a sensory attribute at all, but rather some type of image attribute that lies beyond the scope of products. Another example of this is "real chocolate." No chocolate really ever has enough "real chocolate."

(v) *The JAR values are presented either in terms of the percent data or average data.* When dealing with percents or frequencies, the researcher reports the percent of respondents who feel that the product has too much versus too little of an attribute. There is no sense of the magnitude of the overdelivery or underdelivery. This format of report is similar to the percent top 2 box for purchase intent or acceptance, where the data is summarized in terms of the number of people feeling a certain way. The other way of reporting the data, shown in Table 7.10, is the average magnitude of the overdelivery or underdelivery, a measure that gives a sense of the level of correction that needs to be made.

The consistency of data from the directional or JAR scale

With the increasing popular of JAR scales, one of the questions that continues to arise is whether the results are consistent with liking scales or sensory scales. That is, is there a basic agreement? Would problems that surface in sensory liking scales make themselves known if the respondent uses the JAR scale. Is the JAR scale more sensitive to problems? We already saw from Table 7.10 that some attributes such as "meatiness" are often problematic. Typically meat products don't deliver enough meatiness. Is this really a problem for product acceptance?

The issues with JAR scales are not merely theoretical niceties with which to deal. The JAR is a "workhorse" scale. It is used by many researchers to identify problems with single products, for which there is no other sets of measures. With one prototype and with a variety of JAR scales, the product developer may feel exulted after getting back a perfect "report card" on the product, or may feel dejected and have to work weeks and months to repair problems that emerge from the JAR scales. What are the limits of the JAR scale? Is there anything that we can learn that will allow us to work smarter with the scale? If the scale has problems, perhaps we might be able to better understand the seriousness or lack thereof of departures from "just right."

JAR Question #1: Is 1 JAR unit the same for different sensory attributes or does the size of the JAR unit vary by attribute?

This question gets to the root of what the product developer has to do when faced with the JAR data. Suppose, for example, that the product developer is given a report that the product is "off" by 2 JAR units in sweetness and 2 JAR units in thickness. It is one thing to say that the data suggest problems in a sensory attribute. It is quite another thing to report the magnitude of this, and expect the developer to know what to do with the data. For example, when the JAR is defined on a 0–100-point scale with 50 being "just right," then how much of a change in the product formulation should be made for a JAR score of 55 (namely +5). How about a JAR score of 60 (namely +10)?

One way to address the issue of JAR is by working with a number of related products that are rated on sensory attributes, liking attributes, and JAR attributes, respectively. The data set we deal with here comes from a study of 11 different salsa prototypes. This data set is sufficiently complete to allow for detailed analyses showing the sensory and liking magnitudes corresponding to a unit of JAR. Figure 7.6 shows the relation between the JAR and sensory intensities of aroma (left panel) or liking of aroma (right panel), for the salsa prototypes. The relations are linear, as one might expect. Low intensities correspond to ratings that the aroma is too weak (left panel). When the aroma intensity is "just right" (right panel) the aroma is rated at almost its highest liking. Thus within at least one attribute, aroma, the JAR scale is consistent.

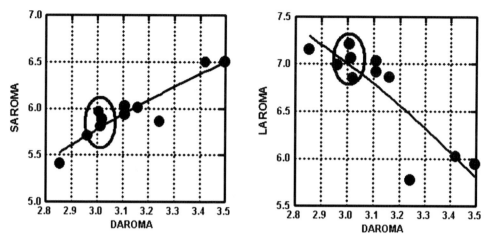

Figure 7.6 How liking of aroma (LAROMA) and strength of aroma (SAROMA) co-vary with the directional or JAR scale (DAROMA). Each filled circle corresponds to one of the 11 salsa products, all rated by the same respondents on a variety of sensory, liking, and directional scales (data courtesy of Peryam & Kroll, Inc.; from the Pangborn, 2003 Workshop on JAR Scales). There is internal consistency of JAR, sensory, and liking scales, but the sensory magnitude of a 1 JAR scale value varies by sensory attribute.

We can go one step further in the analysis by comparing JAR scales across different attributes. We see from Figure 7.6 that there exists a linear relation between JAR value as the independent variable and sensory intensity level as the dependent variable. We can express this equation as: Sensory Level $= k_0 + k_1(JAR)$. Keep in mind that the JAR value is given by DAROMA (short for "directional scale value for aroma"). The question is whether the different sensory attributes each has the same value, k_1, corresponding to 1 JAR value. When the slopes are the same, then this tells us that 1 JAR value for all sensory attributes has the same sensory magnitude. Table 7.11 shows clearly that the magnitude of the JAR scale *is not constant* across the different sensory attributes. Depending upon the particular sensory attribute being used, a 1-JAR value might correspond to a low sensory change of 1.50 units (for tomato flavor), or a much larger sensory change of 2.44 units for firmness. It remains for the researcher to further understand why these different slopes emerge for varying attributes of one product, and to measure these sensory-JAR slopes for other products. Parenthetically, this analysis of the "sensory value" of the JAR unit is reminiscent of the issue of JND or just noticeable difference in experimental psychology. At the start of the twentieth century, when experimental psychologists were thinking of summing JNDs to create a scale, the question circulated as to whether all JNDs were equal in different sensory modalities. For example, would a stimulus of 20 JNDs for bitterness taste as strong as a stimulus of 20 JNDs for sweetness? Issues in sensory science may arise and take on new form, but many of the questions remain fundamentally the same, once the veneer is stripped away, and we get to the essence of the issue.

Implications of the wandering sensory unit in the JAR scale
The importance of the JAR scale in product development means that many decisions are predicated on being "just right." What happens, therefore, when the JAR scale has a wandering unit? One possibility is what we might observe in Figure 7.7.

Table 7.11 Slope relating the JAR or directional scale to the sensory scale.

Sensory attribute	Slope, k_1, corresponding to 1 JAR unit
Tomato flavor	1.50
Flavor intensity	1.65
Amount of herbs	1.75
Heat	1.88
Roast garlic flavor	1.89
Salty	2.07
Tangy	2.17
Thick appearance	2.25
Sweetness	2.31
Firmness	2.44

Note: The slope shows the number of sensory points on a 100-point scale corresponding to 1 JAR or directional unit.

The same JAR value can mean small sensory changes or large sensory changes, depending upon the specific sensory attribute. When the developer doesn't really know the sensory-liking curve, it is quite possible that the directional or JAR data will be meaningless in terms of "how much to change," and may actually lead the developer to overshoot the change and overdeliver where previously he underdelivered an attribute or vice-versa. Figure 7.8 shows the consequences to overall liking of that inability to pin down the sensory magnitude of the JAR.

What should the developer do with a JAR of 2.0?

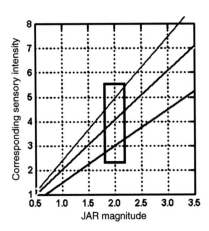

➢ Developer only has one product

➢ And the product gets a JAR of 2.0 in a test on an attribute

➢ Does this signify a big change or a little change?

Figure 7.7 Sensory implications for a JAR having a different value for each sensory attribute. The change in sensory attribute with JAR value will depend upon the slope relating the JAR and the sensory level. High slopes mean that a JAR value, such as 2.0, requires a large sensory change.

The same directional can be interpreted as different sensory levels

.. Leading to different products

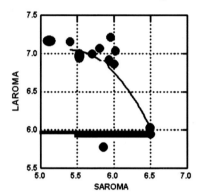

> ➤ We were just told to *decrease* aroma by 1.0 JAR units
> ➤ We don't know the JAR – aroma intensity relation
> ➤ How big a sensory change is .0 JAR units of aroma
>   > ➤   1.0 Unit? 1.5 units? 2.0 Units??
> ➤ How much does liking go up

Figure 7.8   How a given sensory change can affect liking. As a sensory attribute increases, liking changes, often according to a curvilinear function. Depending upon the magnitude of sensory change, liking can change dramatically or hardly at all. LAROMA, liking of aroma; SAROMA, strength of aroma.

JAR Question #2: Does creating a product with JAR's near "all about right" create an optimal product?

Another issue that arises with JAR scales is the ability of the scale to drive product optimization correctly. That is, when the researcher optimizes acceptance, do the JAR scales lie around "just right," as they should. The inverse question can also be asked; when reverse engineering generates a profile with all directional attributes lying near "just right," is the product optimally acceptable?

Let's look at the data from the 11 salsa products. The data were analyzed by the category appraisal specified in Table 7.3. The sensory attributes for the sauce generated a set of principal components. These principal components became the independent variables for equations. The set of equations generated a product model for the salsa. Each individual attribute, sensory, liking and directional (JAR), generated its own equation. The product model then allowed for two analyses, both of which reaffirmed that the JAR scale provides consistent information:

(i) *Optimization*: The product model was used to optimize overall liking, subject to explicit constraints placed on the factor scores and the sensory attributes. The constraints required that the optimal product lie within both the factor score range and the sensory levels achieved by the 11 salsas. The optimal set of factor scores was then used to estimate the full profile of the product on all attributes.

(ii) *Reverse engineering*: The product model was used to identify the factor scores that corresponded to a product whose directional JAR values were all "just right" (3.0 on the 5-point scale; 30 on the 50-point scale). The factor scores were then used to estimate the full profile of the product on all attributes.

(iii) *These two approaches generated profiles that agreed with each other*: Figure 7.9 shows a pair of scatter plots. The left panel shows how the levels of the 13 sensory attributes

Optimizing liking vs. reverse engineering (on target JAR) Agreement of sensory levels, directional (JAR) values

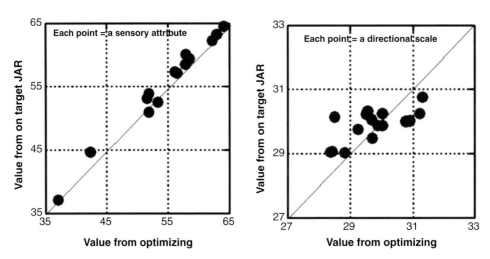

Figure 7.9 How the sensory attributes and the directional (JAR) attributes agree when the sauce product model is optimized in two ways. Each point is a different attribute. The abscissa corresponds to estimated values obtained by optimizing acceptance. The ordinate corresponds to estimated values obtained by reverse engineering. All ratings were multiplied by 10 before being used in the product model.

agree with each other from the two methods of optimizing. This means that despite the wandering unit, the optimization direction obtained from maximizing liking is similar to the optimization directional obtained from generating a product that is just right on all directionals. The right panel shows the expected values of the 15 directional scales. Again the results show a strong agreement. This means that the same information and guidance emerged from the two different ways of maximizing acceptance; direct maximization versus ensuring that no product is "too far away from the desired level in either direction," respectively.

THE SELF-DESIGNED IDEAL AS A VARIANT OF THE JAR SCALE

Another research strategy that works in the same spirit as directional ratings instructs the respondent profile the self-designed ideal. By profiling the ideal the researcher believes that the respondent somehow "knows" what the product should be, and can provide this information to the experimenter. This approach is similar to the JAR scale, in that we assume that the respondent somehow "knows" the level of self-designed ideal. Three issues arise:

(i) *Can the respondent actually identify the self-designed ideal?* This sounds easier than it really is. Respondents tend to describe sensory attributes one at a time, and tend not to verbalize the fact that there are pairwise and higher order interactions among products. Thus when a respondent profiles the self-designed ideal, all too often this ideal is described as if it constituted a set of attributes that do not interact.

(ii) *Consistency*: Once the self designed ideal is profiled, can the researcher create this ideal by formulation, or are there inconsistent attributes in the ideal that create a less than optimally acceptable product? Again the respondent is not constrained to lie within the limits of the products tested, and cannot be aware that it may be physically impossible to create the products. Parenthetically, the same consideration applies to the JAR scales, which are rated one at a time.

(iii) *Anchoring*: The self-designed ideal must be anchored against actual products. Otherwise the researcher has no idea to what specific sensory magnitude a profile value corresponds.

SENSORY PREFERENCE SEGMENTATION

People like different foods and beverages, as can be seen by a simple walk down the aisles of a supermarket. Whether the differences are in the products themselves, the different flavors, forms, health variations, and so forth, the differences are pervasive.

For many years, researchers in psychophysics treated the individual differences as intractable forms of variability that were best averaged by working with many respondents, or in the case of many sensory professionals, suppressed by training. Interindividual variability in hedonics is a simple fact of life, however, so pervasive as to demand recognition and protocols to deal with it (Ekman & Akesson, 1964; Pangborn, 1970). The recognition led to the need to divide people into groups showing similar preferences. At first the conventional thinking was that the segmentation might be somehow traceable to the more conventional aspects of an individual such as geodemographics. This did not work. Although researchers felt that perhaps culture and age would determine preferences, the results disappointed. Certainly there were some obvious effects such as the change in sweetness preference with age, or the higher acceptance of strong flavors by Latinos versus others. These preference differences were interesting, stimulated some new experiments, but could not explain the large differences to be seen on the supermarket shelf. Somehow the marketers and their associated R&D product developers were cognizant of individual differences at intuitive level, but this recognition did not lead to systematic approaches to understanding the nature of person-to-person differences in preferences.

We might find more "pay dirt" by describing sensory preference segments rather than explaining them away by causal mechanisms. That is, a profitable way to understand these interindividual differences is to accept the fact that people differ, and then look for ways to standardize the way we discover and use these differences. We're talking here of industrial-strength approaches to segmentation, rather than one-off small scientific experiments that demonstrate the segmentation but then leave product development opportunities unaddressed.

Researchers have developed different ways to segment consumers, but they all do one thing, namely divide the respondents by a well-defined criterion. We can get a sense of output from segmentation by looking at the results from a study on coffee (Schlich, 1997). In the specific study, the respondents all evaluated eight different samples of the coffee. This means that the data set follows a simple rectangular format, in which there is complete acceptance information for each product by each respondent. The actual study was quite simple—respondents evaluated each product on liking, using an anchored 1–9 scale. The study comprised data from 420 respondents in five countries. The top of Table 7.12 shows the average rating for each product for each country, as well as the averages and standard deviations (StDev) for each product across several countries, and for each country across products. The bottom of Table 7.5 shows a similar type of analysis, but based on sensory preference segments.

Table 7.12 Summary data for eight coffees rating on liking, for five countries.

Summary data for eight coffees—by country

Coffee	Denmark	France	Germany	Poland	England	Average	StDev
15	4.4	4.7	5.1	5.6	6.0	5.2	0.6
08	4.8	4.7	4.9	5.1	5.5	5.0	0.3
02	5.0	4.3	5.4	3.7	5.3	4.7	0.7
12	4.2	4.8	4.7	4.5	4.1	4.5	0.3
10	3.9	3.4	3.9	3.8	5.6	4.2	0.9
13	3.5	3.0	3.7	4.4	4.5	3.9	0.7
11	3.6	3.0	4.3	2.7	4.7	3.6	0.8
04	2.2	2.1	2.7	4.3	4.4	3.2	1.1
Average	3.6	3.4	4.0	3.9	4.5	3.9	0.6
StDev	0.9	1.0	0.9	0.9	0.7	0.7	0.3

Summary data for eight coffees—by sensory segment

Prod	LSeg1	LSeg2	LSeg3	LTot	StDev
15	5.2	5.6	4.9	5.2	0.4
08	4.2	5.4	5.0	5.0	0.6
02	3.5	5.2	4.9	4.7	0.9
12	3.4	3.0	6.7	4.5	2.0
10	3.9	4.4	4.0	4.2	0.3
13	5.8	3.5	3.3	3.9	1.4
11	2.9	3.9	3.8	3.6	0.6
04	6.7	2.4	2.3	3.2	2.5
Average	4.1	3.8	4.0	3.9	1.0
StDev	1.3	1.2	1.3	0.7	0.8

StDev, standard deviation.

Let us now look at different strategies by which to divide these respondents. We will consider three of them, all using the same data, that is, the data shown in Table 7.11:

(i) *Divide people by the magnitude of their reaction to one product.* This is the conventional approach. *This approach does not look for patterns.* It looks at the response of many people to one stimulus, divides the responses into likers and dislikers, and then attempts to identify underlying patterns that co-vary with the membership in these two (or more) groups. This analytical strategy approach can be traced to psychophysicists who worked with thresholds, and defined each respondent in a yes/no way—was the stimulus above threshold for that person so the person detected, or was the stimulus below threshold. This classification of someone into one of two groups was then carried forward into hedonics, with the notion of likers or dislikers (or occasion three groups, i.e., the indifferent group). The classification of a person into liker/disliker has been done by researchers working in the sense of smell, who wanted to classify specific odorants as being primarily liked or disliked (Moncrieff, 1966).

When it comes to applying this first strategy to the data, the researcher is faced with the simple question—which of the samples should be used as the criterion? It might sound obvious at first—one should divide the respondents by acceptors versus rejectors, but what cutoff should be made with a highly acceptable product such as pizza? Almost everyone will be classified as an acceptor. One conventional work-around this problem is to use the average rating, which at least transcends the problem of the product to choose as the determinant of acceptance.

(ii) *Divide respondents by the pattern of liking ratings assigned to a set of products.* The approach does not divide people based upon a single cutoff value, but rather divides people based upon the pattern of their responses. This second approach has proved fairly popular because it uses responses to a set of products, rather than to just one product.

(iii) *Divide respondents by the sensory level corresponding to their maximum achievable liking rating (sensory-preference segmentation).* In the foregoing section on drivers of liking, the point was made that the relation between a sensory attribute and liking often follows a quadratic function. This curvature leads to the possibility that the individual differences among individuals may represent different sensory-liking functions (Moskowitz, Jacobs, & Lazar, 1985). Indeed, when the respondent rates liking for a set of products, and if for these products one knows the different sensory levels on some scale, then one can relate the individual liking ratings to the common sensory scale, one sensory attribute at a time, one individual at a time, respectively. As long as the sensory scale was the same for each person, one may create the individual sensory-liking curve, and identify the sensory level on that curve where the person's liking rating reaches its peak. Doing that exercise generates a distribution of sensory levels, one per person per attribute, where liking peaks. *It really doesn't matter how high the liking rating gets when it peaks. Rather it matters where on the sensory scale the person's liking rating peaks.*

Figure 7.10 gives an example of the variation of these sensory liking curves. From the average data shown on the left-hand panel we see a curvilinear relation between sensory attribute level and liking. The optimum liking corresponds to a sensory level of approximately 45 on a 0–100-point scale. However, the right-hand panel shows that this simple quadratic curve emerges from averaging the data from several different respondents, each person generating his own curve.

Each sensory-liking curve for each respondent generates one optimum sensory level, within the sensory range tested. The input data comprise, therefore, a matrix. The rows of the matrix correspond to the 420 respondents. Each column of the matrix corresponds to a sensory attribute. The data in the body of the table are the optimum levels for the liking

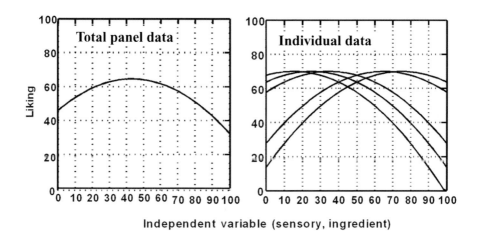

Figure 7.10 How liking co-varies with sensory attribute level or formula level, for total panel and for individual respondents that make up the total panel.

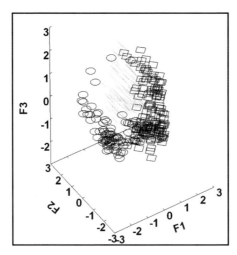

Figure 7.11 Location of respondents in three of the four dimensions for coffee. Each point in the graph corresponds to a particular respondent. The 400 respondents are clustered based upon how close they are in the map. Each of three clusters has a different symbol.

from each respondent on each sensory attribute. With 30 such attributes for coffee (some taken from consumer sensory ratings, some from expert panel ratings), there are 400×30 or 12,000 ratings. Many of the sensory attributes correlate with each other, and are redundant. A principal components factor analysis reduced these 30 attributes to four factors, of which three are shown in Figure 7.11.

The power of sensory segmentation to identify different sensory preferences appears most strongly when we look at the relation between sensory level and liking. Figure 7.12 shows a striking example from the coffee study, for bitterness (from the consumer panel) versus liking. The sensory-liking curves are fitted. The countries show similar patterns. Although no one really describes an ideal coffee as "bitter," in actuality as bitterness increases liking

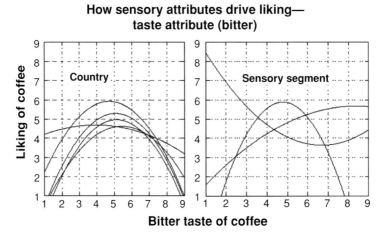

Figure 7.12 How liking co-varies with perceived bitterness for coffee, based upon patterns from five European countries, and from three sensory segments created from respondents across these five countries.

first increases, maximizes at the optimum, and then drops down. The picture is entirely different, however, when we look at the sensory preference segments (right hand side). The segmentation reveals three groups, all groups present in each of the five countries. One group dislikes bitter, one group loves bitter, and the third group finds the optimum acceptance at an intermediate level of bitter. It is this type of segmentation that begins to unravel the differences in preferences that exist among people far more than they exist across geodemographic, or even attitudinal divisions commonly made in consumer research.

THE VALUE OF THE LABEL OR BRAND

This section is labeled "value of the label or brand" rather than "testing products blind versus branded" for a reason. The reason is that the testing itself is only secondary. What is important to discover is whether the brand adds anything to the ratings of the product.

We know that there are expectations about a product. Simply labeling the product can change what people expect it to be. For example, when we call a beverage a health/refreshment drink after exercise we may find people to expect less sweet taste, more metallic/salt taste, and in general a less acceptable product. If we call the same beverage a refreshment drink, we may expect it to taste better, perhaps be carbonated, and so forth. We can demonstrate that effect simply by giving the respondent two samples; one labeled a "Refreshment Drink," and the other labeled an "After Exercise Replenishment Drink." The difference in the ratings shows the effect of the different labels.

Let us take this example one step further, and work with actual brands in the market. The study design is similar to what was just described. The researcher tests the same products twice. The first time the respondents evaluate the products without the benefit of the brand. The products are simply given code numbers to identify them. The respondents rate the products on a wide variety of sensory, liking, image, and even directional attributes, generating a database. The second time that the respondents rate the same products, this time with the products branded. The brand can be the product label presented together with the product. The respondents rate the same products on the same attributes, but of course the products have different identification numbers. The products should be of the same type, but can be distinguishable from each other even by visual inspection alone. Ideally, the same respondent tests the same products, to remove any effect of respondent variability, and to allow the brand effect to come through more clearly.

The foregoing exercise produces a database. That database comprises the same products, blind and branded, profiled on a variety of attributes by respondents who are presumably users of the category. Table 7.13 shows what these data look like for two soups, one vegetable soup with meat, and the other vegetarian vegetable soup without meat. Both products are currently in the market and enjoy widespread distribution. The respondents evaluated the product "blind," without any identification, and then branded, with a labeled can of soup in front of them. The mean ratings from 150 respondents who rated both soups blind and then branded appear in Table 7.12:

Six key questions to keep in mind when inspecting data of this type are:

(i) *What are the differences in the ratings between blind and branded?*
(ii) *Is either of the products more susceptible to branding?* That is, the magnitude of the branding effect may be larger for some brands, such as well-known ones, and smaller for others.

Table 7.13 Blind and branded ratings for two soups: vegetable soup (with meat) and vegetarian vegetable soup (without meat).

	Condensed vegetable soup			Condensed vegetarian vegetable soup		
	Bl	Br	Br–Bl	Bl	Br	Br–Bl
Overall ratings						
Overall liking	65	71	6	58	74	16
Top 2 box eat frequency (5-point scale)	35	46	11	37	46	9
Top 2 box purchase intent (5-point scale)	54	68	14	45	64	19
Appearance						
Like appearance	59	63	5	50	62	12
Clear vs. cloudy	52	45	−7	41	43	2
Broth color	56	54	−2	39	47	8
Broth thickness by appearance	21	18	−3	19	17	−2
Oily	54	56	2	64	55	−9
Amount of pasta	41	49	9	38	50	11
Size of pasta	25	28	3	24	27	4
Amount of meat pieces	3	13	10	40	12	−27
Amount of vegetables	59	64	5	59	65	5
Ratio of liquid to solid	−14	−17	−4	−16	−17	−1
Aroma						
Like aroma	57	66	8	51	57	6
Aroma strength	50	54	3	43	55	13
Taste/flavor						
Like taste	65	72	8	59	74	15
Like aftertaste	49	66	17	48	56	8
Flavor strength	61	66	5	56	67	11
Vegetable flavor	58	66	7	55	61	6
Saltiness	45	44	−1	46	43	−3
Seasoned	56	54	−2	50	49	−1
Meat flavor	6	4	−3	4	5	2
Aftertaste	42	44	2	42	39	−3
Texture/mouthfeel						
Broth oiliness	51	48	−3	60	50	−9
Broth thickness	20	15	−5	13	19	6
Amount of solid ingredients	54	54	1	56	61	4
Vegetable hardness	28	25	−3	31	28	−3
Meat tenderness	46	42	−4	25	17	−9
Image						
Nutritious	61	62	2	57	63	6
Hearty	53	55	2	42	54	12
Home made	40	41	1	28	41	13
High quality	52	59	8	49	58	9
Appropriate accompaniment to a meal	27	32	5	25	32	8
Appropriate for children/adults	39	46	7	39	48	9
Appropriate for males/females	47	51	4	47	49	2
Appropriate when hungry	41	49	8	37	46	10
Appropriate to accompany a sandwich	80	82	2	71	83	12
Substitute for a sandwich	43	46	3	34	47	12

(Continued)

Table 7.13   (*Continued*)

	Condensed vegetable soup			Condensed vegetarian vegetable soup		
	Bl	Br	Br–Bl	Bl	Br	Br–Bl
Appropriate for lunch by itself	45	52	6	39	49	10
Appropriate as accompaniment for dinner	59	57	−1	43	61	18
Appropriate for dinner by itself	26	30	3	26	31	5
Appropriate when eating alone	62	64	2	57	68	11
Appropriate for eating with family	45	52	7	45	48	3
Sophisticated/gourmet	17	18	0	11	23	12
Unique	17	19	2	14	24	11
Would tire of it quickly	50	39	−11	44	36	−8

Bl, blind; Br, branded.
Note: Each of the four soups was rated by the same 150 respondents. Unless otherwise noted, the attributes were rated on an anchored 0–100-point scale.

(iii) *Does branding influence ratings of sensory attributes or just ratings of liking attributes?* We usually think of sensory as being more stable than liking attributes, and hard to affect by anything other than formulation changes. Yet, the data may show that there is an effect on sensory attributes that is hard to explain. Nonetheless, this branding effect does occur.

(iv) *Standing back from the data, and inspecting the liking ratings of these two products and several others (not shown), do we see the same general effect of branding?* For example, if the products are branded and they are all popular, then does this branding contract the liking range, so that the products tend to be more equally performing when tested branded, but more different when tested blind?

(v) *Does the brand effect co-vary with the brand used by the respondent?* That is, are respondents uprating their own brand? This hypothesis can be tested by looking at the brand effect by total panel and by brand used most often. When the respondent "uprates" the brand he uses to a greater degree, this is *prima facie* evidence that brand usage affects product perception.

(vi) *Does brand effect appear to correlate with advertising spending or market share?* This type of analysis requires the evaluation of several different products, and the advertising spending for those products.

AN OVERVIEW

Product testing constitutes the key area for sensory analysts. This chapter has covered product testing from the point of view of application; specifically what the product developer and marketer need. There is a very large literature on product testing. The practical aspects of product testing are handled in a number of publications by the American Society for Testing and Materials) and various texts (e.g., Lawless & Heymann, 1998). For the most part, many of the books and pamphlets are simply "how to" and "best practices." What is important, and what this chapter has stressed, is the business use of the approaches. The researcher must always keep in mind the why of the test—why has it been commissioned, what will it do for the business. If those considerations are maintained and consulted in the selection of the test procedure, we may expect to see far more impactful, useful, and business-promoting product testing in the future.

REFERENCES

Amerine, M.A., R.M. Pangborn, & E.T. Roessler. 1965. *Principles of Sensory Evaluation of Food*. New York: Academic Press.

ASTM, Committee E-18. 1973. Standard recommended practice for establishing conditions for laboratory sensory evalution of foods and beverages, E480. In: *1975 Book of ASTM Standards*, Part 46, p. 266. West Conshohocken, PA: American Society for Testing and Materials.

Beebe-Center, J.G. 1932. *The Psychology of Pleasantness and Unpleasantness*. New York: Van Nostrand Reinhold.

BMD. 1973. *Biomedical Computer Programs*. Berkeley, CA: University of California Press.

Boring, E.G. 1929. Sensation and perception. In: *The History Of Experimental Psychology*. New York: Appleton Century Crofts.

Box, G.E.P., J. Hunter, & S. Hunter. 1978. *Statistics For Experimenters*. New York: John Wiley & Sons.

Brandt, M.A., E.A. Skinner, & J.A. Coleman. 1963. Texture profile method. *Journal of Food Science* 28:404–409.

Cairncross, S.E. & L.B. Sjostrom. 1950. Flavor profiles—a new approach to flavor problems." *Food Technology* 4:308–311.

Caul, J.F. 1957. The profile method of flavor analysis." *Advances in Food Research* 1–40.

Civille, G. & B.G. Lyon, eds. 1995. *FlavLex, Version 1.15*. Copyright Softex. West Conshohocken, PA: American Society for Testing and Materials.

Clapperton, J., C.E. Dagliesh, & M.C. Meilgaard. 1975. Progress towards an international system of beer flavor terminology." *Master Brewers Association of America Technical Journal* 12:273–280.

Crocker, E.C. & L.F. Henderson. 1927. Analysis and classification of odors. *American Perfumer and Essential Oil Review* 22:3–11.

Ekman, G. & C.A. Akesson. 1964. Saltiness, sweetness and preference. A study of quantitative relations in individual subjects. Report 177. Stockholm, Sweden: Psychology Laboratories, University of Stockholm.

Engel, R. 1928. Experimentelle untersuchungen uber die abhangigkeit der lust und unlust von der reizstarke beim geschmacksinn. *Pfluegers Archiv fur die Gesamte Physiologie*, 64:1–36.

Fechner, G.T. 1860. *Elemente der Psychophysik*. Leipzig, Germany: Breitkopf und Hartel.

Gordon, J. 1965. Evaluation of sugar-acid-sweetness relationships in orange juice by a response surface approach. *Journal of Food Science* 39:903–907.

Lawless, H.T. & H. Heymann. 1998. *Sensory Evaluation of Food: Principles and Practices*, 2nd edn. London, UK: Chapman and Hall.

Meilgaard, M., G.V. Civille, & B.T Carr. 1987. *Sensory Evaluation Techniques*. Boca Raton, FL: CRC Press.

Harper, R., E.C. Bate-Smith, & D.G. Land. 1968. *Odour Description and Odour Classification*. New York: American Elsevier.

Meiselman, H.L. 1978. Scales for measuring food preference. In: *Encyclopedia of Food Science*, eds. M.S. Petersen & A.H. Johnson, pp. 675–678. Westport: AVI.

Moskowitz, H.R. 1981. Relative importance of perceptual factors to consumer acceptance: linear versus quadratic analysis. *Journal of Food Science* 46:244–248.

Moskowitz, H.R. 1996. Experts vs. consumers. *Journal of Sensory Studies* 11:19–38.

Moskowitz, H.R. 1997. Base size in product testing: A psychophysical viewpoint and analysis. *Food Quality and Preference* 8:247–256.

Moskowitz, H.R. 1999. Improving the "Actionability" of product tests: Understanding and using relations among liking, sensory, and directional attributes. *Canadian Journal of Marketing Research* 18: 31–45.

Moskowitz, H.R., B.E. Jacobs, & N. Lazar. 1985. Product response segmentation and the analysis of individual differences in liking. *Journal of Food Quality* 8:168–191.

Pangborn, R.M. 1970. Individual variations in affective responses to taste stimuli. *Psychonomic Science* 21:125–128.

Schlich, P. 1997. Personal communication. Data provided from the Coffee Study, courtesy of the European Sensory Network.

Systat 7.0. 1997. Statistics. p. 751. Chicago: SPSS Inc.

Systat. 2004. *Systat, the system for statistics*. User Manual for Version 11. Evanston, IL: Systat, Division of SPSS.

Szczesniak, A.S., B.J. Loew, & E.Z. Skinner. 1975. Consumer texture profile technique. *Journal of Food Science* 40:1253–1256.

Thurstone, L.L. 1927. A law of comparative judgment. *Psychological Review* 34:273–286.

8 So what can sensory do for me (or for my company)?

INTRODUCTION

The primary role of sensory analysis is to provide information about how the sensory characteristics of products, ingredients, or other related information and services relate to perceived quality characteristics and to consumer liking. This chapter discusses what happens in the early stages of new product development, the types of decisions that will be affected by sensory, how sensory evaluation activities help to arrive at outcomes for each milestone in product development, and finally how sensory provides insight into a myriad of business decisions that surround the many activities in new product development.

In the not-too-distant past, product development was conducted in an "information vacuum" (Moskowitz, 2000) where the prevailing attitude was that experts or top management knew what was best for consumers. It was all too often the case that the food product development process simply combined management dictates with creative intuition of the product developer. This vacuum and isolationist attitude worked for many years when there was not an overabundance of food products on supermarket shelves, consumers did not have many alternative products to choose from, and there was not so much competition for consumer dollars. Consumers purchased whatever manufacturers offered. Today, the consumer is deluged with thousands of new product introductions and line extensions yearly. Manufacturers have to compete for valuable shelf space. The theme of a choice economy, an abundance that is no longer a blessing to the manufacturer, continues throughout this book as a pace-setting rhythm that informs what sensory and consumer should do now (many are doing it already) and must do in the coming years.

The current process for developing new food products is seriously flawed and has been so for decades (Rudolph, 1995). Too many new food product introductions fall short of the developer's expectations, visions, and hopes (Lord, 2000). Lots of products get to market, obtain early distribution, and before long distribution wanes and the product is being "swapped out" for another new product (Lord, 2000). It has been estimated that 80–90% of new food products introduced fail within 1 year of introduction. The failure cost to the US food industry is estimated at 20 billion dollars and includes missed sales

Sensory and Consumer Research in Food Product Design and Development, Second Edition.
Howard R. Moskowitz, Jacqueline H. Beckley, and Anna V.A. Resurreccion.
© 2012 Blackwell Publishing Ltd. Published 2012 by Blackwell Publishing Ltd.

targets, lost revenues, and postponed profits in addition to wasted development resources (Morris, 1993).

Successful product launches involve a number of different activities. A disciplined product development process with defined stages and gates such as the Arthur D. Little (ADL) process (Rudolph, 2000) provides the basis for objectivity in decision-making because each stage operationally defines deliverables and metrics (Lord, 2000). Cooper (1996), the originator of the Stage-Gate™ process, notes the importance of the quality of execution of technological activities such as preliminary technical assessment, product development, in-house product or prototype testing, trial or pilot production, and production start-up.

WHY USE SENSORY FOR BUSINESS DECISIONS?

Three major types of business issues arise in companies for which product developers and product marketers seek out sensory scientists. These three business issues are: (1) for advice in designing studies, (2) for assistance in collecting data from experts and consumers, and (3) for guidance in interpreting results (Moskowitz, Munoz, & Gacula, 2003). Among the many steps in product development are the often monumental tasks of measuring the target consumers' response to the concepts, prototypes, pilot plant outputs, initial production line efforts, and ongoing packaged products. Each phase requires different information to be provided to formulators, refiners, optimizers, marketers, and their associated team members (Moskowitz, 2000).

HOW DO THESE BUSINESS ISSUES GET FORMULATED AND BY WHOM?

Cross-functional new product development teams that include product managers, product developers, marketers, and sensory practitioners promote communication among the various functions, with the speed and cooperation in turn enhancing speed to market. The inter-action among team members who are co-located and share a common set of objectives, facilitate the projects for food product development, increasing the success potential of these projects. Cross-functional teams make it easier to assemble, coordinate, and guide necessary resources from the many functions that are required in order to successfully launch a new product.

Pillsbury® used the team approach to develop and launch low-fat Häagen Dazs® (Dwyer, 1998). Twelve people worked on two integrated groups, a business team and a project team. At Nabisco, cross-functional teams comprise eight to twelve members coming from areas as diverse as commercialization, marketing, product development, engineering, quality assurance, packaging, sensory services, logistics, and finance. In some cases, an outside supplier representative may participate at the very beginning of the project (Lord, 2000).

The corporate core competencies are increasingly being focused on consumer marketing and product innovation. As a result, although most medium- and large-sized companies have at least the minimal in-house capabilities to generate new products, few companies today possess all of the necessary skills in order to efficiently develop the new products that today's market demands. This situation creates the need to outsource certain new product development activities (Lord, 2000).

THE SYNERGISTIC INTERACTION BETWEEN SENSORY AND CONSUMER RESEARCH DEPARTMENTS

The specialization by sensory scientists in product testing often fits well with the emerging corporate needs in consumer research, although as previous chapters have noted there is occasionally a dynamic tension between sensory and consumer research. Product developers and product marketers work with sensory scientists in various stages in the product development process, starting from the very early stages where they need advice in designing consumer studies. The sensory practitioner member of the new product development team provides assistance in implementing the studies to collect data from consumers as well as trained panelists. Finally, the sensory practitioners provide invaluable guidance in interpreting results (Moskowitz, Munoz, & Gacula, 2003). The role of the sensory practitioner is thus structurally very important.

Many forward-thinking corporations actively foster communication between sensory scientists and consumer researchers. In cases where the sensory scientists and consumer researchers report to one vice president, turf fights for corporate resources and recognition are effectively eliminated. There is cross-fertilization of advanced techniques. Advances in sensory methods for product development such as response surface methodology (RSM), once reserved for the sensory statistical analyst, and consumer research procedures such as the category appraisal method to assess drivers of liking are now widely used by the entire integrated R&D department or function (Moskowitz, Munoz, & Gacula, 2003).

CONSUMER SENSORY TESTS AND MARKET RESEARCH: THEIR SIMILARITIES AND DIFFERENCES

Sensory consumer tests and market research tests have many similarities as well as differences:

(i) *Who participates?* Both involve testing with consumers and subsequent collection of data on consumers' opinions. In companies where consumer sensory testing and market research coexist, the consumer sensory tests are conducted by the sensory evaluation department for R&D, which is the end user of the information, whereas consumer market research is conducted by the marketing department and marketing is the end user of the information.

(ii) *What types of stimuli are tested?* A difference between the two tests is that consumer sensory tests are conducted on blind samples, labeled to disguise the product origin (e.g., by a three-digit code), under experimentally controlled conditions, and generally with minimal concept information in order to avoid signals that are market-related, rather than stimulus-related. In contrast, consumer market research tests include concept testing and labeled samples. It is important to note that in the past few years sensory researchers in companies have begun to adopt some of the market research practices in the types of stimuli they test, so that the distinction between sensory and consumer research may be blurring somewhat.

(iii) *Where do participants come from?* Another major difference is that consumer sensory test panelists are recruited and screened on the basis of their use of the product—they should be current users. In many consumer tests run by market research the respondents may be first given a concept to read. Only if they are positive to the concept (e.g., definitely or probably would buy) or positive to a specific brand would they then be recruited to participate in the consumer research study (Lawless & Heymann, 1998). The assumption by consumer market researchers is that they want to fine-tune the product to those individuals who they believe are predisposed to the product idea. Such predisposition is measured by accepting the introductory or set-up concept.

Consumer sensory evaluation tests appear to be evolving to the point where they are fairly similar to the types of consumer tests conducted in marketing research. Because of this apparently similar function, a recent trend of merging sensory departments that do field testing with their market research departments has occurred. The mergers may have an unfortunate consequence, due to the nature of the corporate structure. Consumer market research generally reports to the corporate marketing function, where it gets its budget and thus holds its loyalty. The R&D department, just as needful of consumer feedback, albeit for each stage of development, could lose its test link to consumers through the merger of the research functions. The R&D department would thus be deprived of important technical support (Lawless & Heymann, 1998). When sensory testing is unavailable, R&D is deprived of ultimate validation; it is uncertain whether the new product's sensory goals were reached and how consumers perceive product performance. For example, when the consumer sensory test is not conducted and the product fails, the reasons for failure remain unknown—if only a concept test had been performed by market research. Ambiguity is brought about by not being able to determine if it was the sensory properties of the product or failure of the marketplace to respond to the product, as predicted. When product and concept testing are confounded, inferences cannot be conclusive about the reasons for failure.

WHAT DO THE SENSORY PROFESSIONALS DO TO "FORMALIZE" OR "OPERATIONALIZE" THEIR TEST OBJECTIVES?

The sensory scientist's specialization in product testing fits well with emerging corporate needs in consumer research. Product developers and marketers seek the sensory researcher's assistance in designing studies and in collecting data from experts and consumers. The primary role of sensory analysis is to provide information about how the sensory characteristics of products, ingredients, or other related information and services relate to perceived quality characteristics and consumer liking.

Sensory scientists are becoming more educated in basic science and the applications of sensory methods. The variety of data to be generated from sensory tests involving consumers means that a range of test protocols should be employed in several venues: laboratory, office, central location, consumers' homes, and so forth. All consumer-oriented sensory testing is done in order to predict sensory-based consumer behavior. With the proper direction, sensory evaluation can be a powerful tool to guide new food product development (Moskowitz, 2000).

Sensory and acceptance tests are instruments, only as good as the sensory practitioner who uses them, the validity of the test design, the appropriateness of the questionnaire, and the representative nature of panelists selected to participate in the tests. The principle for all sensory tests is that the test method should be selected on the basis of the objectives for the study. The question that must be asked is: What does the test measure? If the objective is to quantify consumer acceptance, then a *consumer affective test* is appropriate; if the objective is to describe the characteristics of a product, then the test should include a *descriptive analysis of product attributes*.

In new product development, much of the success or failure of a food product in the marketplace results from consumer perception of sensory quality. Therefore, new product development should involve, as the very first step, the identification of the product, then the identification of critical quality attributes of the product by regular consumers of a product. Once the quality attributes important to consumers have been determined, appropriate designs for systematic product optimization can be developed.

Over the past decades, researchers have divided tests into two forms, qualitative and quantitative, although as we saw in previous chapters, the emergence of new research methods is blurring some of these two methods. Nonetheless, the division of the research world into these two general parts is very useful because it sets a framework for communicating problem statements and expectations about research outcomes:

(i) *Qualitative research*: Traditionally, qualitative research in the form of focus groups is used to identify the critical quality attributes. Focus groups comprise eight to ten consumers recruited to fit specific demographic, attitudinal, and usage characteristics; they are conducted by a trained moderator who helps to stimulate and direct the discussion (ASTM, 1979; Sokolow, 1988). The focus group is particularly valuable in language generation and very early stage product development (Sokolow, 1988). With this information, formulation or process variables can be systematically modified in order to increase the likelihood of acceptance of the final product by consumers. Focus groups are an effective means to identify those attributes that are important in the product, and that should be included and maximized in the product. Focus groups are just as important to identify characteristics that are undesirable to consumers and that should be minimized or eliminated from the product. Focus group results can be used to assist in questionnaire design of the succeeding consumer test. Quantification occurs after the critical quality attributes have been identified. Rarely do developers simply "run" with the results of focus groups or other qualitative procedures. The culture of testing is such that generally the qualitative portion precedes other parts of the research process.

(ii) *Quantitative research*: Consumer affective tests are conducted to quantify overall acceptance or consumer responses to critical attributes, defined as those attributes that determine a product's preference or acceptance. Consumer testing of new product prototypes or market candidates can provide valuable information to product developers (Lawless & Heymann, 1998). Often descriptive analysis and physicochemical measurements characterize and establish quantitative limits for the critical attributes that determine consumer acceptance. In some industrial situations, much of the product research relies on the technical knowledge of the product specialist. Some traditional approaches to quantification have their origin in quality grading systems such as those developed for dairy products, meat, fruit, and so forth. Unfortunately, quality grades do not always agree with consumer acceptance data (Lawless & Heymann, 1998).

In R&D-oriented sensory evaluation tests, products are mostly evaluated "blind," that is, without brand identification. "Blind-labeled" consumer sensory tests can be used in at least six situations to generate necessary information about the product:

(i) *Pure acceptance*: Measure the level of consumer acceptability, on a sensory basis, without the conceptual claims that normally appear in advertising or in packaging (Lawless & Heymann, 1998).

(ii) *Selection*: Screen multiple formulations for the most promising candidates—nonperformers can be dropped.

(iii) *Development guidance*: Provide direction for formulating new products or reformulating current products. Tests using current products without brand name may be "tricky" because the brand name may lead the respondent to expect a certain sensory profile (witness the New Coke® fiasco in the early 1980s).

(iv) *Early warning*: Facilitate diagnosis of problems that may be perceived by consumers before more expensive market research studies. Problems can be uncovered that cannot be detected in laboratory or in "more tightly controlled central location tests."

(v) *Insurance*: Serve as a final step before multicity market research tests or launching the product in the marketplace.

(vi) *Legal documentation*: Provide documentation for claim substantiation—defending challenges from competitors.

Chapter 9 discusses how consumer sensory tests are conducted. There is substantial literature on consumer marketing studies, but not on consumer sensory testing techniques. It is difficult to find academic courses offered on consumer testing methods, and details of consumer sensory testing of food products. Lawless and Heymann (1998) indicate that novices are quite frequently trained through "shadowing" an experienced researcher in industrial practice. There are a few general guides to consumer sensory testing. Among these are a manual from the American Society for Testing and Materials (ASTM, 1979) and a reference by Resurreccion (1998) on consumer sensory testing for product development.

HOW DOES THE SENSORY PROFESSIONAL THINK IN THE CONTEXT OF THE DEVELOPMENT TASK?

The consumer in product development

Unsatisfied consumer needs represent potential product opportunities. Therefore, effective and successful new product development ought to start with the consumer (Lord, 2000). Consumers are the center of the product development process. The critical importance of consumer response "as a key to product success" has made early stage product testing the focus of attention by corporate management, marketing, and the R&D function. Product developers need to understand consumer behavior and food choice as well as the relation between individual products and the consumer (Earle & Earle, 2001).

In order to increase the chances for commercial success for almost any product, it is imperative to solicit consumer feedback obtained throughout the different phases and steps in the product development process. This is becoming more and more of a truism every day. It is no longer sufficient for a product to be acceptable—a product needs to *delight* consumers to be successful. The consumer of the new food product gives the ultimate decision on the new product. Therefore, it makes sense for the consumer to be involved in evaluations throughout the product development process. Even more important is to include the consumer in the creative steps in the new product development process. Conversely, it is erroneous for company personnel involved in new product development to assume that they know what consumers want and need.

There are three reasons for obtaining consumer input at the early stage of product development. These were listed by Moskowitz (2000):

(i)   Realization by food manufacturers that it is consumer acceptance, first and foremost, that determines product success.
(ii) Consumer input shortens development time.
(iii) Cost of product testing is minimized.

The sensory evaluation literature has exploded in the number of articles on consumer evaluation of products featuring consumer-guided product development and consumer insight. Only two decades ago, early stage product testing was mostly limited to the evaluation of flavor or texture of food. The studies did not have consumers or consumer panels but purposely and primarily involved trained panelists. Today, management recognizes the importance of consumer acceptance testing and the need to obtain a measure of product quality. Guidance research has become increasingly important to management to maintain product quality, and to help product developers understand what consumers want and how they react to the prototypes that are created. All of the reasons for the increased use of guidance research point to one thing—increased business success by creating better products.

Reasons for consumer testing

Ultimately, consumer acceptance tests are the most important tests in the product development process. All other measurements are merely "surrogates" for consumer acceptance, or attempts to look at factors that would influence acceptance by consumers. When research budgets are tight, all other forms of sensory testing can be cut except for consumer testing. Without such data the corporation "flies blind." Perhaps the lucky company can continue with insights gathered years before, with product information gleaned from the popular press, and with the information provided by experts. Such luck doesn't last forever. At the end of the day, it is the ongoing practice of proper consumer testing that gives vision, foresight, and increased opportunities to the companies that practice it.

Thus far this chapter has dealt with product acceptance. Speed to market is also extremely vital in product development. Speed to market is where consumer testing makes its mark because it can be done quickly, with only modest investment. In contrast, characterizing product attributes and sensory profiling require trained panels that take time. Panelist recruitment and training take an inordinate amount of time. Recruiting, training, and maintaining a trained panel can be costly. To minimize the cost of evaluation, testing involving trained panels is often cut in the new product development process. Product developers increase the efficiency of the process by saving time and going directly to consumers. Today, trained panels are considered more appropriate for sensory-based quality assurance (Munoz, Civille, & Carr, 1992) than they are for ongoing research guidance.

PRODUCT DEVELOPERS AND MARKETERS ACTIVELY SEEK THE SENSORY RESEARCHER'S GUIDANCE IN INTERPRETING RESULTS

Sensory scientists have become educated in basic science, sophisticated in their applications of sensory tests, and smarter in their interpretation of results. Statistical analyses provide information as to whether the results of a study can be traced to the experimental treatment or result from the ever-present chance variation. Statistics provides information to be used in the interpretation of the data, and they find a home in both up-front design and subsequent data analyses.

It was categorically stated by Gacula (Moskowitz, Munoz, & Gacula, 2003) that the role of statistics in sensory science is secondary to the competence of the sensory scientist. Although statistics are necessary, they only serve to complement good sensory projects and research. The application of statistical techniques is conducted only after one has ascertained that the best data have been collected. Gacula further stated that the simplest statistical method that provides adequate results is the best choice; the result of this choice is generally easy for the client to comprehend. Statistics becomes a problem when it clouds the results and when good solid thinking is sacrificed in the name of the newest statistical procedure.

Today's sensory scientists have access to statistical packages on personal computers that enable researchers to better understand relations among stimuli or among variables. They lead to new insights that make the results more meaningful and valuable in new product development. Although there are many statistical programs in any single statistics package, the set of approaches used by today's sensory researcher fall into four major classes:

(i) *Tests between and among samples*: The first set of techniques comprises powerful statistical methods such as analysis of variance (ANOVA), factor analysis (FA), discriminant function analysis (DA), etc. These methods look for differences among samples, and rules by which these differences emerge.

(ii) *Mapping methods*: Multidimensional scaling (MDS) techniques map or locate a product, panelist, or attribute or combinations of such in a geometrical space. Points on the map that are located in closer distance to each other are more "similar." Conversely, those points on the map that are further away from each other are "dissimilar."

(iii) *Functional relations among variables*: A third set of techniques involves the relation of two or more sets of variables. These techniques include partial least squares (PLS) and reverse engineering.

(iv) *Methods that search for a goal or satisfy an objective*: The fourth group of techniques includes modeling and optimization. The objectives of these techniques are to identify formulation and/or processing variables leading to a high level of consumer satisfaction (they delight consumers).

These statistical techniques, maps, and models provide significant value to both product developer and sensory scientist in the process of creating products that fulfill consumers' requirements and expectations.

SENSORY RESEARCH AND THE PRODUCT DEVELOPMENT PROCESS

Sensory research fits into the different stages of the product development process in various guises. To understand where the sensory research actually applies, it is probably best to review the process itself. Although product development processes may vary from company to company, generally there are a limited number of "real stages." The four main stages in the product development process as described by Earle and Earle (2001) and Rudolph (2000) are shown in Table 8.1 as follows:

(i) *Product strategy and definition*: Stage I comprises the strategic plan, consumer needs and market opportunity assessment, and product concept and design specification. This corresponds to Phase I of both Earle and Earle (2001) and Rudolph (2000).

(ii) *Product implementation and marketing*: Stage II includes product and process development. This corresponds to Phases II, III, and IV of Earle and Earle (2001) and Phase II of Rudolph (2000) except for scale-up and trial production.

(iii) *Product commercialization*: This includes scale-up and trial production and product launch preparation. This corresponds to Phase V of Earle and Earle (2001) and the last step of Phase II of Rudolph (2000).

(iv) *Product launch and evaluation*: This includes full-scale introduction, final evaluation, and product support. This corresponds to Phases VI and VII of Earle and Earle (2001) and Phase III of Rudolph (2000).

Given this division of the product development into stages, the key thing to keep in mind is the pervasiveness of consumer feedback. In product development, for the most part, the developers and marketers rely on a combination of insight and hope. Consumer sensory feedback in the food and beverage industry is vital. Unlike consumer electronics, where the latest technologies can emerge, grow rapid, peak, and then drop down to irrelevance in 6 months to a year, the food and beverage industries work with slow-moving trends. The different types of sensory and consumer information provide a guidance method for a situation in which otherwise the development cycle might go completely astray. The feedback from the marketplace is simply too slow, although such feedback can be brutal after the product is launched, if it doesn't move off the shelves. Few companies today can afford such reality based testing, so enter the consumer feedback system as a vital metric in virtually every part of the development cycle.

PRODUCT DEVELOPMENT MILESTONES AND THE CONTRIBUTIONS OF SENSORY RESEARCH

Rudolph (2000) discussed the ADL comprehensive philosophy to guide food product development activities. This is based on establishing consistent milestones for the entire development process and identifying the required deliverables by each of the functional groups that contribute to product development in the company (Rudolph, 2000).

In their product development process, milestones for each phase, such as the strategic plan, market opportunity assessment, product business plan and development, prototype development, market strategy and testing, scale-up and trial production, and product introduction and product support, are viewed as an opportunity to monitor progress against a planned set of goals, to review the next task and anticipate problems, and to initiate program changes. It is critically important that each functional group, namely management, marketing, R&D, manufacturing, sales, distribution, and product support, provide input to all of the milestones throughout each phase of the process.

Barclay (1992a, 1992b) put it succinctly: "The food product development process needs to be linked with corporate objectives and to the external environment to allow new ideas

Table 8.1 Steps in the product development process and the appropriate sensory/market test.

Phase in the product development process	Stage in the product development process (according to Earle & Earle, 2001)	Phase in the product development process (according to Rudolph, 2000)	Appropriate sensory/ market test	Nature of the output
I. Product strategy and definition	I. Business strategy	I. Product definition		
	Idea generation and screening	Strategic plan	Consumer focus groups	Vision of the company's direction
		Market opportunity assessment		Product report, identification of new consumer needs
		Product business plan		A business plan
	Product concepts and design specifications	Product definition		Product concept
II. Product implementation and marketing	II. Product and process development	II. Product implementation		
	Product formulation Process development	Prototype development	Consumer tests; Descriptive analysis tests	Prototype product
		Benchmarking	Consumer surveys; Consumer tests; Descriptive analysis tests	Comparison with other products
	III. Product testing	Product optimization	Consumer tests; Descriptive analysis tests	Optimized product
	IV. Market testing	Market strategy and testing	Market tests	
III. Product commercialization	V. Product launch preparation	Scale-up and trial production	Consumer tests	Commercial feasibility report; commercial product
				Production capacity built; ready sales force and distribution
IV. Product launch and evaluation	VI. Product launch	III. Product introduction		Full-scale introduction
	VII. Postlaunch evaluation	Product support		Final evaluation report

into the organization." The ADL milestone-driven product development process recognizes that a good process should be both flexible to deal with current situations and continually evolve to deal with the future (Rudolph, 2000). With this in mind, let's look more closely at the phases of the product development process to see whether sensory research (really sensory-relevant feedback) fits in, and if so, then in what way.

Phase I: Product strategy and definition

Strategic plan

The strategic plan articulates the issues that the company must consider as it decides how to define overall technology strategy, set projects and priorities, allocate resources among R&D efforts, balance the R&D portfolio, measure results, and finally evaluate progress. Corporate management of business and R&D must act together, as one mind, to incorporate business and R&D plans into a single coherent action plan that serves the short-, mid-, and long-term strategies of the company (Rudolph, 2000). A major output of the strategic plan is an articulated vision of the company's direction. It characterizes the markets served and that competitive environment; details regulatory hurdles; and identifies the company's market positioning, core competencies, and profitability targets (Rudolph, 2000). Typically sensory research plays little role here, because there are no stimuli to evaluate.

Market opportunity assessment

The second step characterizes the market opportunity. This means that the market requirement must be defined. Therefore, consumer research needs to be conducted, and "sensory gaps" may need to be identified. Frequently, focus groups at this assessment stage help to identify potential opportunities for new products (Rudolph, 2000). Focus groups have their positives and negatives. The advantages of using a qualitative method such as focus groups (real-time knowledge elicitation) may be offset when misunderstanding of the data occurs because of a strong observer bias. On the other hand, quantitative methods such as conjoint analysis (Rosenau, 1990) were previously limited by the unrealistic nature of the options that can be presented and evaluated by the consumer. As Chapter 6 shows, conjoint analysis can now be done quickly.

Sensory research can play a major role in the assessment of market opportunities, especially in light of its ability to profile the sensory characteristics of in-market products. For example, Rudolph (2000) discusses the ADL approach to conducting *real-time knowledge elicitation,* which combines the power of description using unrestricted choices in free-choice profiling with efficient, statistically oriented collection of quantitative data. The method of real-time knowledge elicitation uses software designed to capture and analyze data during the focus groups. Such data provide feedback to the focus group moderator regarding how well descriptions of products (i.e., the consumer language) differentiate the samples. The moderator can then further probe in order to uncover subtle but important differences among the samples.

The ADL method is effective in explaining the full range of dimensions where individuals are able to discriminate effectively among items. Subsequent statistical analysis further reduces the set of sensory characteristics to the specific set of descriptors that accounts for the greatest amount of variance, that is, those descriptors that are critical to defining the differences of products. As features are defined during real-time knowledge elicitation activities the features, that is, the language, can be translated into product concepts for further testing. This exercise improves the success rates of the concepts at the market opportunity assessment stage.

The business plan

Identifying new consumer needs and then creating product concepts, comprise two work products emerging from the previous steps. These two work products are incorporated into a business plan. The business plan constitutes a document that describes a market opportunity and the program required to exploit that opportunity. Usually the business plan is written for a 12-month period. The plan generally comprises the following components:

   (i) Define the business situation (past, present, and future).
  (ii) Define the opportunities and problems facing the business.
 (iii) Specify business objectives.
 (iv) Describe the marketing strategy and programs needed to accomplish the objectives.
  (v) Assign responsibility for program execution.
 (vi) Create timetables and tracking mechanisms for program execution.
(vii) Translate objectives and programs into forecasts and budgets for planning by other functional groups within the company.

Product definition

Product definition constitutes the last step in the product strategy and definition phase. "The key to product definition is the integration of multiple and often conflicting objectives" (Rudolph, 2000). At this stage the task is to integrate consumer requirements, business objectives, product delivery requirements, and product safety/regulatory issues. In defining the product, questions that must be answered are:

  (i) Who are the consumers?
 (ii) What do the consumers want?
(iii) How will our product deliver those wants?

Phase II: Product implementation and marketing

Prototype development

After the product definition phase, R&D will have received guidance, either formally or informally, allowing them to formulate one or several prototypes to accord with the product definition, as they understand it. In this step, the sensory researcher uses descriptive sensory analysis to demonstrate that the prototype in its "protocept" but more or less final form will meet the technical and business objectives established. Descriptive analysis to support this effort is discussed more fully in Chapter 9.

   No one descriptive system has achieved the status of a universal method. There are several descriptive analysis methods used by sensory practitioners, as well as the ever-increasing number of hybrids of the different descriptive analysis methods that have been invented in order to deal with specific problems faced by an individual, a company, or even an industry (Einstein, 1991). Descriptive analysis constitutes an objective method of sensory analysis that requires the use of trained panels to detect, describe, and quantify sensory properties/attributes of a food or food product. Data from descriptive analysis tests provide a profile or complete description of all attributes of the product.

Benchmarking

Benchmarking is defined by Gacula as the establishment of a *reference point* for product categories or services (Moskowitz, Munoz, & Gacula, 2003). The reference point can be a company's product, the leading product in the marketplace, or all products within a category or across all relevant categories. The definition of benchmarking is fluid because benchmarking finds many uses, with each use defining what the benchmarks should be.

Benchmarking becomes of exceptionally prime importance in the new product development process, because it shows in many ways whether or not the prototype meets product objectives. Often these objectives are defined with reference to other products, either explicitly or implicitly. For example, when developing a reduced fat or fat-free product, it is necessary to determine whether this new product provides the sensory properties similar to the full-fat product. Descriptive analysis is the most useful tool for this purpose. Which products and how many products to use for benchmarking is a critical part of the benchmarking process. This decision lies within the team made up by the sensory scientist, the product developer, and the market researcher. Benchmarking is discussed in greater detail in a later part of this chapter.

Product optimization

Optimization comprises a set of techniques that can improve the efficiency of the research and development activity on new products (Schutz, 1983). According to Sidel and Stone (1983), optimization is a procedure used for developing the best possible product in its class. The process involves measuring the opinion or response for the most liked/preferred product. A widespread method used for optimization is RSM, a useful statistical tool for investigation of the impact of formulation and processing variables on response (dependent) variables such as flavor, texture, and other sensory attributes, and/or a product's acceptance ratings (Chu & Resurreccion, 2004). RSM examines the effects on the response variables of several factors acting in concert (formulation or process variables or both). In the best of cases product optimization time greatly reduces compared to the time for traditional step-by-step improvement involving a single ingredient or process variable. In the worst of cases product optimization teaches a great deal about the dynamics of the product, which learning eventually leads to a better product anyway. Chapter 9 discusses the optimization approaches in greater detail.

Market strategy and testing

After the optimization step, either formal with RSM or informal with trial and error, the company will have invested time and money in product development, from the initial concept through product optimization. Up to this point, the product definition phase is based on models that are reasonably accurate representations of market response, but not of reality. Failed national introductions of New Coke® and Milky Way II® (25% fewer calories and 50% fewer calories from fat) indicate that things can go wrong, even when market forecasts look good. Long-run sales, based on product trials and repeat purchase behavior, can be forecasted if market test analyses can predict the number of consumers who will try the product and who will become repeat users.

Various models are employed by marketing groups to forecast cumulative trial and repeat purchases, such as stochastic or random models and a combination of trial/repeat and attitude models (Rudolph, 2000). At this point formal sensory research gives way to simulated test markets, which may include sensory assessment by the product, along with responses to concepts. However, by this time the product is assumed to be as good as it could be, although in many cases companies shorten the development cycle, eliminate the previous steps, and simply "hope" that the results from the market testing bear out their (often unreasonable) expectations.

Phase III: Scale-up and trial production

The optimized product must be reproducible at a commercial plant, on a commercial scale, in order to make "business sense." Involving the manufacturing function early in the product development process helps avoid problems when engineering constraints change the product in a way that affects consumer expectations. Successful commercialization of the manufacturing process results from the compromise between R&D and manufacturing functions. Implicit in scale-up and trial production of the new food product is a total quality program that continuously identifies, analyzes, and controls risk. A hazard analysis critical control points (HACCP) matrix is useful in identifying and prioritizing hazards that may affect product quality. At this level sensory research may play a role, albeit an informal one, especially when it's necessary to modify the formulation somewhat to accord with engineering constraints.

Phase IV: Product launch and evaluation

In this phase, field trials have been conducted and the product is designed to meet the needs and expectations of the consumer. The new product is packaged and priced appropriately to convey the message of quality and value. Packaging for transport has been tested and the product has been distributed in a timely and correct fashion so that it flows through the distribution chain without any impediments (Rudolph, 2000).

Product introduction is led by the sales function but supported through all functional areas, especially marketing and distribution. This is the phase that all functional groups have been waiting for. Consumers can now see the product in its entirety and their response tells us the potential for product success or failure.

Product support

A final milestone to building product success and repeat business is product support. In reality, even with the most thorough prelaunch research program, marketers inevitably face many more unknown than known obstacles when they launch a new product. In order to provide product support, a program to assess the new product's performance in the marketplace is required. This program will tell the team and company management: (1) how well the product is doing, and more important, (2) where its performance needs improvement if the sales prognosis is not good (Olson, 1996).

Assessment methodologies for product support include the following:

(i) *Sales figures*: Actual sales are an important measure. Initial consumer sales reflect early consumer trials of the new product. These numbers can mislead because they can be overestimated by short-term promotions, thus things appear "rosier" than they really should. The sales velocity of most new products peaks in the first few months then falls off as trial flattens. Understanding the month-by-month sales growth and decline curves is essential to projecting sales accurately (Olson, 1996).

(ii) *Leading consumer behavior indicators*: Consumer research including consumer panel and tracking data can ascertain the magnitude of consumer response in the initial months following launch. The consumer's buying behavior is studied: who is first buying and who is rebuying, how much are they buying at each purchase, and what is the time between purchases. This information is typically used as input in sales forecasting models in order to provide accurate estimates of eventual long-term sales. Data can also be used to provide key diagnostic learning regarding the product's strengths and weaknesses. Tracking studies by telephone interview, after consumers have just bought the product, often measure consumer awareness, advertising awareness, trial, reasons for purchase or nonpurchase, repurchase, repeat intention, and product likes and dislikes. Face-to-face interviews can likewise be conducted outside the supermarket when they have purchased the product. Panel data from individual households can be obtained by in-store scanners or through pencil and paper diaries. These provide better data than telephone tracking studies but do not obtain attitudinal data that tracking studies provide. Therefore, if panels are used, additional research is needed to flesh out the problems and point to corrective actions (Olson, 1996; Earle, Earle, & Anderson, 2001).

(iii) *Consumer perceptions and evaluations*: There is a need for information regarding consumers' perceptions of the new product and the product values and attributes. It is important to test the product attributes on which the product was designed, and to determine whether consumers identify new product attributes, which influence their acceptance of the product. In order to maintain product quality on the shelf after product launch, after reformulation due to returns, or after processing changes, sensory affective tests should be conducted. Furthermore, quality audits should be conducted after product launch and at various points in the distribution chain, using trained panelists and descriptive analysis. However, descriptive measurements corresponding to levels of consumer acceptance must be established on the basis of company specifications for product quality (Figure 8.1). Otherwise, consumer affective tests will need to be conducted each time a descriptive analysis test is carried out (Resurreccion, 1998; Earle, Earle, & Anderson, 2001).

(iv) *Other performance measurements*: Other measurements, such as customer inserts with mail-back questionnaires, monitoring of 1-800 customer complaints or comments, and interviews with field reps and consumers about product returns, help to provide insight regarding why a new product performs as it does. Furthermore, qualitative research, such as focus groups, is conducted among purchasers and nonpurchasers in order to discover barriers to success. However, qualitative research should be followed with quantitative research, so that these insights can be quantified in terms of seriousness.

These product assessment methods generate valuable information for other functional areas within the company. That information, in turn, may lead the way to business opportunities, such as line extensions and product upgrades.

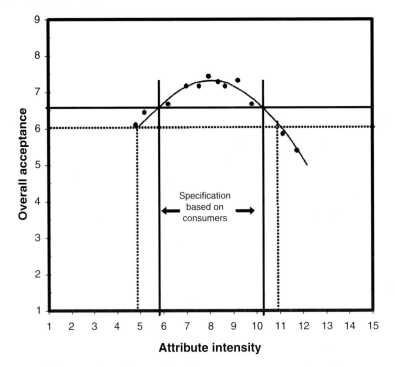

Figure 8.1   Establishment of specifications based on consumer response. Overall acceptance was measured using a consumer panel that rated product on a 9-point hedonic scale (1 = "dislike extremely"; 5 = "neither like nor dislike"; 9 = "like extremely"). Attribute intensity was determined from mean ratings by a descriptive panel using a 15-point (150-mm) line scale.

SYMPTOMS OF A "BROKEN" PROCESS

Not all processes work, no matter how hard the participants desire it. Four signs of what Rudolph (2000) calls a "broken" new product development process are:

(i) Longer development time compared to competitors.
(ii) Missed target introduction dates.
(iii) Significant number of "crash projects."
(iv) A succession of stop/go decisions.

These signs are beyond the issues dealt with in this book, and pertain more to the nature of the corporation and business strategy than to the expertise in product, concept, and package research.

EXPLORING BENCHMARKING IN MORE DETAIL: A KEY OPPORTUNITY FOR SENSORY RESEARCHERS

The widespread use of benchmarking in business is a good reason to present the logic in detail, outside of its important position in the product development process. This section

Table 8.2   Changes in market conditions that require benchmarking.

Changes in market conditions	Examples of specific changes
Formula, process, and packaging changes	Product reformulation Ingredient change Ingredient supplier change Product improvement Packaging innovation
Market condition changes	Marketing of new/special/unique products, innovations Major product reformulations to meet market needs (new flavors, features, benefits, etc.)
Consumer changes	Attitude change Behavioral change Change in consumer expectations

explores the different aspects and uses of benchmarking in more detail, focusing on the way a sensory researcher might use the data.

The so-called benchmark is not a permanent reference point. Markets change, situations change, products change, and indeed nothing, including the benchmark is permanent, carved in stone. Experience shows that benchmarking is a continuing process, not a one-time event. Benchmarking is likewise necessary because market conditions in the food and other consumer product industries are in a constant flux. Changes are brought about as a result of new market conditions, as indicated in Table 8.2. Gacula (Moskowitz, Munoz, & Gacula, 2003) stated: "Products that are not continually benchmarked are likely to fail over time."

Changing market conditions, changing products, changing consumers

Changes in product characteristics are important to track. When product characteristics are what is to be evaluated, descriptive analysis is the most useful tool for this purpose. It is important to track the complete product category. This entails measuring how the company's product performs in that new/modified category, and how new product introductions fare. The results provide a key picture of what's happening in the category.

Good research practice dictates that benchmark products should be tested along with new products. Heavily advertised products should be tested, even when these products they do not command a large market share at the time, for the simple reason that they may gain in popularity. If there is a rapid increase in the purchase and consumption of a particular product or set of products, then it is advisable to repeat part of the old study and incorporate the new data into the database.

Despite the normative value presented by benchmarking (i.e., tests with supposed constant products as references), benchmarking does not necessarily produce totally consistent results. Some variation in benchmark scores occurs because the consumer panelist changes, often in ways that can be traced to market conditions but also often in ways that cannot be easily understood. Moskowitz, Munoz, and Gacula (2003) listed examples of changing market conditions that affect sensory scientists. The panelists, themselves, change their internal reference standard over time, due to experience, to what they read, to maturation, or to a host of other reasons that cannot necessarily be traced. Panelists may become sensitive to salt or sweet foods. Thus over time, foods that have the same salt or sweetener concentration may be rated higher in salty or sweetness intensities. It is a challenge for the sensory practitioner to keep up with this change in panelists' internal sensory references. As products in the marketplace change, the consumer's frame of reference changes as well. Such changes present many problems to sensory researchers, who often use the competitive frame of products as stimuli for their research. Researchers want to know what a highly rated product is/does and, conversely, what a low-rated product is/does. When product quality continues to change, the norms used in one year will not be consistent for the next year.

A good example of the effect of changing market conditions comes from the business history of jarred pasta sauce. In the 1980s consumers of pasta sauce could, by their preference patterns, be easily divided into three segments: those who preferred (1) low flavor impact, thin and smooth sauces, (2) high flavor impact and spicy sauces, and (3) chunky sauces. Over time, with introduction of different flavored products, the segmentation changed from a simple and straightforward segmentation to a broader, more complex flavor-based segmentation. These changing market conditions brought on by the competitive frame and the evolution over time of the products available mean that databases must be updated on a regular basis.

Changes in consumer attitudes are also important to benchmark

Consumer attitudes may be important to track, in order to measure changing consumer response to the marketplace. Consumer awareness of brands and their product usage are important. A knowledge of the changes in consumer expectations, wants, and needs in a product category is necessary to be able to develop and reformulate products that meet those needs.

Historically, the sensory researcher has typically not been part of the tracking of consumer attitudes, even toward the sensory characteristics of products as they are "thought of" in the consumer mind, or as they are responded to when put into the form of a concept instead of an actual product. Two of the authors (Beckley and Moskowitz) have created a series of tracking studies for particular food products, using conjoint analyses to measure the impact or utility of brand, use situations, and product description (Beckley & Moskowitz, 2002). Their approach, described in Chapter 6, provides a way to track the consumer "mind-set" about what sensory characteristics are desirable in a product, based upon descriptions of the product, rather than upon the product itself.

Business issues in benchmarking: cost and frequency

The cost of benchmarking is an important consideration in the planning and design of the study. To reduce the cost of benchmarking, historical and/or descriptive analysis results can be statistically clustered (i.e., a bi-plot of principal components allowing the researcher to select two or more products from each cluster for inclusion in the benchmark study). This clustering strategy is reasonable when the products in the cluster share a great deal of sensory similarity, but the strategy can provide possibly misleading results if the cluster contains products that are radically different.

The frequency with which the category should be benchmarked varies with external considerations. Issues to be taken into account include new product introductions, improvements, or other changes in products in the marketplace, or new and technology and social developments such as those related to safety and other health reasons. There is no clear guideline on the number of years for which a database can be considered valid. A reasonable recommendation is to retest the product category every 2–3 years, in order to ensure the data remain current.

Procedures used in benchmarking

As in every other aspect of sensory research, there is no single standardized way to benchmark products. Companies that produce and market products use a number of different methods. Furthermore, marketing research and sensory research companies offer their own general as well as tailored methods specific to a product category. One of the techniques used is a consumer survey dealing with the sensory properties of the named products or product categories—no actual samples are evaluated. A descriptive analysis of the products of interest is conducted next, followed by a consumer test. The relations among the three test results are explored in order to establish benchmark patterns to aid in decision-making. However, even when looking for these relations among the data sets, researchers will differ, depending upon their objectives, their statistical proclivities, and their intellectual heritages.

The three tests that are used in benchmarking are:

(i) *Consumer surveys*: These provide information on product characteristics that consumers look for in a product that can lead to an assessment of preferences and acceptance, negatives and positives, and norms. MacFie and Thomson (1994) and Meiselman (1996) provide technical information to guide the development of a consumer survey questionnaire.

(ii) *Descriptive analyses*: These methods provide information on the intensity of the sensory attributes in a product and are used to monitor changes and document perceived sensory attributes of existing and new products. The descriptive test is usually conducted by the sensory group. Applications of descriptive analysis are found in Gacula (1997) and in the ASTM publication by Hootman (1992).

(iii) *Consumer tests*: These provide information on the degree of liking and consumer perception of intensities of product attributes for product improvement, new findings, preferences, and other acceptability measurements. For details on consumer testing for product development, consult Resurreccion (1998). In some situations, the consumer survey is not necessary. This is true when benchmarking established products with fully defined sensory properties that are understood. In all cases, however, the consumer tests and descriptive analysis tests are required.

Documenting change

The documentation of changes is best approached using a continuous monitoring system, and collecting data in order to establish a database. When a continuous monitoring system has not been implemented and there is no database. The alternative turns out to be less than optimal; one reacts by making product comparisons when market conditions change. The product comparisons constitute evaluations of the new product along with the company's other products. The sensory profiles can be compared, either qualitatively or quantitatively, to specify the nature and the magnitude of the change.

The ideal for benchmarking, and a good way to finish this chapter is to describe briefly how the company might produce a powerful benchmarking system, and how the sensory researcher could contribute to the system. The goal is to create a continuous monitoring system to ensure that the product is the same over time, or when the product changes, to document the nature of the change and quickly estimate the impact of that change.

The continuous benchmark begins with the creation of useful databases of information on the product category and the company's brands. Within this database new product introductions by the company and its competitors could be easily included, along with the information about competition. The database would further include methods for comparing two products, and norms to determine when two products are *meaningfully* as well as significantly different, and on which particular attributes. The word "meaningfully" is important, because not every departure from the gold standard product is relevant. Some can be safely ignored—but the question is which departures, in particular.

The ideal benchmarking database should include physical and chemical measurements on the same product, sensory profiles, as well as sales and marketing data. New product evaluations are made and results are compared with the database. This information comprises the sensory profile of the new product, information on the position of the new product in the category's sensory space (differences and similarities between existing and new products in the category), and the new sensory dimensions and direction established by the new product.

Can this database be established? The answer is an unqualified *yes*. The limiting factor is not money, samples, or willing participants and contributors. The answer is simply lack of shared vision—the difficulty of different members in the company to work together to put this database in the system for a year or two. Databasing the characteristics of products is critical, especially in a time of intense competition and the search for new product opportunities. Benchmarking is often left to isolated efforts, such as the structured development of a single product. A benchmark database, however, would transcend one time period and provide a valuable corporate resource in which the sensory researcher would play a major role.

REFERENCES

ASTM. 1979. Committee E18. *Manual on Consumer Sensory Evaluation*, ed. E.E. Schaefer. ASTM Special Technical Publication 682.

Barclay, I. 1992a. The new product development process: Part I. Past evidence and future practical applications. *R&D Management* 22(3):255–263.

Barclay, I. 1992b. The new product development process: Part 2. Improving the process of new product development. *R&D Management* 22(4):307–317.

Beckley, J. & H.R. Moskowitz. 2002. Databasing the consumer mind: The Crave It!™ Drink It!™ Buy It!™ & Healthy You!™ Databases. Presented at the Institute of Food Technologists Conference, Anaheim, CA.

Chu, C.A. & A.V.A. Resurreccion. 2004. Optimization of a chocolate peanut spread using response surface methodology. *Journal of Sensory Studies* 19:237–260.

Cooper, R.G. 1996. New products: what separates the winners from the losers. In: *The PDMA Handbook of New Product Development*, eds M.D. Rosenau, A. Griffin, G.A. Castellion, & N.F. Anschuetz, pp. 3–18. New York: John Wiley & Sons.

Dwyer, S. 1998. Inextricably linked. *Prepared Foods* 167(2):21–22.

Earle, M.D. & R.L. Earle. 2001. *Creating New Foods: The Product Developer's Guide*. London: Chadwick House Group Limited.

Earle, M.D., R.L. Earle, & A. Anderson. 2001. *Food Product Development*. Cambridge, England: Woodhead Publishing Limited.

Einstein, M.A. 1991. Descriptive techniques and their hybridization. In: *Sensory Science Theory and Applications in Foods*, eds H.T. Lawless & B.P. Klein, pp. 317–338. New York: Marcel Dekker.

Gacula, M.C., Jr., ed. 1997. *Descriptive Sensory Analysis in Practice*. Trumbull, CT: Food & Nutrition Press.

Hootman, R.C., ed. 1992. *Manual on Descriptive Analysis Testing for Sensory Evaluation*. West Conshohocken, PA: ASTM.

Lawless, H.T. & H. Heymann. 1998. *Sensory Evaluation of Food: Principles and Practices*. New York: Chapman & Hall.

Lord, J.B. 2000. New product failure and success. In: *Developing New Food Products for a Changing Marketplace*, eds A.L. Brody & J.B. Lord, pp. 55–86. Lancaster, PA: Technomic Publishing Co.

MacFie, H.J.H. & D.M.H. Thomson. 1994. *Measurement of Food Preferences*. London: Chapman & Hall.

Meiselman, H.L. 1996. The contextual basis for food acceptance, food choice and food intake: the food, the situation and the individual. In: *Food Choice Acceptance and Consumption*, eds H.L. Meiselman & H.J.H. Macfie, pp. 239–292. London: Chapman & Hall.

Morris, C.E. 1993. Why new products fail. *Food Engineering* 65(6):132–136.

Moskowitz, H.R. 2000. R&D-driven product evaluation in the early stage of development. In: *Developing New Food Products for a Changing Marketplace*, eds A.L. Brody & J.B. Lord, pp. 277–328. Lancaster, PA: Technomic Publishing Co.

Moskowitz, H.R., A.M. Munoz, & M.C. Gacula, Jr. 2003. *Viewpoints and Controversies in Sensory Science and Consumer Product Testing*. Trumbull, CT: Food & Nutrition Press.

Munoz, A.M., G.V. Civille, & B.T. Carr. 1992. *Sensory Evaluation in Quality Control*. New York: Van Nostrand Reinhold.

Olson, D.W. 1996. Postlaunch evaluation for consumer goods. In: *The PDMA Handbook of New Product Development*, eds M.D. Rosenau, A. Griffin, A. Castellion, & N.F. Anschuetz, pp. 395–410. New York: John Wiley & Sons.

Resurreccion, A.V.A. 1998. *Consumer Sensory Testing for Product Development*. Gaithersburg, MD: Aspen Publishers.

Rosenau, M.D. 1990. *Faster New Product Development*. New York: American Management Association.

Rudolph, M.J. 1995. The food product development process. *British Food Journal* 97(3):3–11.

Rudolph, M.J. 2000. The food product development process. In: *Developing New Food Products for a Changing Marketplace*, eds A.L. Brody & J.B. Lord, pp. 87–102. Lancaster, PA: Technomic Publishing Co.

Schutz, H.G. 1983. Multiple regression approach to optimization. *Food Technology* 37(11):46–48, 62.

Sidel, J.L. & H. Stone. 1983. Introduction to optimization research—definitions and objectives. *Food Technology* 37(11):36–38.

Sokolow, H. 1988. Qualitative methods for language development. In: *Applied Sensory Analysis of Foods*, vol. 1, ed. H.R. Moskowitz, pp. 4–20. Boca Raton, FL: CRC Press.

9   What types of tests do sensory researchers do to measure sensory response to the product? and... why do they do them?

CONSUMER ACCEPTANCE TESTS VERSUS MARKET RESEARCH TESTS

At the very start, it is important to emphasize that the consumer acceptance test is neither a substitute nor a competitive alternate for the standard large-scale market research test. In other words, consumer acceptance tests should not be used as inexpensive substitutes, in place of large-scale market research tests when the latter are needed. Each type of test is conducted by a different functional group within the company and one type of test does not replace the other. A comparison between consumer sensory acceptance tests and market research tests is shown in Table 9.1.

The consumer acceptance test is a small-panel test usually involving from 50 (Meilgaard, Civille, & Carr, 2007) to 100 panelists (Chambers & Wolf, 1996), whereas the large-scale market test usually requires at least a hundred or more panelists from each of several strategically selected geographic locations. The emphasis of the consumer acceptance test is usually in the earlier stages, such as selecting the best product for subsequent work. In contrast, the large-scale market test is generally more of a report card for go/no-go decisions.

Recently, Hough et al. (2006) recommended using at least 112 consumers for acceptability tests based on a root mean square error divided by the scale length (RMSL) of 0.10. However, they indicated that means of RMSL over 108 measurements in five countries had an overall RMSL of 0.23. Thus for an RMSL of 0.2, only 29 panelists are needed for an α of 0.05 and a β of 0.10.

Consumer acceptance tests focus on the product and are designed for obtaining direction for technical guidance, if needed. Issues addressed in the consumer test include overall preference or liking of a product, as well as a detailed assessment of the product's sensory properties such as appearance including color, flavor, and texture (Meilgaard, Civille, & Carr, 2007). It is worth restating that the output of each of the pieces of information in the consumer acceptance test is important for the product developer, who looks at the results with an eye toward identifying shortcomings and changing the product. In contrast, market

Sensory and Consumer Research in Food Product Design and Development, Second Edition.
Howard R. Moskowitz, Jacqueline H. Beckley, and Anna V.A. Resurreccion.
© 2012 Blackwell Publishing Ltd. Published 2012 by Blackwell Publishing Ltd.

Table 9.1 Comparison of consumer sensory acceptance tests and market research tests.

	Consumer sensory acceptance tests	**Market research**
Scale	**Small panel**	**Large panel**
Number of panelists	50–112	≥112
Purpose	Determine overall preference or liking of a product, or a product's sensory properties such as appearance including color, flavor, and texture	Usually focuses on consumer populations and identifying the consumers to whom the product will appeal, and developing the means to reach those consumers

research tests on the same product typically focus on consumer populations and identifying the consumers to whom the product will appeal, and developing the means to reach those consumers (Stone & Sidel, 2004).

When looking at the tests, therefore, we see that sensory research and market research each work with consumers, but the former uses consumers to help with a sensory assay of the product, whereas the latter looks at the product to do an assay of the consumer himself. The two types of studies complement one another and rely on different testing procedures. Both are vital in the development of new products, because they tap the product and the person, respectively, two radically different worlds.

Acceptance and preference tests

Consumer sensory affective tests are conducted throughout the different phases in the product development process. There are two approaches to consumer sensory acceptance testing: the measurement of preference and the measurement of acceptance (Jellinek, 1964). They sound similar, and are often used interchangeably. Yet, these two approaches entail radically different executions, mean quite different things when the data are interpreted, and come from different intellectual histories.

Acceptance tests measure consumer acceptance or liking of a product. The consumer acts as a measuring instrument, assigning numbers or words from a scale. In contrast, preference tests measure the appeal of one food or food product over another (Stone & Sidel, 2004). The consumer selects one product in a set.

The methods most frequently used by sensory practitioners to determine preference and quantify acceptance are the paired preference tests for preference (selection of one product over another) and tests employing the 9-point hedonic scale, respectively. Acceptance measurements can be made on single products and do not require comparison to another product. The questions asked during acceptance tests are, "How much do you like the product?" (Stone & Sidel, 2004) or "How acceptable is the product?" (Stone & Sidel, 2004; Meilgaard, Civille, & Carr, 2007).

There are many different scales one might use to measure the magnitude of acceptance, because once the consumer acts as a measuring instrument the entire world of "scaling" opens up. Ratio scaling was proposed by Moskowitz (1974) as a method to quantify acceptance and preference but is not as frequently used by sensory practitioners as the 9-point scale, even though the magnitude estimation method had been validated by experimental psychologists for many years before and is a scaling method of choice in many research laboratories.

The 9-point hedonic scale has been used for about 60 years, beginning with acceptance research conducted for the US Army. Such extensive use has brought with it exploration and validation in the scientific literature (Stone & Sidel, 2004). Nine-point scales, however, may not be optimal for scaling purposes, especially when working with children. Adaptations to the hedonic scale may work better with younger consumers, such as the 9-point facial scale for children 8 years or older (Kroll, 1990; Kimmel, Sigman-Grant, & Guinard, 1994). The 3-, 5-, and 7-point facial hedonic scales were found to be more useful when working with 3-, 4-, and 5-year-old children, respectively (Chen, Resurreccion, & Paguio, 1996). The key thing to remember, however, is that the scales require the respondent to act as a measuring instrument that shows magnitude of acceptance.

Acceptance tests

What exactly does it mean to measure acceptance? Are we talking about accepting the idea of a product, or are we talking about accepting the product itself? Both aspects are generally subsumed in the word "acceptance." Consumer acceptance of a food may be defined as (1) an experience, or feature of experience, characterized by a positive attitude toward the food; and/or (2) actual utilization (such as purchase or eating) of food by consumers.

The importance of consumer acceptance means that the acceptance tests are used to answer a variety of questions and are appropriate in different business scenarios. Examples and caveats in their use are:

(i) *Overall performance*: This is degree of liking for a product(s), or the percent of "target" respondents who say that they prefer one product to another. The word "target," used by consumer researchers, denotes the fact that the participants in the study match the ultimate consumer for whom the product is intended.

(ii) *Diagnostics*: Liking ratings or preference (vs. control) for a product's sensory properties including: appearance, flavor, and texture, which are generally known as "diagnostics" show performance of specific aspects of the product.

(iii) *Relations among variables*: Relate consumer responses to the profile generated by descriptive analysis results and/or physical and chemical measurements.

(iv) *Caveat*: The consumer acceptance test estimates of product acceptance a based on the product's sensory properties. This measure of acceptance does not guarantee success in the marketplace because other factors may exert their own individual influences as well. These factors are marketing variables such as packaging, price, and advertising. Nonetheless, a well-executed consumer test will reveal the basic magnitude of consumer liking, a measure that is critical for the product developer and sensory researcher.

Preference tests

Preference tests refer to all affective tests based on a measurement of preference (i.e., choice) or a measurement from which relative preference may be determined (IFT/SED, 1981). Preference in its most classical treatment by experts in the food industry was defined in three different ways: (1) an expression of higher degree of liking, (2) choice of one object over others, and (3) psychological continuum of affectivity (pleasantness/unpleasantness) upon which such choices are based (Amerine *et al.*, 1965). The term "preference" may include the choice of one sample over another, a ranked order of liking, or an expression of opinion on a hedonic (like/dislike) scale. With the proliferation of sensory research procedures, however, it is probably most productive to limit preference to choice-based studies, where respondents choose the sample that they prefer, using the criteria imposed by the researcher.

Preference tests measure the appeal of one food or food product over another (Stone & Sidel, 2004). The panelist has a choice of products, and one product is to be chosen over another. The questions asked in preference tests are unambiguous: "Which sample do you prefer?" or "Which sample do you like better?" (Meilgaard, Civille, & Carr, 2007). Preference tests are useful when one product is compared directly against another, such as in determining preference of the new improved product over the existing one, or assessing preference for a new product against competing products.

Scaling methods allow us to directly measure degree of liking and to infer preferences from these data. The inferential method might be as simple as *winner take all* (the product chosen is given a liking value of, say, 100 and the nonchosen products given liking values of 0). The inferential method might be more complicated, such as estimating the scale differences between two products on the "theoretical underlying liking scale" by processing the proportion of times one product is preferred to another, across the entire population of respondents. The measurement of preference from paired comparison or ranking tests is direct, whereas preference from hedonic ratings is implied. There is an obvious and direct relation between measuring product liking/acceptance and preference (Stone & Sidel, 2004).

Preference methods can be used to determine differences in preference, but not differences of products; discrimination tests should be used for this purpose. A preference test is occasionally used to determine whether the company's "gold standard" or target product is preferred over products from the different plants, or prototype made with ingredients from different suppliers. However, when the objective of the test is to determine whether the gold standard or target product differs from the product from different plants or suppliers, the preference test is probably not as good as the discrimination test.

Methods used in acceptance and preference testing

A number of consumer sensory evaluation methods are well known, established, and validated. The three principal types of tests used in consumer acceptance testing are (1) the paired preference, (2) ranking, and (3) rating tests. However, two of the most frequently used tests to measure consumer preference and acceptance are the paired preference test, and ratings of acceptance using one or another form of the hedonic scale (e.g., good to bad, palatable to unpalatable, etc.).

Paired preference tests

In order to execute this test, the researcher presents the judge or panelist with two samples, typically simultaneously but occasionally sequentially. Sequential presentations are appropriate when the samples have characteristics that would make simultaneous presentation either impossible or create some artifacts. The question is easy—which of the samples does the panelist prefer. The preference test is easy and straightforward to execute, and it works well even when the consumer panelists have little reading or comprehension skills. The test instructions may or may not force the panelist to make a decision. When a forced choice is imposed, the panelist may not give a "no-preference" response (ASTM, 1996) and must indicate a preference for one sample over another. The paired comparison method to elicit preferences may also be used for multiple paired preferences within a sample series, such

---

PAIRED PREFERENCE TEST

Name _____   Session Code _____   Date _____

Please rinse your mouth with water before starting.

In front of you are two samples. Beginning with the sample on the left, taste each one and **circle the sample that you prefer.** You must choose one. You may re-taste as often as you need to.

Figure 9.1   Example of a ballot for the paired preference test. In this particular ballot the "no-preference" option is not shown, so that the panelist must choose between the two samples. Reproduced from Resurreccion, 1998. Fig. 2.1, p. 12. Copyright 1998 by Aspen Publications, with kind permission of Springer Science and Business Media.

as a standard product versus each of several experimental products. One or more sample pairs of products may be tested in each panel session. The upper limit for the number of sample pairs that can be tested is determined by physiological and psychological constraints (Resurreccion, 1998).

The paired preference test is simple and relatively easy to organize and conduct. The two coded products are served together to the panelist, who is instructed to sample the products and then indicate which sample is preferred using a specific criterion that the researcher indicates. When the researcher does not specify a criterion such as "prefer on flavor," then we assume that the preference criterion is overall acceptance. Figure 9.1 shows a sample ballot. It is advisable to include the "no-preference" and "dislike both equally" choices among the responses, and to use an adequate sample size when conducting this test. Samples are generally masked as to their identities, typically using a 3-digit random number. From time to time the preference study may use identified or even branded samples, depending upon the objectives of the research. The serving order is always balanced, with each product appearing first in half the trials, to minimize order bias (e.g., the first sample generally being preferred—called the *tried first bias*).

Executional issues also emerge in preference tests, especially when the panelist evaluates more than two products. In a multiproduct test, there will be considerable interaction because of flavor carry-over from one product sample to the next. In addition, memory can be a confounding variable if a sequential presentation is selected for the paired preference test.

Advantages of the paired preference test
Simplicity and ease are the two major advantages of the paired preference tests. The paired preference tests are easy to organize and to implement. Only two orders of presentation are possible, A–B and B–A. However, the order of presentation must be balanced across panelists, or else position bias is likely to occur. One panelist usually evaluates only one

pair of products in a test without a replication. When the "no-preference" choice is used, the number of consumers answering a preference for either sample will be smaller because some of the panelists will choose the "no-preference" option. Therefore, this option should be introduced only when there are a larger number of consumers participating in the test, for example, a hundred or more. Furthermore, if the researcher chooses to include the option of "dislike both equally" and/or no-preference, then the analysis should be performed only on those responses that show a definite preference for one of the two samples.

The special case of no-preference
The issue of "no-preference" is a perennial problem in preference tests, as it has been in experimental psychology where it originated. Stone and Sidel (1985) suggested that if at least 5% of participants select the "dislike both equally," then it is necessary to seriously question the appropriateness of the products being tested or the sample of consumer panelists recruited to participate in the product evaluations. In their opinion the "dislike both equally" option raises a red flag, indicating that most likely the product prototype did not incorporate the critical attributes defined by the consumers, or that the product prototype did not match product definitions and more work needs to be conducted. Selecting an inappropriate respondent sample may likewise generate a large number of responses in the "dislike both equally" category. For example, when testing products such as a soft drink developed for teenagers, preference panel comprising a high proportion of adult or elderly company employees in the consumer test would likely generate in a high "dislike both equally" category. Similarly, using a sample of consumers who do not eat shrimp when testing a new microwavable frozen breaded shrimp product would likely result in a high response rate for the "dislike both equally" category.

Disadvantages of the paired preference test
From the standpoint of sensory research, the paired preference test provides less information than a rating test because pure preferences themselves give no direct measure of the *magnitude of preference*. Thus, the paired preference test is less efficient than a rating test because only one response per product pair is obtained as opposed to one response per product when a rating test is used. Furthermore, when the "no-preference" category is not used, it is virtually impossible to determine whether or not both products were disliked. Because preference testing asks the respondent to evaluate the product on an overall basis, the testing may be especially susceptible to unintentional biases that can arise, such as the effect of slight differences in placement on the serving tray, serving temperatures, sample volume, and variability of the sample from the manufacturing process, among others (ASTM, 1996). On the other hand, if magnitude of preference is important the researcher can instruct the panelist to rate magnitude of preference, along with indicating the preference itself. However, to determine whether both products are disliked requires classification or rating of the products to show that they belong in the class of disliked products. Preference measures cannot do that, and preference measurement cannot be adjusted to indicate membership in the category of "disliked."

Data analyses appropriate for the paired preference test
In the analysis of the paired preference test, the probability of randomly selecting one of the two products is 50%, or one out of two. The null hypothesis states that when the consumer has

no preference on the key attribute, the consumer will pick each product an equal number of times; the probability of the null hypothesis is $P = 0.5$. If the underlying population does not have a "real" preference for one sample over the other then the probability of choosing product A is equal to that of choosing product B. This may be written as $H_0: P(A) = P(B) = 1/2$. The alternative hypothesis for the paired preference test is that the underlying population has a true preference for one product over the other. In the study, the preferred product would thus be selected more frequently than the other product. This may be written as $H_0: P(A) = P(B)$.

When analyzing the results of the paired preference test, the conservative sensory researcher uses a two-tailed test because there is no prior expectation regarding which of the two products will be preferred by the consumer population. In other instances, such as new product prototypes, there is an ingoing expectation that the new product will be preferred. The appropriate test is then a one-tailed test (Stone & Sidel, 2004). Analysis of the paired preference tests is accomplished using any of the following statistical methods: binomial, chi-square, or normal distributions or use of tables. Basic sensory evaluation texts (O'Mahony, 1986; Resurreccion, 1998; Lawless & Heymann, 1999a; Stone & Sidel, 2004) detail all of the analyses methods used for paired preference data. For the most part preference testing methods and statistical analyses have been well documented in the literature, perhaps because in the minds of many individuals much of sensory analysis sprung from the question, "Which product is better?"

A potpourri of statistical issues and approaches in preference and acceptance testing

Binomial distribution and tables

The published binomial probability calculations are manageable for small panels, but as the number of observations becomes large, the binomial distribution begins to resemble the normal distribution (Lawless & Heymann, 1999a). Before calculators and computers, these calculations became so cumbersome as to become unworkable. However, Roessler *et al.* (1978) published tables that used the binomial expansion in order to calculate the number of correct judgments and their probability of occurrence. The compilations such as that shown in Table 9.2 simplify the procedure of determining whether or not a preference for one sample over another is statistically significant in paired preference tests.

To illustrate how these compiled tables simplify decision-making, consider a paired preference test involving two formulations of a new powdered raspberry soft drink, with the study involving 50 consumers. A total of 30 of the 50 consumers preferred drink B, and 20 of the 50 preferred drink A. According to Table 9.2, a formulation would have to be preferred by 33 or more of the total of 50 panelists in order to be significantly preferred at the 5% level, or to be preferred by at least 35 of the 50 panelists in order to be significantly preferred at the 1% level of significance. Because one sample was preferred by only 30 of the 50, we conclude that the panel of consumers did not significantly prefer one raspberry soft drink formulation over the other. If 34 or 37 of the panelists had shown a preference for formulation A, in the two aforementioned cases, then the conclusion would be different. We would conclude that the panel prefers raspberry drink A over raspberry drink B at the 5% level of significance. In turn, we could elaborate this conclusion by saying that if we would repeat the study 100 times, more frequently than not the panel would show a preference for A. Indeed we could go further and say that 95% of the studies would show more people preferring drink A and fewer people preferring drink B.

Table 9.2   Minimum numbers of agreeing judgments necessary to establish significance at various levels for the paired-preference test (two-tailed, $p = .05$)[a].

Number of trials (n)	Probability levels		Number of trials (n)	Probability levels	
	0.5	0.01		0.5	0.01
7	7		32	23	24
8	8	8	33	23	25
9	8	9	34	24	25
10	9	10	35	24	26
11	10	11	36	25	27
12	10	11	37	25	27
13	11	12	38	26	28
14	12	13	39	27	28
15	12	13	40	27	29
16	13	14	41	28	30
17	13	15	42	28	30
18	14	15	43	29	31
19	15	16	44	29	31
20	15	17	45	30	32
21	16	17	46	31	33
22	17	18	47	31	33
23	17	19	48	32	34
24	18	19	49	32	34
25	18	20	50	33	35
26	19	20	60	39	41
27	20	21	70	44	47
28	20	22	80	50	52
29	21	22	90	55	58
30	21	23	100	61	64
31	22	24			

Source: Resurreccion. 1998. Table 2.1, p. 242. Copyright 1998 by Aspen Publications. Reproduced with kind permission of Springer Science and Business Media.
[a]Value (x) not appearing in table may be derived from $x = Z[(n + n + 1)/2]$, where n = number of trials, x = minimum number of correct judgments, if $x = 1.96$ at a probability $(\alpha) = 5\%$, and $Z = 2.58$ at probability $(\alpha) = 1\%$.

The chi-square (X^2) test
Typically a binomial test is used to test whether a panel of consumers prefers raspberry soft drink mix A to raspberry soft drink mix B. However, researchers often use the chi-square test in order to confirm hypotheses about frequency of occurrence. The chi-square may be used to obtain the same type of information as the binomial test, but it has the advantage that it can be used to test hypotheses when the response falls into one of several categories, not just two categories. Calculation of the chi-square statistic can be obtained from standard texts such as O'Mahony (1986). As in the binomial test, the alternative hypothesis determines whether a one- or a two-tailed test is used. A two-tailed test is used because we have no knowledge of which raspberry drink mix sample, A or B, the consumer will prefer.

Dealing with cases where preference is not forced
The data analyses described in the preceding text are based on the so-called forced choice paired preference test. The panelist must choose one of the two raspberry beverages, even when both are disliked, or even when both are perceived to be identical. Sometimes, it is important to work in situations where a vote of "no-preference" is meaningful. If the

---

PAIRED PREFERENCE TEST

Name _____   Session Code _____   Date _____

Please rinse your mouth with water before starting.

In front of you are two samples. Beginning with the sample on the left, test each
one and **circle the sample that you prefer or circle no preference if that is feeling
about the samples.**

You may re-taste as often as you need to.

Figure 9.2 An example of a ballot to be used in a paired preference test with "no-preference" option.
Reproduced from Resurreccion, 1998. Fig. 2.2, p. 19. Copyright 1998 by Aspen Publications, with kind
permission of Springer Science and Business Media.

project leader and requester jointly decide on use of the "no-preference" option, then the
no-preference option might provide useful information regarding the product. Figure 9.2
shows an example of the ballot that incorporates "no-preference".

When the "no-preference" option is used the analysis has to change to accommodate this
new, possibly troublesome, but equally informative option. Researchers have developed at
least three alternative methods to handle "no-preference" methods that depend upon the way
the researcher thinks about what the panelist is trying to convey. These are:

(i) *Ignore the response*: The first option ignores the "no-preference" responses. Ignoring
these results will decrease in the number of responses and decrease the power of the
test.

(ii) *Split the "no-preference" responses equally*: This mistakenly assumes that the panelists
would have selected one of the product samples over the other, were they have been
forced to make a choice. The sample size is not decreased and there appears to be no
effect on the power of the test. This procedure actually dilutes the signal-to-noise ratio
by assuming that panelists who express no choice would respond randomly (Lawless
& Heymann, 1999a). This conservative approach protects from a false positive result
while still running the risk of missing a significant preference.

(iii) *Split the "no-preference" responses proportionately, corresponding to ratio of preference between the two product samples* (Odesky, 1967): This proportional split is not a
common practice, and the risks associated with this procedure are not known. When
the panel size is more than 100 and fewer than 20% respond "no-preference," the confidence intervals for the multinomial distribution can be calculated. If the confidence
intervals of the proportions of those expressing a preference do not overlap, then one
can test whether one product is preferred over the other.

Misusing the paired preference test

The paired preference test is often combined with other tests. The combination occasionally creates unforeseen biases:

(i) For example, the paired preference test should not be combined with discrimination tasks. Asking preference questions should not follow a difference question or vice versa.
(ii) The consumers who are recruited to participate in the preference tests must be naive users of the product; they would not qualify as panelists in discrimination tests. Furthermore, they should not be asked to focus on differences and then on preference.
(iii) Panelists used in a discrimination test are not recruited to represent the target population of a food product sample.
(iv) The discrimination test is an analytical test, whereas the paired preference tests determine whether consumer affective responses to one sample are higher compared to another sample.

The four key venues for affective tests: laboratory, CLT, mobile CLT, and HUT

Consumer responses needed for the quantification of acceptance or preference can be conducted in a sensory laboratory setting, in central location tests (CLTs), or in home use tests (HUTs), which are also known as home placement tests. A specialized form of CLT that has been used is one that is conducted in a mobile laboratory (Resurreccion, 1998).

Sensory laboratory tests

Sensory laboratory tests, as the name implies, are conducted in the sensory laboratory of a food manufacturing company, consulting firm, or research organization. This type of test is the most frequently used venue by R&D when the objective is to measure consumer acceptance. The panel generally comprises consumers who are recruited from a list or consumer database. The database, in turn, comprises prerecruited consumers screened for eligibility to participate in the tests, who have agreed to participate. Usually 25–50 consumers participate in a test; however, 50–100 responses are considered adequate (IFT/SED, 1981), so the sensory laboratory test may run over several sessions. The recommended number of products per sitting is two to five.

Laboratory tests—advantages

The major advantage of using a sensory laboratory to conduct a consumer affective test is the convenience of the location of the testing facility to the researchers. It is particularly appealing to manufacturing companies because of the accessibility of the laboratory to a large number of employees. Any number of employees can be recruited to participate in the tests on short notice. Although the sensory laboratory provides a convenient location for the research team and is accessible to employees, it is less convenient for local residents or other consumer panelists who will participate. In instances when the consumer panelists used are employees of the company who have participated previously in consumer tests, the researcher often can dispense with the orientation, thus saving time.

Among the other consumer tests, the laboratory test allows for the greatest control over the sample preparation and testing conditions, including lighting and environmental conditions. In a sensory laboratory, the researchers are able to control all conditions of the test such as the product preparation and the product evaluation environment. Control of the product evaluation environment includes control of the testing environment such as lighting, noise, and other distractions, and conduct of the test in individual partitioned booths that isolate panelists from each other. Lighting can mask color differences and other appearance factors so that panelists can focus on either other sensory attributes or on acceptance without visual cues. Of course, for the most part the lighting used in the booths attempts to approximate the daylight spectrum.

Most specially constructed sensory laboratories are set up to be adjacent to a fully equipped kitchen. The adjacency works well to control many of the key factors in sensory analysis; sample preparation steps including recipe formulation can be standardized, and the duration and temperature of cooking, holding and reheating, slicing (portion sizes of serving), and serving (with bread or crackers, dishes on which food is served) can be also be carefully controlled. Often these well-controlled laboratory tests are designed for a limited number of panelists—for example, a total of 25–50. The presumption is that the control over the test conditions and sample quality is so great that there is no need for a large base of consumer panelists to "average out the noise."

One of the advantages of conducting a consumer test in the sensory laboratory is the rapid turn-around time for results to be obtained because of the proximity to the data processing facilities. When a computerized sensory data entry and analysis system is employed, feedback of results is almost instantaneous and available shortly after the last consumer completes the test.

Laboratory tests—disadvantages

The disadvantages of conducting consumer tests in the sensory laboratory generally pertain to the nature of the panel that is used. Having a sensory laboratory also means having certain types of limitations on the consumer population. By putting the laboratory in the corporate complex there is the great temptation to use corporate employees as panelists, simply on the basis of cost. It presumably costs "less" to work with an employee than to recruit an outside consumer panelist, although a true accounting of the costs would probably reveal that the corporate employee actually costs most—it's just that the costs don't appear as line items and "extras."

The sensory researcher needs to remember the nature of the risks associated with using employees in product maintenance tests; employees should not be used in product development, product improvement, or product optimization tests. When a decision to use employees in product maintenance is made, care must be taken to recruit only those employees who are not familiar with the production, testing, or marketing of the product; in addition, employees' rating patterns must be compared to that of a representative panel.

Comparative methods (Lawless & Heymann, 1999a; Stone & Sidel, 2004) to determine whether an employee panel would present great risk and should not be used are discussed in Chapter 10. When consumers screened from a prerecruited consumer database are used, then those consumers who would be most willing to participate are, for the most part, those individuals whose homes or place of employment is close enough to the sensory laboratory. On the other hand, local residents who participate in the tests may be biased due to the belief that the products originate from a specific company or plant. If care is not exercised to recruit an appropriate panel, then the demographic characteristics of the panel may be skewed on characteristics such as income, education, race, and any other factors that may have an influence on usage patterns for the food.

Another disadvantage of laboratory tests is that a sensory testing booth is very different from a real eating environment and the realism of the laboratory test can be questioned (Hersleth *et al.*, 2005). In some studies, laboratory measurements of food preference were shown to be poor predictors of consumption (Cardello *et al.*, 2000; Kozlowska *et al.*, 2003). An example of the limited information that may be obtained from a laboratory test is the absence of normal consumption patterns, such as a sip versus drinking from a full glass. These limited procedures may influence detection of positive or negative attributes (Stone & Sidel, 2004; Meilgaard, Civille, & Carr, 2007). Furthermore, when the food is prepared or tested in the laboratory, product performance may differ from the product performance that the researcher would observe in the more natural home use situations. A final disadvantage of the laboratory test is the limited amount of time that the consumer is exposed to the product, compared to a home use test. The very nature of the laboratory test makes it a test with a short exposure to the stimuli.

Laboratory tests—panel size

The consumer laboratory panel comprises consumers who are recruited and screened for eligibility to participate in the tests. Typically the consumer panelists are either prerecruited consumers in the neighborhood (nonemployees) or company employees. The appropriate base size is not fixed. Good practice usually works with 25–50 responses per product, with suggestions of at least 40 by one set of experts (Stone & Sidel, 2004) and about twice that, 50–100, by another set (IFT/SED, 1981). These numbers are not "laws" *per se*, but merely good research practices. The reason is that there are issues of variability and representation in the data. People differ from each other, so it is important to have a sufficient number of ratings so the uncorrelated variation in the data cancels itself out. Furthermore, in consumer tests by sampling lots of different people the underlying hope is that the test will gather opinions from the different types of mindsets in the population, making a study that better represents the array of consumers.

In a consumer test comprising only 24 panelists, it may be difficult to establish a statistically significant difference in a test with that small number of panelists. However, even with the small-panel sizes it is still possible for the sensory professional to identify trends in the responses and to provide direction to the product development and marketing functions involved in developing the new product. With 50 panelists, statistical significance of differences has a far greater chance to emerge. Stone and Sidel (2004) provide a guide for selecting panel size based on two factors: the expected degree of difference in ratings from the 9-point hedonic scale, and the size of the standard deviations. They warn their readers that different products have their own requirements. Smaller panels can generate results that are highly significant when the intrinsic variability within a sample is small so that sample-to-sample differences can easily appear. In contrast, when the intrinsic variability within a single product is high from sample to sample, larger panels are needed as variability of the products increases.

Laboratory tests—panelist recruitment to participate in studies

Preference or acceptance tests require different selection criteria for panelists compared to the criteria required for discrimination or descriptive tests. The composition of the preference and acceptance panels must be defined by the marketing and sensory functions of the company. In acceptance tests, the characteristics of the panelists should match those of the target market for the product being tested. One approach to recruitment develops a database of consumer households with consumers who may be available for testing. In many cases, the demographic characteristics of the individuals in the database are known and updated on a regular basis. Recruitment of consumer respondents may be done through a market research agency or by designated recruiters in the project team who recruit by telephone calls, referral, or personal contacts. Other methods often used to recruit panelists are random selection from a telephone directory, random digit dialing, posters in retail stores, organization mailing lists, assorted consumer databases, and intercepts at malls, shopping areas, or restaurants. A combination of these methods is generally necessary and thus frequently used when developing the initial database. Figure 9.3 shows an example of the information collected for a potential participant in a database. This information will help the recruitment for subsequent studies.

Laboratory tests—employees versus nonemployees
Employees should not be used in affective tests unless sufficient tests have been conducted to ensure that the employees exhibit food consumption and preference patterns similar to that of the target market for the product. Furthermore, over time employees may develop certain unconscious attitudes toward the sensory attributes of the corporate products, and in many cases learn to recognize the "sensory signature" of the product, even without explicit training. At this point, the employee panel has changed from being surrogates of the consumer to a new type of being, not necessarily consumer and not necessarily expert. Certainly the long-term employee is biased and may accept characteristics in a product simply because these otherwise unacceptable characteristics have become familiar and expected.

Laboratory tests—using local residents
One efficient method of recruitment brings local residents into the laboratory. This approach is growing in popularity. Local residents as panelists are convenient for the sensory staff, making it easier for the latter to schedule tests. The availability of motivated local panelists allows tests to be completed more rapidly, reduces morale issues due to interruptions on employees to participate, and reduces the very great but hidden costs associated with using employees as panelists. However, the method of recruiting local residents involves prerecruitment, database management, scheduling, and budgeting activities; and accessibility and security issues. A system must be established to contact and schedule local residents for a test. A budget and means of providing incentives or honoraria to panelists is required; this is not needed in tests involving employees. Access of the laboratory to panelists should be limited so that they do not wander in restricted areas and all panelists should be escorted at all times for security reasons. The advantages of using local residents are that participants are usually highly motivated; they will show up promptly for each test and are willing to provide considerable product information.

Laboratory tests—other options and outsourcing the testing function
Another option for a company is to develop and maintain a satellite or off-premises laboratory test facility. The off-premise solution might well eliminate many problems associated with bringing local residents to the company premises. This option will, however, be more costly than bringing people to the sensory laboratory. Another alternative will be to contract with a sensory evaluation company, assuming the contracted company has the necessary sensory

For Office use only
Panelist Code #

1 2 3 4 5 6 7 8 9 0

Panelist Code #_________

Please answer all the questions. Your name is not on the questionnaire and will not be identified with your answers.

Q-1 We would like to begin by asking if you are the person who buys and cooks the food in your household most of the time?

Yes ❑ No ❑

Q-2 In your household, the person who buys food and cooks most of the time is the:

Husband ❑ Wife ❑ Child ❑ Self ❑ Other ❑

Q-3 We want to ask you a few questions about yourself. This information will help us compare opinions of people with different backgrounds.

What is your date of birth?_____________: What is your age group?

18-25 ❑ 25-34 ❑ 35-44 ❑ 45-54 ❑ 55-64 ❑ >64 ❑

What is your sex?

Male ❑ Female ❑

Which do you consider yourself to be? (Check one).

White ❑ Black ❑ Spanish/ Hispanic ❑ Other, Specify______ ❑

What is your marital status? (Check one).

Never married ❑ Married ❑ Separated ❑ Divorced ❑ Widowed ❑

Please mark how many people in each age group live in your household. Include yourself in the count.

	0	1	2	3	4	5	6	7	8	>8
under 6 yrs. old	❑	❑	❑	❑	❑	❑	❑	❑	❑	❑
7-12 yrs.	❑	❑	❑	❑	❑	❑	❑	❑	❑	❑
13-18 yrs.	❑	❑	❑	❑	❑	❑	❑	❑	❑	❑
19-24 yrs.	❑	❑	❑	❑	❑	❑	❑	❑	❑	❑
25-64 yrs.	❑	❑	❑	❑	❑	❑	❑	❑	❑	❑
over 64 yrs.	❑	❑	❑	❑	❑	❑	❑	❑	❑	❑
Total Number of people in household	❑	❑	❑	❑	❑	❑	❑	❑	❑	❑

Figure 9.3 Example of a demographic questionnaire for consumer database participants. Reproduced from Resurreccion, 1998. Fig. 4.1, pp. 84–88, Copyright 1998 by Aspen Publications, with kind permission of Springer Science and Business Media.

test resources. Both options generate above-line costs that become public. When budgets are tight and corporations opt to work inside, the cost of running the testing can be absorbed into other functions. These alternative options reflect the reality that testing is a corporate expense and needs to be treated as such.

In each case, the product development team will need to assess its current acceptance test resources and anticipated workload before determining which, if any, of these options is feasible. There is no question that consumer acceptance tests need to be conducted in new

If your household has any children under 19, how many parents are living in it/

One Parent	Both Parents	No Parents
❏	❏	❏

How far did you go in school? (Check one).
- ❏ Less than seven years of school
- ❏ Junior high school
- ❏ Some high school
- ❏ Completed high school or equivalent
- ❏ Some college
- ❏ Completed college
- ❏ Graduate or professional school

Please check the one which best applies to you:
- ❏ Employed full time
- ❏ Employed part time
- ❏ Home maker
- ❏ Student
- ❏ Retired
- ❏ Unemployed
- ❏ Disabled

How many people in your household contribute to the household income?

	0	1	2	3	4
Number of people employed full time (check one):	❏	❏	❏	❏	❏
Number of people employed part-time (check one):	❏	❏	❏	❏	❏
Number of people with other sources of income (pensions, social security etc). check one:	❏	❏	❏	❏	❏

What was the approximate income level of your household last year? (Check one).

under $9.999	$10.000-$19.999	$20.000-$29,999	$30.000-$39,999	$40.000-$49,999	$50.000-$59,999	$60.000-$69,999	$70.000 and over
❏	❏	❏	❏	❏	❏	❏	❏

Check the item that best describes the job of the head of your household.
(If retired or not now working, describe the former or usual job).
If you are the head of the household, Check the item that describes your job.

- ❏ Executive or proprietor of large concern
- ❏ Manager or proprietor of medium concern
- ❏ Adminstrative personnel of large concern or owner of small independent business
- ❏ Owner of little business establishment, clerical, sales work or technician
- ❏ Skilled worker
- ❏ Semi-skilled worker
- ❏ Unskilled worker

If not sure, describe the job:

Title_____

Kind of work_____

For Office use only
Panelist Code #

Figure 9.3 *(Continued)*

product development. The question is where it will be done, who will be the test subjects, and what will be the relative costs of the different alternatives (Stone & Sidel, 2004).

Laboratory tests—screening panelists to join the panel and to participate in specific tests
For a given test, consumer panelists should qualify according to predetermined criteria that describe the target consumer for the product. A recruitment screener should be developed by

DM103

For Office use only
Panelist Code #

1 2 3 4 5 6 7 8 9 0

Do you own or rent the place where you live?

- ❑ own (or buying)
- ❑ Rent
- ❑ Other

In what type of home do you live?

- ❑ One family house
- ❑ Mobile home
- ❑ Apartment, condominium, duplex, or townhouse

How many years have you lived in Georgia?________years

For Office use only
years

1 2 3 4 5 6 7 8 9 0

In Georgia, people live in a city and a county. Please give the name of your city__________

For Office use only
cities name

1 2 3 4 5 6 7 8 9 0

Please give the name of your county__________

For Office use only
country name

1 2 3 4 5 6 7 8 9 0

Q-4 We want to ask you some questions about your shopping and eating habits.
How many times per month do you shop for groceries?__________

For Office use only
cimes per month

1 2 3 4 5 6 7 8 9 0

How many different grocery stores do you regularly shop at (at least once per month)?__________

For Office use only
stores per month

1 2 3 4 5 6 7 8 9 0

Figure 9.3 *(Continued)*

the sensory practitioner in collaboration with the "client," that is, the person who requests the test. The recruitment screener should ensure validity of the screening process (i.e., checks on the screening) and allow for intermediate and final tallies of participants to ensure that quota requirements are being met (e.g., half users of the product, half nonusers, etc.).

The selection of prospective participants should be conducted as rigorously as possible. Consumer panelist recruitment should be monitored, whether the job of recruitment is done by the sensory function or contracted to a firm that handles consumer recruitment. One ought to discourage the undesirable, albeit frequent practice of using those consumer panelists who can be most conveniently contacted, such as those people who live close to the facility. Similarly, the use of friends and relatives of project staff should be avoided. These individuals

Figure 9.3 (*Continued*)

may bias the results of the discussion and should not be used. Demographic criteria such as age, gender, frequency of product use, availability during the test date, and other criteria such as employment with the sponsoring company or similar business concerns and other security screening criteria are often used. In addition, ethnic or cultural background, occupation, education, family income, experience, and the last date of participation in a consumer affective test may be used. Frequency of participation may also be important, with declining response rates recruiters often opt to work with a limited set of individuals who are willing participants in tests, and as a consequence participate more frequently than one desires.

Besides, the conventional issues of appropriateness to participate based on product usage or demographics, it is important to recognize that in the food world there are allergies. It is vital to screen panelists to ensure they don't have certain food allergies. Furthermore,

For Office use only
Panelist Code #

How many pounds do you want to lose or gain? _____ 1bs

For Office use only
take-out

Check the statement that comes closest to describing you at the present time. (Check one only)

- ❏ I am on a diet right now
- ❏ I diet from time to time not now
- ❏ I don't diet but I never eat certain fattening foods
- ❏ I cut down somewhat on fattening foods
- ❏ I sometimes cut down on fattening foods for a few days but in general pay no attention to my weight
- ❏ I eat all I want and am not concerned about putting on weight

Figure 9.3 (Continued)

it is generally important to work with panelists who do not have an in-depth knowledge of a product. All persons who have in-depth knowledge of the product, or those who have specific knowledge of the samples and the variables being tested should not be included in the test unless the test is designed specifically to solicit the response of these "expertized" individuals.

Laboratory tests—attendance and no-shows
Tests cannot be conducted without participants. A continuing, occasionally very major problem is "no-shows" at the test session, where a panelist agrees to participate but something comes up at the last minute and the panelist either phones in or just as frequently fails to show up. This occurs for in-house panels as well as for consumer panels. The problem of "no-shows" can be reduced by a number of steps, some more radical than others. It may be necessary to overbook according to a historical "no-show" rate, obtained from previous tests. A rate of 20% may be high for a panel of consumers recruited from a database; however, a higher rate possibly close to 50% or higher would be expected when the panel is prerecruited through store intercepts and invited, for their first time, to participate in laboratory tests.

To ensure participation, select participants who live within a 30-minute traveling radius to the test facility, give participants clear directions and a map to help them find the location, and plan test dates that do not conflict with major community or school events. Reminder letters, phone calls, or mail-outs of brightly colored postcards to post in a visible location—such as the home refrigerator; adequate honoraria, rewards, or incentives; and overbooking for the test do increase the attendance rate. The most important practice is to have participants commit to the process; to make participants understand the importance of their attendance and promptness and the value of their participation.

Laboratory tests—incentives to participate
When the sensory profession began the conventional approach was not to pay participants. Free participation in studies also characterized market research as well, and indeed it might be correctly said that the vast majority of the participants gave free data. Today, with time pressures and the use of external panelists, participants in laboratory tests are usually paid for their effort. The amount of incentive paid to participants may differ according to many factors including: length of the test, location of the test and associated travel expenses, and incidence rates of qualified participants. Payment ensures motivated panelists and allows the panelist to set aside a chunk of time to participate, allowing more and richer data to be collected.

Payment is provided after the session is completed. Certain organizations may have restrictions on the type of incentive that may be awarded to panelists; therefore, these vary considerably. Incentives for participation may be provided in the form of a cash honorarium, product coupons, election from a gift catalog, gift certificates and tickets to special functions such as ball games or concerts, or donation to charity or a nonprofit organization.

Laboratory tests—location and design of the test facility
The location of the sensory laboratory is important because location may determine how accessible it is to panelists. The laboratory should be located so that it is convenient for the majority of the test respondents. Laboratories located in places inconvenient to panelists will not only reduce the number of consumers that will want to participate, but also limit the type of panelists that can be recruited for the tests. When the sensory laboratory is located in a company facility, it should preferably not be situated near a noisy hallway, lobby, or cafeteria because of the possibility of disturbance during the test (ASTM, 1996). The panelists should not be able to hear the telephones nor hear/see other offices, food production or laboratory equipment, company employees, visitors walking or conversing, and so forth. When the sensory laboratory is located in such areas to increase accessibility to panelists, the laboratory should be equipped with special sound-proofing features. At the same time, when the sensory laboratory is located in a remote area within the company, the very remoteness may be a negative factor.

The sensory laboratory should be planned for efficient physical operation. Furthermore, the laboratory should ensure a minimum amount of distraction of panelists caused by laboratory equipment and personnel, and often emerging from the interactions between panelists themselves (ASTM, 1996). The laboratory should comprise separate food preparation and testing areas. These areas must be adequately separated to minimize interference, during testing, resulting from food preparation and serving operations. The sample preparation area should not be visible to the panel. Individual partitioned booths are essential to avoid distraction between panelists; however, these should be designed so that they do not elicit the feeling of isolation from the rest of the panel.

The physical environment must be pleasant. It is important to have a reception area, separate from the testing and food preparation areas, where panelists can register, fill out demographic, honorarium, and consent forms, and be oriented on testing procedures before and after a test, without disturbing those who are doing the test. This area, away from the test itself, encourages social interaction and allows for the administrative tasks to be done in a comfortable environment, each leading to a more pleasant experience for the panelists. In a world where the consumer is king and the researcher is buying the consumer's time, this pleasant experience will be helpful in both the short and long runs, to keep the test facility functioning well.

Sanitation and surroundings are important. The test area must be kept as free from extraneous odors as possible. Adequate illumination is required in the evaluation areas for reading, writing, and examination and evaluation of food samples. Special light effects may be used either to emphasize or hide irrelevant differences in color and other aspects of appearance. To emphasize color differences, different techniques are used. Examples are spotlights, changes in the spectral illumination by changing the source of light from incandescent to fluorescent, changing the types of fluorescent bulbs used, or changing the position of distance of the light source. To de-emphasize or hide differences, the researcher may use a very low level of illumination, although this can make the experience less than pleasant. Special lighting such as sodium lights, colored bulbs, or using color filters over standard lights may likewise be used.

Laboratory tests—number of products per sitting
For tests involving employees who cannot spend too much time away from their work assignment, the recommended number of products per sitting is 2–6. Nonemployees are available for longer periods of time and can evaluate more products. The number of products that can be evaluated depends on the panelist motivation, which will make itself known in complaints, such as "inability to test more products," or overt displays of boredom. More samples can be tested if a controlled time interval between products is allowed and only acceptance is measured. Kamen *et al*. (1969) observed that under such conditions, consumers could evaluate as many as 12 products.

The use of complete block experimental designs for consumer testing is recommended even when panelists have to return to complete the entire set of test samples. The complete block design requires that each panelist acts as his own control. When a complete block design is not feasible, incomplete block designs may be used. In such cases it is recommended that incomplete cells per panelist be kept to a minimum of one-third or less of the total number of samples being evaluated by the panelist (Stone & Sidel, 2004).

CLT—the central location intercept test

One of the important venues in obtaining guidance for maximizing product acceptance is the central location, with its eponymous CLT. CLTs are frequently used for consumer research by both sensory and market researchers. CLTs can handle a variety of techniques and procedures. The tests are conducted in one or, more frequently, several locations away from the sensory laboratory, in a public location, accessible to a large number of consumers who can be intercepted to participate in the tests. These tests are usually conducted in a shopping mall or grocery store, school, church or hotel, food service establishment including cafeterias, or similar type of location that is accessible to a large numbers of potential consumer panelists (Resurreccion, 1998).

The major advantage of the CLT is the capability to recruit a large number of consumer participants and thus obtain a large number of responses, which allows for stronger statistical analyses. In addition, only consumers or potential consumers of the products may be recruited to participate in the tests; no company employees are used. The well-conducted CLT, properly recruited and executed, will have considerable impact and validity because actual consumers of the product or product category are used. This type of testing enables the collection of information from groups of consumers under conditions of reasonably good control (ASTM, 1979) compared to a home use test. Furthermore, several products may be tested, a researcher goal harder to achieve in the less well-supervised home use test.

The major disadvantage of the CLT is the distance of the test site from the company. An example is a mall location, often rented by the researcher for the purposes of a single test. The company sensory practitioner, sensory staff, and company employees from other functional groups have to travel to the off-site central location. Often, the central location will have limited facilities, equipment, and resources necessary for food preparation and conduct of the test. Examples of food preparation and testing facilities that may be lacking in a central location are space for a registration area, suitable sample preparation areas, and individual partitioned booths for the sensory test. However, suitable equipment and testing space can be made available and a small number of trained personnel may be used so that preparation of samples and serving of products can be controlled. When the test product samples require extensive sample preparation and product handling, the benefits of the lower cost of mall intercepts may be offset and reduced by the lack of food preparation facilities and the increased time and cost of testing associated with the preparation of a large amount of sample or its transport in the prepared state to the test site.

Sensory research often requires special facilities to accomplish the test objectives (e.g., food preparation from the raw materials, controlled temperatures, etc.). The inadequacy of facilities limits the type of tasks that can be performed by the consumer panelists, and of course limits the ability of the consumers to evaluate the product under the highly controlled experimental conditions of a laboratory test, or under the actual use conditions provided in a home use test. In a CLT, the potential for distraction is high; "walk-outs" (panelists just getting up and leaving) should be taken into account when planning the test, no matter how unpleasant that thought may be. The large number of panelists that can be recruited for this type of test is a disadvantage just as it is an advantage. Although the location is ideal for recruitment of a large panel, problems that plague studies with large panel sizes include the time and manpower requirements needed to conduct the test. Data are usually collected by trained interviewers rather than by self-administered questionnaire, adding to the number of personnel needed for the test, and increasing the time required to collect data from a given number of respondents. Logistics and choreography of the test session play an increasingly important role as the number of panelists in a test session increases.

In the CLT, the products are tested under conditions that are significantly more artificial than the conditions encountered at one's home or restaurant, where the product is typically consumed. The artificiality limits the nature of information. In the short CLT, the researcher can only ask only a limited number of questions. Examples of the limited information available from the data are preferences of consumers, along with some cursory demographics such as age and socioeconomic status.

When using mall or store intercepts, panelists are, in general, not paid for their participation, although they may receive a token reimbursement in the form of a very small gift. By not paying the panelists the researcher ends up with possibly less than optimal data. For example, intercepting shoppers in a mall will generate a panel of people who are busy with their shopping activity, and who are not willing to spend more than a few minutes on the interview before becoming visibly irritated. Supermarket shoppers have been observed to have a greater time constraint, because they need to return home to store perishables, than retail store shoppers. This short interview limits one's exposure to the food sample(s) being tested, and thus limits the information. In contrast, a home use test (HUT) allows a lot longer exposure to the product and permits more data to be gathered about the consumer's feelings toward the product (Stone & Sidel, 2004). Similarly, those laboratory tests wherein the consumers are paid for several hours generate more data. The consumers are paid for their participation, expected to participate in a longer test, and can do more complex tasks.

When the CLT is used primarily in tests to screen products and/or to define what type of consumer accepts the product or prefers it to a competitor, then the longer exposure time to the product is often unnecessary. However, when testing a product that may change over the normal recommended home storage period, the CLT would not provide information on consumer acceptance during actual home use conditions. In this case, the home use test would provide more realistic data.

The consumer panel occasionally comprises individuals who are recruited from a database consisting of prerecruited consumers (so-called prerecruited panelists). More frequently, consumers are intercepted to participate in the tests, screened for eligibility, and when qualified, immediately recruited to participate in the tests. Several central locations may be used in the evaluation of a product to determine regional or demographic (socioeconomic status) effects, as well as to ensure that the study has a minimum base size, often set around 100 or more. The recommended number of products to be evaluated per sitting in a CLT is 1–4, especially when the panelist is unpaid, and has just been intercept to participate.

CLT—how many panelists are really needed in a study, and why?
Panelists cost most, whether they are paid or not. Thus, it is vital to optimize the information that one can obtain from a panelist, and at the same time reduce the panel size to a feasible number whose costs to acquire the data are not too high. In a CLT, usually 100 (Stone & Sidel, 2004) or more consumers (responses per product) are obtained, but the number may range from 50 to 300 (Meilgaard, Civille, & Carr, 2007), especially when consumer segmentation is anticipated (Stone & Sidel, 2004). The increase in the number of consumers in a CLT compared to the laboratory test is necessary to counterbalance the expected increase in variability due to the inexperience of the consumer participants and the "novelty of the situation" (Stone & Sidel, 2004). Several central locations may be used in the evaluation of a product. Tests that use "real" consumers have considerable validity and credibility.

CLT—shopping mall intercepts
When the test is conducted at a store or shopping mall, consumer intercepts are considered to be the best method of recruitment. Interviewers use visual screening for gender, approximate age, and race to select likely looking prospects from the traffic flow. A person who appears to be appropriate is greeted by the interviewer and asked the necessary screening questions. When a person qualifies as a panelist, the test is explained and the person is immediately invited to come to the test center to take the test. Usually, a small reward in the form of money, coupons, or otherwise is offered. This method of recruiting has become more popular and used more often during recent years.

CLT—affinity groups: panelists from civic, social, or religious organizations
Recruitment of panelists may be done through civic, social, or religious organizations such as clubs of different types, church groups, chambers of commerce, and school associations. For having their members participate these groups receive a cash award for participation. These groups may provide homogeneous groups of participants, a condition that often helps the otherwise more difficult recruitment process. Often, the affinity group allows the researcher to conduct the interviews in facilities provided by the organization, such as a church or school hall, meeting room, and so forth. The organization may agree to provide a number of panelists of a certain type such as elderly consumers, homemakers with young children, or teens at intervals to fit the testing schedule. In other cases, the organization may provide the project

personnel with lists of members who may be called, screened, and qualified ahead of time and scheduled. A warning about the use of these groups is that they represent a narrow segment of the population (Stone & Sidel, 2004) and more often than not possess characteristics that are similar. In fact, many of them know each other or may be related to one another. Their demographic responses as well as their food preference patterns may, therefore, be skewed. With these affinity groups economic considerations often provide powerful arguments for their use, despite any bias.

CLT—other sources of consumer panelists

Often researchers set up a booth or temporary facility at a convention, fair, industrial show, or similar event, where crowds of people are likely to congregate. Visitors or passers-by are invited to participate. This is usually limited to brief tests such as a "taste test" between two samples (ASTM, 1979). Additional sources of panelists for CLTs are newspaper, radio, and TV advertisements and flyers at community centers, grocery stores, and other business establishments; referrals from current panelists; letters to local businesses requesting their employees to become panelists; purchased mailing lists of consumers in a geographic location or telephone directories; and random digit dialing.

CLT—screening for consumer panelists

The screening procedure for consumer panelists should be unbiased. The screening criteria for acceptance and preference tests are very important to establish, because when the wrong group of consumers participates the results of the test may be misleading at worst or less than meaningful at best. A thorough understanding of study objectives along with discussions with the client or requester generally suffices to identify the nature of who should participate. When screening the test participants it is important to ask the questions in an unambiguous fashion, to be sure that the consumers can properly identify themselves as the relevant target. However, the actual screening criteria should be undisclosed to participants so that the consumers will answer with actual information about themselves and not what recruiters want to hear.

There are a number of issues to keep in mind when briefing the staff who will do the recruiting, especially when those recruiting consumer panelists *on the floor* at a shopping mall. Thirteen of these appear below; there are often more, depending upon the specific study, the nature of the products, and the nature of the participants:

(i) The central location should be selected on the basis of matching the demographic profile of shoppers to the target consumer of the test product.

(ii) Recruitment should be scheduled during the peak traffic hours for the desired participant. For example, if full-time employees are required, then recruitment should be scheduled before or after regular working hours or during the lunch break. Otherwise the cost of recruitment during regular working hours would be high.

(iii) It may not be a wise practice to recruit panelists who are active panelists for another company. Indeed, many screeners have the so-called security-check questions to ensure that the panelist does not work for a competitor company, an advertising agency, a market research company, etc.

(iv) Individuals or immediate household members of individuals whose occupation is related to the manufacture or sales of the test product, or engaged in market research, should be disqualified from participating.

(v) So-called professional panelists should be avoided. These individuals may no longer perform as naive panelists depending on the degree or frequency of their past participation in consumer tests.

(vi) The frequency of participation of a panelist in tests within a product category may also be one of the screening criteria. For instance, many companies mandate that the panelist not have participated in a similar study for the past 2 months. It is getting harder to fulfill this requirement.

(vii) Frequency of use of the product may be an important consideration.

(viii) Income level may be included as one of the criteria for screening potential panelists.

(ix) Other factors such as age, marital status, gender, and education may be used as screening criteria.

(x) In some cases, the number of persons in a household, whether these persons are adults, adolescents, or children may be important.

(xi) Whether the individual is the primary purchaser or preparer of food in the household may likewise be important.

(xii) Availability and interest of the panelist will determine whether or not they can be scheduled.

(xiii) Prospective panelists who are allergic to the product should immediately be disqualified, as would individuals who are ill or pregnant.

CLT—orienting panelists about what they are to do

The orientation of consumer panelists should be limited to the information absolutely required. Generally, this information describes the mechanics of the test. The orientation may deal with how to use the features of the booth area, such as explanations about the sample pass-through door, signal lights, and so forth. Other parts of the orientation may deal with the use of data acquisition devices, such as computerized ballots, and instructions on using a light pen, mouse, or keyboard. The orientation must be carefully planned to avoid any possibility to alter the panelists' attitudes toward any of the food samples to be evaluated.

CLT—preparations and serving considerations

The sensory practitioner will need to plan for preparation and presentation of samples, the duration and temperature of holding after cooking until serving, portion size of serving, and the method of serving. The use of a carrier should be considered, including dishes on which food is served and whether or not to include bread, crackers, or water for rinsing. The control of testing environment (red light, etc.) should likewise be considered.

CLT—number of panelists in a single test session

The number of consumers to be handled at one time depends on a number of factors that include the product type, the capacity of the testing facility, and the number of technicians available to conduct the test. In some situations, only one or two panelists at a time may be recruited to participate, whereas in other cases a larger group of panelists may be handled at one time. Too many panelists in one area will encourage inattention or interference and if the panelist to technician ratio is too high then panelists' mistakes or questions that arise during the course of the evaluation may go unnoticed. The author (Resurreccion) has observed that with more than 12 people per session, evaluating products using a self-administered questionnaire in an open area under the supervision of one trained technician leads to problems in the session. The practice of administering a test simultaneously to a roomful of people where the ratio of panelists to technicians is over 1:12, at any given time, would likely result in loss of control and would not be recommended.

CLT—number of products to be evaluated by a single panelist in a standard central location test

The number of products to be evaluated by a panelist per session at a CLT should be four, with fewer samples being even better, at least from the quality of data, (but not from the cost per rating, which is greater with fewer samples). The sensory practitioner should consider the number of samples that can be ideally presented at a mall location, and in fact in any test that employs the intercept method of recruiting panelists. In these instances, the panelists are more likely to walk out in the middle of the test if the interview is lengthy, complicated, unpleasant, or boring.

CLT—what the panelist does

Any of the standard affective test methods can be used in CLTs. The decision on what test to use depends on the objectives, the product, and the panelists, respectively. Usually, the information is collected through a self-administered questionnaire. In such cases, the questionnaire should be easy to understand and not take too long to answer. The attention span of the panelists must be considered as well. For example, if there are one or two samples then it is permissible to ask several questions about each. However, as the sample number increases, the number of questions that must be answered should be decreased accordingly.

Panelists intercepted at a central location will often not agree to a test that will take longer than 10–15 minutes. Most panelists' interest and cooperation may be maintained for this length of time. Longer tests should be avoided unless there are special circumstances, such as giving the panelists a substantial cash incentive. Prerecruited panelists who are asked to come to a central location may be told beforehand how long the test will last. For example, in such cases, the panelist may be told that the test would take an hour.

On the test dates, panelists are assembled, singly or in small groups, in the test area where trained personnel conduct the test. The number of personnel assisting will vary with the size of the group being handled, the stimulus control requirements in any given case, and the test procedure. For example, one-on-one interviewing may be required, and in other cases, it may be sufficient for one person to handle 4–5 or as many as 12 people responding to a self-administered questionnaire at one time.

The panelists are given a brief orientation in the form of written or oral instructions to assure adequate understanding of the test, including the number and type of test product samples to be tested, presentation of samples, waiting periods between samples, and other details. The panelists are informed about the type of information to be collected by a brief review of the questionnaire.

CLT—proper handling of test samples

The handling of test samples for a CLT is more involved than a laboratory test because these samples must be shipped from the laboratory to the test site. Sufficient sample should be provided for the test. Sample containers need to be clearly identified with appropriate sample identifying codes and inventoried. Sample preparation instructions need to be written in detail, as clearly as possible. The preparation instructions should cover holding time and temperature conditions.

The timing of sample preparation is an important consideration. Timing must be planned so that sample preparation coincides with serving times. A dry run of the entire procedure is often necessary to ensure that the schedules are correctly timed. When preparation is going to be done in the test site, all necessary equipment for food preparation should be available. Temperature holding devices such as warming lamps or steam tables should be available if needed. Training of personnel in sample preparation and equipment use may be necessary. Sample serving specifications should be outlined. Appropriate serving utensils such as plates, plastic cups with or without lids, scoops, knives, or forks should be specified.

CLT—test facilities

Facilities for a CLT test vary widely. In general, testing should be conducted at a location that will be convenient for the participants to travel to, with ample parking spaces. Access to public transportation may also be an important consideration. Much of the testing may occur in settings that are less than optimal. Two general requirements for test facilities are agreed upon: (1) the facilities should allow adequate space and equipment for preparation of the product and for presentation of the product to panelists in a controlled environment, and (2) the facilities should provide proper control of the "physiological and psychological test environment" and include adequate lighting, temperature and humidity control for the general comfort of panelists, freedom from distractions such as odors and noise, and elimination of interference from outsiders and between subjects (ASTM, 1979).

Central location testing can be contracted out to a marketing research company that operates its own testing facility. A CLT facility may be purchased, constructed, or leased. In such cases, proper control can be maintained in the test area by constructing a sample preparation laboratory, panel booths, and adequate ventilation systems.

CLTs using groups are often conducted in public or private buildings such as churches, schools, firehouses, and the like. This type of test offers the least amount of control. Testing in these sites is usually done without booths, controlled lighting, or ventilation. Noise and odors may likewise pose a problem. Sample preparation must be done in advance and test samples and equipment will need to be transported to the test facility. The problem of clean-up and waste disposal may involve hiring of a custodian when the test is conducted in these facilities.

Ideally, panelists should be in individual partitioned booths. Isolating the panelists eliminates distractions from other panelists or test personnel. When booths are not available, panelists should be isolated from each other as well as possible in order to minimize distractions and provide privacy. Seating should be comfortable and at an appropriate height for the table or counter where testing will take place.

When using a test site for the first time, it should be examined at least 1 week prior to the test in order to ensure that the test site will "work" properly in terms of running the test. When the test site has been used previously, evaluation may take place a day or two before the test date. Final details may be arranged then. Equipment should be tested to ensure that it is in good working order. For example, ovens should be calibrated prior to the test day. Food samples, ingredients, and other supplies should be examined for quality and set up for the test. Storage areas and containers for samples and supplies, requested in advance, should be examined. Whenever possible, a separate dry run of the sample preparation procedure using the available equipment should be made.

Samples should be prepared out of the sight of the panelists and served on uniform plates or sample cups and glasses. The serving area should be convenient to both the test personnel and the sample preparation area. The serving area should have the serving scheme posted for test personnel to mark as they obtain samples for each panelist.

Large containers for waste disposal need to be positioned near the sample preparation and serving areas. Arrangements need to be made for waste disposal from the test site after completion of the study. Samples that were not used should be handled according to the client's directions. Instructions regarding the handling of the product should be agreed upon beforehand and specified in writing.

Panelists need a comfortable place to wait when they arrive for the test. When planning for a reception area, sufficient space should be available for panelist registration, orientation and waiting for the test, payment of panelists, and treats and social interaction after a test. In some cases, an area for childcare is designated to allow panelists with children to take turns babysitting while they take turns testing. It may be helpful to establish rules regarding children and to post these rules so that fewer problems will arise.

In a briefing session before the actual test, it is essential that the project leader review the complete test protocol to determine whether the test personnel are familiar with their assigned tasks, the procedures, and the serving instructions. It is a common practice for an agency to conduct a briefing with their personnel prior to the test. The briefings are a good opportunity to review the instructions for the study and explain any special requirements. If a script will be used, a dry run of the reading of the script is conducted to identify any problems. The importance of reading scripts verbatim is emphasized. The serving scheme for test samples is explained at this time.

Mobile test laboratories

During the past two decades, a new form of central location facility has been developed, best described as being a mobile test laboratory. A fully equipped mobile laboratory with complete facilities for food preparation and sensory testing is driven to, and parked at, a central location. Tests are conducted within the mobile laboratory. In this type of test, the panel comprises consumers who are intercepted at the site in order to participate in the tests. There is usually no prerecruitment or screening except for age. Usually 75–100 or more responses are obtained in one location. Often, the test finishes in one location and the mobile laboratory is then driven to other locations where additional tests are conducted. Several central locations may be used in the evaluation of a product. Data are collected by trained interviewers rather than by self-administered questionnaires, adding to the number of personnel needed for the test and the time required to collect data from a given number of respondents. The recommended number of products to be evaluated is 2–4.

The major advantage of using the mobile laboratory for a consumer affective test is the ability to recruit a large number of "actual" consumers for the test, in test facilities similar to a laboratory, which maintains experimentally controlled and environmental conditions conducive for food preparation and testing. The disadvantages of the mobile laboratory test are the expense of maintaining the mobile laboratory and logistical arrangements that have to be made prior to the test for parking and power supply.

HUTS: home use tests

The home use test is also referred to as a home placement or in-home placement test. This test, as the name implies, requires that the test be conducted in the participants' own homes. They provide testing conditions that are not researcher-controlled; therefore, they could yield the most variable results. The HUT is used to assess product attributes, acceptance/preference, and performance under actual home use conditions. The food product samples are tested under normal use conditions. The HUT thus provides additional and valuable information regarding the product that may not be obtained in any other type of test. From the product developer's point of view, HUTs provide information about the sensory characteristics of a product under uncontrolled conditions of preparation, serving, and consumption.

HUTs are valuable in obtaining measurements about products that are difficult to obtain in a CLT or a laboratory setting. The HUT involves preference, acceptance, and performance (intensity and marketing information). The HUT questionnaire can also get information about product preparation and attitudes, and ask other relevant questions specific to the product, or even attitudes to the product category or to the different brands/companies in the category, respectively. An example of this is to determine consumer acceptance and use of a new chicken product, irradiated for increased food safety, wherein the company needs to know not only how much the consumer likes the irradiated chicken samples, but also the performance of the packaging and how the consumer will cook and serve the irradiated chicken. Since products can fail due to packaging or product use, a home use test would likewise be able to test the performance of the packaging early in the development phases of the product. Knowing how consumers will cook and serve the irradiated chicken will provide information on product use by the consumer.

HUT—advantages
The major advantage of the home use test is that the products are tested in the actual home environment under actual normal home use conditions. Testing in the consumer's home is considered to be more optimal compared to laboratory and CLTs, in regard to realism during consumption of a product. Another advantage of conducting HUTs is that more information is available from this test method because one may obtain the responses of the entire household on usage of the product. Responses may be obtained not only from the participant, who is usually the major shopper and purchaser of food in the household but also from the other members of the entire household. In many instances, other pieces of information may be obtained such as the types of competitive products found in the home during the test, usage patterns of all household members, and other information that would be useful in marketing the product.

The HUT method can be used early in the product formulation phase where it is used not only to test a product for acceptance or preference but also for product performance (Hashim, Resurreccion, & McWatters, 1995). In addition, marketing information can be obtained, such as the types of competitive products found in the home during the test and usage patterns, as well as any other information that would be useful in marketing the product. Information regarding repeat purchase intent obtained from the consumer participating in this test would be more useful than the corresponding information one could obtain from a sensory laboratory test or a consumer CLT.

Participants in a home use test should be selected to represent the target population. If the participants of the tests are prerecruited from an existing database and screened, then typically the participants are aware of their role and the importance of the data collected. Their awareness and ongoing motivation will likely generate a high response rate. Finally, HUTs lend themselves to product delivery and return of ballots by mail. When using the mail to conduct a home use test, the researcher can achieve speed, economy, and broad coverage factors that improve the efficiency of the test (Resurreccion & Heaton, 1987).

HUT—disadvantages

The main disadvantages of the home use test are that it requires a considerable amount of time to implement, to distribute samples to participants, and to collect participants' responses. It often takes at least 1–4 weeks to complete a HUT (Meilgaard, Civille, & Carr, 2007). Lack of control is another disadvantage; little can be done to exert any control over the testing conditions, once the product is in the home of the respondent, and the potential for misuse is high (Stone & Sidel, 2004). This lack of control may result in large variability of consumer responses. In consumers' homes there is generally a lot of variation, home to home, for the same food when it comes to preparing the food, time of day when the sample is consumed, and variation in the products consumed along with the test sample.

The test design must be as simple and unambiguous as possible—it is best to conduct a home use test on only one or two samples. Otherwise the test situation would be too complex for most respondents. Therefore, the home use test is not appropriate or at least not well suited for a test where one participant evaluates several different samples.

The home use test is the most expensive test in terms of actual product cost because the larger sample sizes of participants need to test a product, and the bigger amounts of product to be evaluated. Others might, however, argue that the bigger amounts of served test product allow the food to be tested under actual home conditions, such as drinking rather than sipping a beverage, which would be the case in a laboratory, central location, or mobile laboratory test.

Distribution costs may become a major expense associated with this test. There is also the cost of producing the samples for the test. Depending upon the nature of the samples and the number of test participants, the test samples may be produced "on the bench," or in some larger cases, may be require more expensive "scale-up" production facilities, which runs into far more money.

The cost of the products placed in participants' homes will add to the cost of the test. On the other hand, if the consumer panel size used is lowered because of these costs, then the information one would obtain from the HUT is correspondingly limited. Furthermore, since placement of products is by mail, perishable and nonmailable products cannot be tested.

When participants have not been prerecruited and screened from an existing consumer database, the participants will likely be less aware of the importance of their role and the importance of the data being collected. In such cases, response rates will be lower than expected.

Questionnaires must be self-explanatory, clear, and concise because there is no opportunity to explain and elaborate as one would be able to do in a personal interview. Visual aids, if needed, should be limited to graphics on a questionnaire. Finally, the researcher needs to realize that the consumer respondent will be able to read all the items on the questionnaire beforehand, thereby limiting the impact of sequencing of questions.

HUT—panel size

The sample size for the HUT should be large enough but not unduly large, so that the researcher can get the necessary number of respondents within a reasonable length of time. Most research agencies guarantee a 70% response rate. The factors of cooperation and nonresponse may influence the results of these samplings (ASTM, 1979). The assumptions are that the nonresponders will react similarly to those who respond—fortunately, there are instances where mail panel results closely approximate probability sampling results. Most of the time, participants of HUTs conducted by mail are represented as being quota samples that match US Census distribution. The apparent superiority of sampling of respondents may be at least partly an illusion. People who volunteer for such activity and sustain their interest may be different from the normal user of the product type. Also, it is often alleged "though seldom proven" (ASTM, 1979) that panel members who are used repeatedly may develop special ways of responding that are not typical of consumers.

HUT—recruitment and placement

In HUTs, product samples are tested in the consumers' own homes. There are various methods to locate and recruit participants, and to deliver the test product to their home. We talked about the mail previously. One method is to prerecruit from a consumer database and screen for panelists' eligibility to participate in the tests. However, the practice in some cases is to involve employees of a company who have little or no responsibility for production, testing, or marketing of the product. There is some risk associated with employee panels in product maintenance tests. Employees should not be used in product development, improvement, and optimization studies.

Due to the uncontrolled conditions of testing, a larger sample than that required for a laboratory test is recommended for a home use test. Usually 50–100 or even more responses are obtained per product. The number varies with the type of product being tested and with the experience of the respondent in participating in home HUTs. With "test-wise" panelists, Stone and Sidel (2004) recommend that a reduction in panel size could be made. Because these panelists know how the test is conducted and feel more comfortable in the test situation, the testing will be less susceptible to error and the results of psychological variables associated with being a subject. In multicity tests, 75–300 responses are obtained per city in three or four cities (Meilgaard, Civille, & Carr, 2007). The number of products to be tested should be limited to 1–2.

One approach uses the telephone for preliminary questioning to establish qualifications and solicit cooperation. Calling may be random within a given area or from lists of members provided by cooperating organizations. This approach has an advantage in that the recruiting is less expensive when qualified respondents are expected to occur with low frequency. On the other hand, it is probably harder to gain cooperation over the telephone and the problem of delivering products and test instructions has to be solved. Furthermore, more effort may be required in recontacting the participants because they may be more geographically dispersed.

Another approach conducts a door-to-door household survey in areas selected on the basis of having a high probability of producing desired subjects. An interviewer asks questions to establish qualification and to obtain background information that might be useful in analyzing the results. When a qualified family is found, the product is *immediately* placed in that household, instructions given about the interview, and the interviewer makes arrangements for a recontact or tells the participant how to return the ballot. An advantage of this approach is that it permits distributing the sample as desired in a given territory. This stratagem eliminates the problem of product delivery. Finally, the participants are easier to contact for the final interview. However, the method may be more expensive because much effort is wasted in contacting unqualified people.

Mall intercept or in-store recruitment has been relied on to an increasing extent for many types of market research studies. Mall intercepts, the mainstay of CLT tests, are also useful for HUT recruitment. Interviewers recruit prospective participants in the mall or store and can quickly screen them to determine whether or not they meet requirements. Consumers who agree to participate can be given the products virtually "on the spot," with little delay. Sometimes, bulky or perishable products must be delivered to the respondents' homes at a later date, after the recruitment is arranged at the mall. The advantage of mall recruitment for

HUTs is that it permits screening a large number of prospects relatively quickly and cheaply, which can be particularly helpful when trying to locate users of a low-incidence product type. A serious disadvantage of this recruitment method is that it may be time-consuming and expensive to recontact the respondents for personal interviews since the participants may live in a region far from the mall, or their residences may be dispersed across a wide area.

When HUT panelists are to be reached by mail, databases of mailing lists must be developed, kept updated by rotation and replacement of dropout participants. Developing a database involves initial location of the respondents, contact, gaining cooperation, acquiring necessary information, and maintaining sufficient interest for sustained participation. Contacts can be made through the telephone through random digit dialing, newspaper advertisements, recruiting through organizations, and so forth. Prospective panel members provide information about themselves, which is placed on file and retained for a certain period of time. The frequency with which a panel member may be contacted varies widely. Usually HUT participants are given some incentive for their participation, which keeps them interested in remaining on the database.

The key to obtaining a valid sample of HUT respondents is the initial selection and gathering of participant data, which allows for a "smart" selection of participants. The usual approach is to seek protection in large numbers, with lists containing thousands of names and representing more or less the national census distribution of the population. Information is obtained by means of questionnaires. Such demographic data as location, the number and identity of family members, family income, race, education, and occupation are nearly always available. Beyond this first step, it is a matter of product usage and proper panelist motivation. The information necessary to find and recruit panelists is generally stored in a searchable database. Drawing the actual sample for a given study is a mechanical matter of identifying families with the desired characteristics, then randomly selecting the desired number of participants to contact, of which some percent will agree to participate.

HUT—facilities and procedures
HUTs are conducted in the respondents' own homes. No controls of the environment or the testing conditions can be made, since the HUT is conducted in a location wherein the researcher has no control over the test, other than providing detailed instructions and hoping that the panelists comply. The products are tested under actual use conditions. Responses from one or even several members of the family are usually obtained, either at the time of preparation/consumption or later by telephone "callback" or mail return of a completed questionnaire. Given the lack of control, it is best to keep matters simple. The test design should be as easy to perform and uncomplicated as possible. The focus should be on consumer acceptance, overall, and acceptance of specific attributes.

HUTs require additional considerations and preparatory steps that occur before the test. The relatively simple considerations can produce problems if done incorrectly. Issues to resolve include the nature of containers and labels, product preparation instructions that are not ambiguous and must be complete, product placement—whether mailed or delivered to consumers' homes, and the specific self-administered HUT questionnaire.

HUT—product samples and placement
The product containers to be used in HUTs are often those in which the product will be sold but should be plain and have no graphics. Containers should be labeled with the sample code number and a label containing the preparation instructions, contents, and a telephone number to call in case questions arise.

Preparation instructions must be complete and easily understood. These must describe exactly how the products are to be prepared. Directions as to how to complete the question-naire should be included. The instructions should be unambiguous and should be prepared so that they are clear to the panelist and do not cause confusion. Pretesting the instructions and the questionnaire is important to the success of a home use test and is always a good idea.

The length of time required for each product dictates number of products. The number of products to be tested should be limited to two, primarily because of the length of time needed to evaluate each product. Often HUTs are designed to get information about products in actual use. The use may be once in several days, or may be daily over 4–7 days. The goal of the test is to capture information in "normal consumer use." The participant generally fills out a separate, parallel questionnaire for each product. When additional products are included in the test, the HUT would take longer and the risk of nonresponse would be greater due to one or more reasons such as loss of the questionnaire, trips out of town, or loss of panelist interest.

There are three methods that can be used to get the samples to the HUT participants. These are to: (1) mail the product, (2) deliver the product, or (3) have consumer panelists come to a central location to obtain the product. Providing the participant with all the products during the beginning of the test minimizes the direct cost of the test because the logistics demand only one appearance by the participant, and one appearance by the test coordination group. Lowered direct costs can be achieved by mailing products to the consumers prerecruited by telephone or mail to participate in the test. Mailing products to subjects is generally risky and should be avoided except in rare instances, such as when the products are microbiologically shelf-stable and are of a shape and size that lend themselves to mailing. Examples of such shelf-stable products that could be mailed to participants are: pecan halves, small products that can resist the impact of temperature changes and other rough conditions encountered during the mail-out (Resurreccion & Heaton, 1987). The problems with using the mail to place products with participants are: the time required, the choice of mailers, microbiological safety, and the not-too-infrequent happening that the package may be received by someone other than for whom it was intended so the participant never tests that particular product.

Providing two or more products at the same time is not recommended for the following two reasons: it allows the participants to make direct comparisons, and it increases the probability of recording responses for the wrong product on the questionnaire (Meilgaard, Civille, & Carr, 2007), invalidating the results. For these reasons, Stone and Sidel (2004) recommend that cost saving should not be seriously considered in HUTs when the precision of the results is critical. The researcher might well end up wasting a lot more money because of the steps used to cut costs.

The practice of having consumers, recruited for the HUT, come to a central location to receive products is efficient and attractive to the sensory practitioner planning the test (Hashim, Resurreccion, & McWatters, 1995). When there is a second sample, consumers will come back to the central location, return the completed questionnaire and the now-empty product container, and then receive the second product and a new questionnaire. This method distributes the travel across the participants, and when local residents are used, this HUT strategy for product distribution is more effective and cost-effective than delivering the product to consumers' homes. In certain cases when the method of sampling for a home use test is to intercept and qualify prospective participants, qualified participants may be given the first of the samples immediately after qualifying. The participants then return to the same location to pick up the next sample and questionnaire and return the empty sample container and questionnaire.

HUT—measuring product acceptance

The primary objective in the home use test is to measure overall product acceptance. In order to maintain uniformity and continuity (Stone & Sidel, 2004) between different methods of testing such as the laboratory test, CLT, and home use test, the researcher should use the same acceptance scale for each. The questionnaire can include other measures and tasks besides rating scales for acceptance. These are attribute diagnostic questions, addition of product performance questions, and paired preference questions.

The addition of specific attribute questions to the basic acceptance ratings add more information, but occasionally raise controversy among professionals. In addition to overall acceptance, acceptance of specific product attributes such as appearance, aroma, taste, and texture may be obtained but the results should not be used as one would use results from a sensory analytical test. Objections to using diagnostic questions were raised by Stone and Sidel (2004), who recommended that diagnostic questions be excluded from the test. They based their assertion on their opinion that the researcher cannot directly validate these attribute ratings, although there is no evidence that this is a cogent criticism since the participants are assumed to be able to rate liking. Furthermore, they assert rating overall acceptance requires the evaluation of the product as a whole, without direction to what sensory aspects are considered important, and by calling the participant's attention to a specific set of attributes there is always the possibility that the participants will be biased.

Often, there is interest in learning more than overall acceptance from a home use test. Issues such as product performance (*How easy is it to pour milk from this stand-up pouch?*) or purchase intent (*If available in the market, how often would you purchase this product?*) can be readily addressed.

Some researchers ask a paired preference question on products that are tested during two different time periods. This final question, when asked, relies on the consumer's memory to remember what was tested the week before and to compare it with the product currently being tested. This use of the paired preference question is bias-fraught and thus, perhaps, inappropriate. The 9-point hedonic scale for each product yields more useful information than the final paired preference question. However, management often likes these simpler preference questions, despite bias, and it is to address these management preferences that the researcher often agrees to ask the paired preference question.

HUT—data collection

There are various ways to obtain often critical, postuse information about the respondent's preferences, attitudes, and opinions. These are: the personal interview, use of a self-administered questionnaire that is mailed back, and the telephone interview:

(i) *Interviews*: The most effective approach among the three methods is the personal interview. The personal interview allows the researcher to review procedures with the respondent and determine whether or not the product was used properly. The interview enables the interviewer to answer any questions regarding test procedures or product use conditions. Respondents are usually more motivated and involved when a personal interview occurs at the start of the evaluations. The interviews offer capabilities not available with a self-administered questionnaire. For example, the interviewer can control the sequence of questioning by using visual displays such as concept cards, pictures, or representations of rating scales. Probing for clarification of answers or more detailed information is also possible. The major disadvantage of the interview is its added cost. When personal interviews are selected as the data collection method, it is extremely important that all interviewers undergo rigorous training prior to the interviews. Dry runs should be conducted using actual consumers who are not participants in the HUT.

(ii) *Self-administered questionnaire*: In many cases, the product samples are either mailed, available on the Internet, or given personally to respondents. The instructions are included, along with the self-administered questionnaire to be completed, at times during and always after the products are used. The questionnaire is mailed back to the researcher after it is completed. Self-administered methods require that the questionnaire be designed so that instructions and questions are easily understood without assistance. It is advisable in such cases for the questionnaire to have a telephone number and name of a contact person who can be reached to answer any questions. Self-administered methods are almost always initially less labor intensive than telephone interviews and, when conducted properly, eliminates interviewer bias and reduces the cost of the interview. The disadvantages of the self-administered questionnaire are that cooperation may be poorer and the response rate lower when the instructions are not well understood. The sequence of questioning cannot be controlled as the entire questionnaire can be read prior to starting the test. Furthermore, there is no opportunity to correct errors or probe for more complete information unless the respondent telephones the contact person.

(iii) *Telephone interviews*: With proper planning, data may be collected by telephone interviews after the participant has received the product and evaluated it. Telephone interviews have the advantages of lower costs and greater speed of data collection. Therefore, less time is involved and thus lower costs are incurred. Furthermore, the telephone makes it easier to use procedures appropriate for random sampling. In addition, telephone interviews make it easy to work with participants who are geographically dispersed. The interviewer can conduct the interview in the evening and, thus, diminish the ever-present, annoying, nonresponse rate. Compared to mail interviews, the telephone interview's advantages are; greater speed of data collection, ease of maintaining random samples, lower nonresponse rates, and ease of getting "hard-to-reach" respondents such as employed men and teens.

There are a number of limitations to telephone interviewing. Compared to personal interviews, telephone interviews should be relatively short or the participant may just terminate the telephone call. Thus, rapport with the panelist is not as good as the rapport developed in personal interviews. Telephone questionnaires have several constraints. These constraints include the difficulty to get complete open-ended responses as from personal interviews, and the impossibility of using visual displays. Furthermore, only rudimentary scaling can be used in a telephone interview. Information that can be gathered through observation, such as the race of the respondent or the type of housing they live in, is not possible to obtain. Telephone interviews are more expensive to conduct than mail questionnaires. The telephone interview is often chosen in order to limit costs, particularly in those cases where the respondents may be difficult to contact due to geographic constraints or limited availability during specific time periods.

Telephone interviews may generate poorer cooperation, with a lower response rate than personal interviews. Furthermore, the range of questioning is often limited because visual displays are not possible to use. Probing, when used, may not be as effective because the visual displays are referred to, rather than pointed to.

The interview questionnaire should be pilot-tested with a small sample of respondents, especially those questions that require the use of a scale. It is extremely important to ensure that both supervisors and telephone interviewers undergo extensive training on how to administer the question and how to administer the actual interview. Quality control measures that may be employed are audits of the interviewing process.

Telephone interviews are now often, but not always run in the form of a "factory product mode." The telephone interview may originate from a central location where a bank of telephones is installed, or from the interviewer's own home. The central location permits immediate supervision of interviewers to handle problems as they arise. The telephone lines may be long distance so that interviews may be conducted in any city desired.

Summing up the HUT

All things considered, the home use test is reserved for the later stages of sensory evaluation testing with consumers. HUTs enjoy a high degree of face validity. They provide the researcher with the tool to measure how other members of the family react to the product, under the more normal conditions at home. The only problem is that in some cases the HUT may be more costly than other sensory tests, requires more time to complete, and lacks environmental controls (Stone & Sidel, 2004).

The test location itself may influence the ratings

One might think that the carefully executed tests should generate bias-free information, no matter which testing location has been chosen to conduct the test. Some research shows just this finding—testing location may not influence the rating. For example, Hersleth et al. (2005) showed that changing the environment and the degree of social interaction in the consumer tests exerted no significant effect on hedonic ratings for unbranded popular cheese samples. The authors attribute the high degree of consistency to the absence of a meal context during testing, similar expectations during testing, and a high familiarity with the samples tested. Similarly, Pound and Duizer (2000) measured responses for overall liking of unbranded commercial chocolate in three venues: in a formal sensory laboratory used for teaching, in a central location, and in a home use test. They found similar results from testing in the various locations.

However, other investigators found the opposite—that test venue plays an important role. King et al. (2004) found that introduction of context effects in a CLT can improve the ability to predict actual liking scores in a real-life environment. They reported that the relation between context effect and consumer acceptance may not be consistent within and across meal components. Thus, meal context had the strongest effect on tea; social context has a strong negative effect on pizza; environment had a weak positive effect on pizza and tea and a negative effect on salad; and choice had a positive effect on salad. These results led the authors to conclude that context variables affect product acceptance, but the relation between context effect and consumer acceptance may not be consistent within and across meal components. Consistent with these findings, Hersleth et al. (2003) found that hedonic ratings for Chardonnay wines were higher in a reception room, where some socializing occurred, compared to acceptance measured in partitioned booths in a laboratory. The effects of context factors may be significant when testing certain foods, and should be considered in planning an acceptance test.

Scales: the measuring instrument

Scale ratings indicate the panelists' perceived intensity of a specified attribute under a given set of test conditions. The panelist may scale liking or sensory intensity. The key thing to keep in mind is what specifically is being rated. The "what" may be a specific attribute or an overall impression. It's impression to specific just what is being rated.

As one might expect, the scaling of acceptance is the more important type of measurement, in contrast to the scaling of intensity, just as paired preference to assess which product is preferred has been done far more frequently than paired comparisons of perceived intensity. For business, the measurement of acceptance is paramount because of the very reasonable belief that consumers will buy what they like, and won't buy what they dislike.

Today's array of scales to measure acceptance comprise different types of ballots (text vs. facial), different numbers of scale points, and different language and, indeed, different aspects of liking. The following list and the explication provide a sense of the range of alternative metrics for acceptance facing today's sensory researcher.

Scales—hedonic measurement (simple degree of liking)

Hedonic measurement is the "workhorse" of sensory researchers. Perhaps today, in the middle of the first decade of the twenty-first century, we can point to a scale that has been around more than 50 years—the traditional 9-point hedonic scale (Peryam & Pilgrim, 1957). The scale was originally developed to measure food acceptance in the US military. The 9-point scale constitutes a rather simple, unambiguous rating scale, used for many years to measure the acceptance of a food. The scale has been used to generate a benchmark number with which to compare products, to compare batches, and to assess the level of acceptance of products in a competitive category. The consumers' task is simple; using the scale, record the degree of liking. There are four presumably equally spaced categories for liking, a neutral point, and then a corresponding four presumably equally spaced category for disliking. The hedonic scale is easy to use, intuitive, and makes no pretense of measuring anything other than the panelist's liking for the product (Moskowitz, Munoz, & Gacula 2003). We see an example of the 9-point hedonic scale in Figure 9.4.

A considerable amount of research was conducted to create the scale, in terms of the number of categories and anchors in each category. The 9-point hedonic scale has been

Panelist Code _____

Session No. _____

Take a piece of sample 452 using a fork and place the whole piece into your mouth.

Please answer the following questions by completely filling in the square that best reflects your feelings about this sample.

OVERALL, how would you rate this sample?

Dislike Extremely	Dislike Very Much	Dislike Moderately	Dislike Slightly	Neither Like nor Dislike	Like Slightly	Like Moderately	Like Very Much	Like Extremely
□	□	□	□	□	□	□	□	□

Figure 9.4 An example of a hedonic scale for overall acceptance. Reproduced from Resurreccion, 1998. Fig. 2.3, p. 21. Copyright 1998 by Aspen Publications, with kind permission of Springer Science and Business Media.

validated in the scientific literature in terms of predicting product acceptance in later studies and even in the subsequent market phase (Stone & Sidel, 2004). Furthermore, the scale is reliable, giving similar answers on repeated use, meaning that it can be used for benchmarking, which it is by corporations and by the US military. Of course there are always objections, and the 9-point scale is not free of critics who feel it could be "improved." For example, one objection is that the scale should be unbalanced. The logic is, for the dissenters, rather simple. Most companies test "good products," not poor products, so why waste scale real estimate (i.e., 4 points) to quantify something that would never reach testing? If there are 9 points, why not allocate 6 of those 9, or even more, to show differences among the "liked products," where the business focus will lie? Others feel that it would be best to have more points because more points presumably generate better discrimination. Still others feel that fewer points would be better because the respondent does not use all the points of the scale. Researchers sensitive to some of these issues have used shorter scales, including 5- and 7-point scales. The 3-point scale is not recommended for use with adult consumers since adults tend to avoid using the extreme points of the scale in rating food product samples. Finally, still others do away with the 9 points and use a line scale instead, arguing that the line scale is more sensitive to differences. This group would still use anchors at the two ends, one corresponding to "like extremely" and the other corresponding to "dislike extremely."

Testing with children may not work with verbal scales. In some instances, an adaptation of the 9-point hedonic scale in the form of a 9-point facial scale proves useful with children. The 3-, 5-, and 7-point facial hedonic scales were found to be appropriate for 3-, 4-, and 5-year-old children, respectively (Chen, Resurreccion, & Paguio, 1996). Figure 9.5 shows examples of the facial hedonic scales.

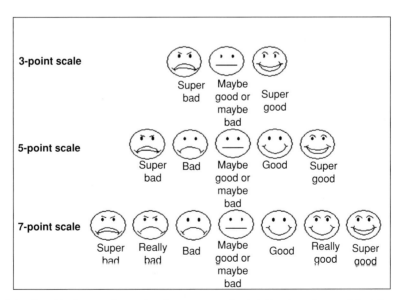

Figure 9.5 Facial hedonic scales used to determine preschool children's preference (Chen, Resurreccion, & Paguio, 1996. Food and Nutrition Press, Inc., reprinted with permission; reproduced from Resurreccion, 1998. Fig. 2.4, p. 22. Copyright 1998 by Aspen Publications, with kind permission of Springer Science and Business Media).

Table 9.3 Descriptors used in the food action rating scale.

I would eat this food at every opportunity I had.
I would eat this very often.
I would frequently eat this.
I like this and would eat it now and then.
I don't like it but would eat it on an occasion.
I would hardly ever eat this.
I would eat this only if there were no other food choices.
I would eat this only if I were forced to.

Source: Adapted from Schutz, 1965; Resurreccion. 1998. Table 2.2, p. 22.
Copyright 1998 by Aspen Publications, with kind permission of Springer
Science and Business Media.

Scales—food action (FACT) rating scale

The FACT rating scale was devised by Schutz (1965) to measure acceptance of a product by a population, but at the same time combine the attitude toward a food with a measure of expected action that the consumer might take (e.g., to consume or not consume the food). The FACT scale comprises nine categories, as shown in Table 9.3. As in every rating system the researcher presents the samples sequentially, generally in a randomized order balanced across the panelists. The panelist selects the appropriate statement. The ratings are converted to numerical scores to facilitate statistical analysis of data (Resurreccion, 1998).

Scales—ranking methods

Ranking is, in effect, an extension of the paired preference test. Many of the advantages of the paired preference test apply to ranking. These include simplicity of instructions to participants, a minimum amount of effort required to conduct, relatively uncomplicated data handling required, and minimal assumptions made about level of measurement because the data are treated as ordinal so only the rank order conveys information, nothing else (Lawless & Heymann, 1999a). Three or more coded samples are presented simultaneously, sufficient in amount so that the panelist can retaste the product. The number of samples tested depends upon the panelist's span of attention and memory as well as on physiological considerations. With untrained or naive panelists, no more than four to six samples should be included in a test (ASTM, 1996). The panelist assigns an order to the samples according to his or her preference. As with the paired preference method, rank order evaluates samples only in relation to one another. An example of a ballot for a ranking test is presented in Figure 9.6.

Scale-based norms for consumer acceptance

Once consumer acceptance for the product has been quantified, the lower boundary of attribute intensities for a given consumer acceptance rating can be defined. This is usually done with input from company management. Some companies that pride themselves on the quality of their products may be unwilling to produce a product that is only "liked slightly" (6) on a 9-point hedonic scale and may opt to accept a limit set at the "like moderately (7) or higher" for the product.

PREFERENCE TEST - RANKING

Name _____   Session Code _____   Date _____

Please rinse your mouth with water before starting, before each sample and any time you need to.

In front of you are five samples. Beginning with the sample on the left, taste each one.

After you taste all samples, you may re-taste as often as you need to.

Rank the samples from most preferred (=1) to least preferred (=5)

Sample   Rank (1 to 5)

          Ties are NOT allowed

643   ____

296   ____

781   ____

528   ____

937   ____

                 Thank you

Figure 9.6   An example of a ballot to be used in a ranking test. Reproduced from Resurreccion. 1998. Fig. 2.5, p. 23. Copyright 1998 by Aspen Publications, with kind permission of Springer Science and Business Media.

Scales—just about right (JAR)

The JAR scale is a bipolar scale and constructed as category scale using boxes (Figure 9.7), or continuous scale using a line (Figure 9.8). In a category scale, responses are limited to the number of points chosen for the scale, whereas, responses for continuous scale are theoretically infinite (Rothman & Parker, 2009). Category scales typically comprise 3, 5, 7, or up to a maximum of 9 points (Rothman & Parker, 2009), anchored on both ends with equal number of categories of either "high or too much" or "low or too little" of a sensory attribute, with the middle category anchored with the phrase "just about right" (Chambers & Wolf, 1996; Gacula *et al.*, 2007; Moskowitz, Munoz, & Gacula, 2008). Although three is

Figure 9.7   Examples of 3-, 5-, and 7-point just-about-right scales using box scales. These scales should not be placed on the same ballot. They are presented here for illustrative purposes only.

the minimum number of points, many researchers are uncomfortable with it because of "end avoidance" which can force respondents to the "just-about-right" scale (Rothman & Parker, 2009). Continuous scales are anchored at the middle of the line with "just about right" and with semantic opposites anchored at each extreme end. When a JAR scale is used, it is assumed that naïve respondents know how to use the scale and understand the attributes (Chambers & Wolf, 1996). Epler *et al.* (1998) found that the form of the scale, whether category or continuous, did not result in differences in the information obtained from the JAR data.

JAR scales have the advantage of providing direction for formulation and reformulation, and they are easily understood by researchers and management (Moskowitz, 2008). As such, JAR scales are championed as a diagnostic tool for consumers tests (Stone & Sidel, 2004). The JAR indicates whether the product is higher or lower relative to the consumer's perceptual ideal level of that attribute (Moskowitz, Munoz, & Gacula, 2008).

JUST-ABOUT-RIGHT SCALING METHOD

Name _____   Session Code _____   Date _____

Please rinse your mouth with water before starting the test.

You are given a coded sample. Taste the sample and **place a vertical mark on the line scale that best describes your reaction to the product**.

Code 259

Aroma

|_____|_____|
Too strong   Just-about-right   Too weak

Thank you.

Figure 9.8 Example of a just-about-right scale using line scale.

Gacula *et al.* (2007) concluded that JAR meanings invoke preference and acceptability. They found that panelists defined JAR as "okay," "very good," "I like the product," "like it very much," "highly favorable," "high acceptability," "desirable; like the product" and "best for the situation." They suggest using descriptive statistics specifically the mean of responses, although skewness and the kurtosis would make the median more appropriate but less discriminating.

The anchors for the JAR scales must be carefully selected (Moskowitz, 2008) to ensure correctness of the scale. Many inappropriate versions of the JAR scale exist because of the misuse of the anchors. Examples of incorrect pairs of end anchors which are not opposites are "much too salty/much too sweet," or "much too soft/much too crispy." True opposite word/terms should be chosen when using a JAR scaling method such as "soft/firm," "light/dark," and "dull/shiny."

Lawless and Heymann (1999b) discussed a problem of the centering bias arising when comparing multiple products in a multiproduct test. The centering bias is a tendency to put a sample having an intermediate intensity in a series of products at "just-right" point which could result in false conclusion that the middle product was "just right," when in fact the true optimum could still be higher or lower.

DISCRIMINATION TESTS

Discrimination testing comprises a class of the sensory analytical tests that address the question regarding whether or not a sensory difference exists between samples. These tests are useful only if there is a perceptible or subtle difference between two products but are not useful when the difference between the samples are very large or obvious (Lawless & Heymann, 1999b). For example, discrimination tests are used to determine whether product reformulations, for example, using low-cost ingredients or changes in the processing, packaging and storage to produce a cheaper product will cause changes or differences in the sensory characteristics of the original formulation. Discrimination tests are also applied during benchmarking and when a product improvement or product optimizations are made; in these cases difference between products is desired.

The major discrimination tests used in research and industry include: duo-trio, triangle, n-alternative forced-choice tests (2-AFC and 3-AFC), and paired comparison tests, where the trained panelists are forced to make a choice or decision. Duo-trio, triangle, and paired comparison tests are typically used to detect overall difference without identifying specific attributes. In contrast, n-AFC tests are designed to determine whether or not the samples differ in a specific attribute, for example, in sweetness (Meilgaard, Civille, & Carr, 2007). In discrimination testing between products that involve the detection of small differences with high confidence, the 2-AFC and 3-AFC methods were reported to be far superior to the duo-trio and triangle test (Ennis, 1990, 1993).

The *following sections* describe each of the major discrimination tests.

Duo-trio test
The duo-trio test is used to determine whether or not overall difference exists between samples without identifying any specific attribute (Meilgaard, Civille, & Carr, 2007). In this test, three samples are presented simultaneously to the panelists; the first is identified as the reference sample (control or standard) and the other two are coded (Stone & Sidel, 2004). One coded sample matches the reference (Lawless & Heymann, 1999b; Meilgaard, Civille, & Carr, 2007) whereas the other is different. The panelists are instructed to identify which of the two coded samples is most similar to the reference (Meilgaard, Civille, & Carr, 2007). An example of a ballot for duo-trio test is shown in Figure 9.9. The panelist should be trained to perform the task described in the score sheet correctly (Lawless & Heymann, 1999b). The two formats for presenting the reference samples—constant reference and balanced reference duo-trio tests can be used (Lawless & Heymann, 1999b; Meilgaard, Civille, & Carr, 2007).

The constant reference duo-trio test presents all panelists the same sample formulation as the reference using two possible orders of presentation, Reference A–AB and Reference A–BA which should be balanced across the panelists, measuring equal number of sequences appear equal number of times. This format is used when the panelists are familiar with the product, such as in a current product formulation (Lawless & Heymann, 1999b; Meilgaard, Civille, & Carr, 2007).

The balanced reference duo-trio test serves one sample formulation as the reference to half of the panelists, and the other sample formulation as the reference to other half of the panelists. Four orders of presentation are possible using this format: Reference A–AB, Reference A–BA, Reference B–AB, and Reference B–BA. This format is chosen when

---

DUO-TRIO TEST

Name _____   Session Code _____   Date _____

Please rinse your mouth with water before starting the test.

You are given three samples, one is marked as R or reference and the other two are coded. One of the coded samples is the same as R while the other is different. Taste the samples from left to right starting with R then proceed with the coded samples.

Circle the code that is different from R.

R   253   179

Figure 9.9   Example of a ballot to be used in a duo-trio test.

panelists are not familiar to both products such as prototype products or when there is insufficient quantity of the more familiar product (Lawless & Heymann, 1999b).

The duo-trio test is simple and easily understood. Compared to the paired comparison test, the advantage is that a reference sample is presented. Using the reference avoids confusion with respect to what constitutes a difference, but introduces a disadvantage because three samples, rather than two, must be tested (Meilgaard, Civille, & Carr, 2007).

When samples have pronounced aftertaste, the duo-trio is less suitable than the paired comparison test (Meilgaard, Civille, & Carr, 2007). Based on the probability of obtaining a correct result by guessing, the duo-trio test is believed to be less efficient than the triangle test because it has 50% or 1 in 2 probability of guessing (Meilgaard, Civille, & Carr, 2007) while in the latter has only 23%. However, the power of the test shows that the duo-trio test will perform better.

Triangle test
The triangle test is well known and has been extensively applied as it was mistakenly believed to be more sensitive than other methods (Stone & Sidel, 2004). The triangle test is used to determine whether or not the samples are different, either in a single sensory attribute or in several, and the nature of the difference between the samples is unknown (ASTM, 2011). In the triangle test, three coded samples are presented to the panelists, two of which are the same and one is different (Stone & Sidel, 2004; Meilgaard, Civille, & Carr, 2007). The samples are served to the panelists either simultaneously or sequentially; the two serving methods are equally valid (Meilgaard, Civille, & Carr, 2007). Sequential presentation is preferable when samples are bulky, leave an aftertaste, or show slight differences in appearance (Meilgaard, Civille, & Carr, 2007).

Before the actual testing, the panelists are trained to understand the format, task and the procedure of evaluation given in the ballot (ASTM, 2011). The panelists are informed that two samples are identical and one is different. For the actual evaluation, the respondents are instructed to identify or guess the different sample by indicating this in the sensory ballot (Rousseau & O'Mahony, 1997). An example of a ballot for triangle test is shown in Figure 9.10. Six orders of presentation are possible: ABB, BAA, AAB, BBA, ABA, and

---

TRIANGLE TEST

Name _____   Session Code _____   Date _____

Please rinse your mouth with water before starting the test.

You are given three coded samples. Two of the samples are the same while the other is different or odd. Taste the samples from left to right.

Circle the code that is different.

527   913   468

Figure 9.10 Example of a ballot to be used in a triangle test.

BAB. The panelist evaluates the samples from left to right and has the option to go back to repeat the evaluation during testing (Meilgaard, Civille, & Carr, 2007).

The odd or different sample presented in the triangle test can either be one that has the stronger or weaker stimuli. Bi (1995) studied the performance or sensitivity of the triangle test as affected by the use of strong stimulus as different from the others, or its dual, the weak stimulus is different from the others. Bi reported that performance on a triangle test with the strong stimulus as different is better than using the triangle test with the weak stimulus as different. Based on these results, it is best to use the strong stimulus as the different sample to obtain the best performance on the triangle test.

The triangle test is suitable only when the products are homogeneous. The triangle test enjoys limited use with products that generate carryover, sensory fatigue, or adaptation (ASTM, 2011), as well as not appropriate with panelists who find testing of three samples too confusing (Meilgaard, Civille, & Carr, 2007).

In general, the triangle test can be said to be a very difficult test because the panelist has to recall the sensory characteristics of the two products before evaluating the third and then make a decision (Stone & Sidel, 2004). Adding to this difficulty is the view that a triangle test is a combination of three paired tests, A-B, A-C, and B-C (Stone & Sidel, 2004). Triangle tests, with chance probability of only 1/3 for correct answers when the answer is given by pure guess, is often believed to be more efficient than the paired comparison and duo-trio tests. This 33% probability and the fewer correct scores required for statistical significance account for the test's claim of greater sensitivity (Stone & Sidel, 2004). However, the sensitivity should not be confused with the power of the test, which is a different issue.

Two alternative forced choice (2-AFC) test
The 2-AFC test is also called as a "directional" paired comparison test (Lawless & Heymann, 1999b). The panelist is required to identify which of the two samples has more of a specified

---

TWO-ALTERNATIVE FORCED CHOICE

Name _____   Session Code _____   Date _____

Please rinse your mouth with water before starting the test.

You are given two coded samples. Taste the samples from left to right.

Which of the samples is sweeter? Circle ONE

713   965

Thank you.

Figure 9.11   Example of a ballot to be used in a two-alternative forced choice test.

sensory attribute, for example, sweetness, or tenderness, without focusing on the overall difference. This test is a simple test and can be used when carry-over effect is present. The 2-AFC test should not be used though when sensory attributes are not easily specified, or not known in advance. This method does not identify the magnitude of the difference for that specific attribute, and neither addresses preference (ASTM, 2010). In the 2-AFC test, the panelist is presented a pair of coded sample, and is informed of the attribute to be evaluated. Training is required before the actual testing, to ensure that panelist recognizes and quantifies the specified sensory attribute, and correctly performs the task described in the sensory ballot (Lawless & Heymann, 1999b). A sample 2-AFC ballot appears in Figure 9.11. Two orders of presentation are possible, A-B and B-A (Stone & Sidel, 2004), which should be presented in balanced order to the panelists.

The 2-AFC is more sensitive than 3-AFC test (Rousseau & O'Mahony, 1997). The 2-AFC test provides a high level of power for small-product differences, and allows subjects relatively low levels of response bias while undertaking it (Hautus *et al.*, 2009).

In the 2-AFC test, the analyst has to know which sample is higher in a specified sensory attribute. The alternative hypothesis states that if the underlying population can discriminate between the samples based on the specified sensory attribute, then the sample higher in the specified dimension, (say sample A) will be chosen more often as higher in intensity of the specified attribute than the other sample (say sample B).

Three alternative forced choice (3-AFC) test
The 3-AFC test is similar to a "directional" triangle test. The panelist is presented with three samples simultaneously and forced to identify the sample that is higher or lower in the specified sensory attribute (Lawless & Heymann, 1999b).

THREE-ALTERNATIVE FORCED CHOICE

Name _____ Session Code _____ Date _____

Please rinse your mouth with water before starting the test.

You are given three coded samples to evaluate. Taste each coded sample in the set in the sequence presented from left to right. Place the entire sample in your mouth. Do not re-taste. Within the row of three, **circle the number of the sweeter sample**. Rinse your mouth with water afterwards and expectorate all sample and water.

219   953   468

Figure 9.12 Example of a ballot to be used in a three-alternative forced choice test.

The 3-AFC method differs from the triangle method because the 3-AFC method specifies the nature of the sensory difference. For example, one product may be sweeter than the other two; the panelist must indicate the sweeter sample, or one product may be less sweet than the other two and the panelist must indicate the less sweet sample (Rousseau & O'Mahony, 1997). Before the test, the panelists must be trained on the identified specific sensory attribute of the samples and on correctly performing the task described in the sensory ballot (Lawless & Heymann, 1999b), an example of which is shown in Figure 9.12. There are six possible orders of presentation: AAB, ABA, BAA or BBA, BAB, ABB that should be balanced when presented to the panelists. The instructions may be altered depending on which sample is different (Rosseau & O'Mahony, 1997). When the odd sample has the higher intensity, the instruction should be that the panelists identify which sample has more of the intensity of the specified attribute, for example, "*Which sample is saltier?*" Otherwise, if the odd sample has the lesser intensity of the specified sensory attribute, the question will be "Which sample is less salty?" The panelists are allowed to retaste the samples as often as they wish provided that they always taste all three samples in the triad in the order specified to maintain the integrity of sequential effects (Rosseau & O'Mahony, 1997).

The performance of a 3-AFC test using strong or weak stimulus as the different sample was studied by Bi (1995). Here is a summary of the findings:

(i)   The performance of 3-AFC with the strong stimulus as the odd or different one is better than 3-AFC with weak stimulus as the different one.
(ii) The 3-AFC with either strong or weak stimulus as different performs better than a triangle test with both strong and weak stimuli as different.
(iii) The performance of 3-AFC with either strong or weak stimulus as different is better than the performance on that part of a triangle test where the weak stimulus is the different one.

 (iv) The performance of 3-AFC with the weak stimulus as different is better than that part of a triangle test with strong stimulus as different only under certain conditions.
 (v) Based on these results, to achieve the best performance of a 3-AFC test, it is best to use the stronger stimulus as the different or odd sample.

The 3-AFC method has a superior performance compared to triangle test due to difference in cognitive strategies (Rosseau & O'Mahony, 1997). During 3-AFC test, the panelist is looking for the highest or lowest intensity on a continuum whereas during a triangle test, the judge compares the sensory distances between samples (Rosseau & O'Mahony, 1997).

Paired comparison test
The paired comparison test is a two-product analysis (Stone & Sidel, 2004). There are two types of paired comparison tests:

 (i) The difference paired comparison (also known as the simple difference or the same/different test).
 (ii) The directional paired comparison or 2-AFC (Lawless & Heymann, 1999b) discussed earlier in this chapter.

Unlike the 2-AFC test where a sensory attribute is specified, the difference paired comparison test determines whether or not the overall difference exists between two products without indication of the sensory attribute on which they differ and direction of difference (Lawless & Heymann, 1999b). The two coded samples are served simultaneously. The panelist is asked only to indicate whether the two samples are the same or different by circling the appropriate word on the sensory ballot (see an example in Figure 9.13). The difference paired comparison test has four possible orders of presentation: A-A, A-B, B-B, and B-A, which should be balanced across panelists (Lawless & Heymann, 1999b).

In terms of organizing and implementing, the difference paired comparison test is relatively easy because it involves only two samples. It is best used to replace the triangle or duo-trio test, when product has a lingering effect or in short supply or the simultaneous presentation of three samples would not be feasible (Lawless & Heymann, 1999b).

In terms of statistical analysis, the difference paired comparison test is one-tailed since the sensory specialist knows the correct answers asked of each of the panelists, that is, whether the two samples are the same or different (Lawless & Heymann, 1999b). The alternative hypothesis for paired comparison test states that the samples are perceptibly different, and that the population will correctly indicate that the samples are the same or different more frequently than 50% of the time (Lawless & Heymann, 1999b).

A not-A test
The A not-A test is another version of the paired difference test where the samples are presented sequentially (Lawless & Heymann, 1999b). This test is used when the objective is to determine whether a sensory difference exists between two products, particularly when these are unsuitable for dual or triple presentation. Examples of such situations are when a researcher cannot make two samples of the same color, or shape, or size (Lawless & Heymann, 1999b), or when the products have strong and/or lingering flavor, are very complex stimuli and are mentally confusing to the panelists (Mailgaard *et al.*, 2007). Since the panelists do not have the samples simultaneously for evaluation, they must mentally compare the two

```
┌─────────────────────────────────────────────────────────────────┐
│                     PAIRED COMPARISON TEST                        │
│                                                                   │
│                                                                   │
│                                                                   │
│   Name _____    Session Code _____   Date _____│
│                                                                   │
│                                                                   │
│   Please rinse your mouth with water before starting the test.    │
│                                                                   │
│   You are given two coded samples. Taste the samples from left to right. Indicate whether │
│   the samples are the same or different by encircling the appropriate word. │
│                                                                   │
│                                                                   │
│   Sample Set # 1                                                  │
│                                                                   │
│                                                                   │
│   The samples are (CIRCLE ONE)                                    │
│                                                                   │
│                                                                   │
│        932–478          Same            Different                 │
│                               Thank you.                          │
│                                                                   │
└─────────────────────────────────────────────────────────────────┘
```

Figure 9.13   Example of a ballot to be used in paired comparison.

samples and decide whether they are similar or different (Lawless & Heymann, 1999b). Prior to testing, 10–50 panelists are trained to familiarize with the "A" and the "not A" samples using 20–50 orders of presentation of each sample in the study (Mailgaard *et al.*, 2007). During testing, each panelist is presented with samples, some of which are product "A" whereas others are product "not A." For each sample, the panelist responds whether the sample is "A" or "not A" and indicate this on the sensory ballot. An example of the ballot for A-not A test is shown in Figure 9.14.

The panelist's ability to discriminate is determined by comparing the correct identifications with the incorrect ones using chi-square test (Mailgaard *et al.*, 2007). The calculated chi-square value should be equal or greater than the tabular value to establish significant difference between samples.

COMPARING THE DIFFERENT DISCRIMINATION TEST METHODS

Table 9.4 summarizes the comparison of the different test methods based on the number of samples presented, number and orders of presentation, method of sample presentation to the panelists, expected results, probability of chance of guessing, power of the test, and proportion of correct responses for given d' values.

```
┌─────────────────────────────────────────────────────────────────────┐
│                            A-NOT A TEST                              │
│                                                                       │
│                                                                       │
│                                                                       │
│   Name _____      Session Code _____    Date _____│
│                                                                       │
│                                                                       │
│   Before starting the test, familiarize yourself with the flavor of   │
│   samples "A" "Not A".                                                 │
│                                                                       │
│                                                                       │
│   Rinse your mouth with water before starting the test.  You are given│
│   10 coded samples. **Taste each sample one at a time starting from    │
│   left to right.**  After tasting each sample, record your response    │
│   below by checking the appropriate box. Spit out the sample and rinse │
│   your mouth with water. Taste the next sample and do the evaluation   │
│   and rinsing as the first sample. Do the same for the rest of the     │
│   samples. Thank you.                                                  │
│                                                                       │
└─────────────────────────────────────────────────────────────────────┘
```

Figure 9.14   Example of a ballot to be used in A-not A test.

All of these discrimination tests are similar in some aspects:

(i)   Determine whether or not a difference exists between two samples, A and B.
(ii) Require training of panelists prior to actual testing to ensure that they are trained on the correct performance of the tasks indicated in the ballots.
(iii) Require that the number and order of sample presentations should be balanced across the panelists.

However, the tests differ in nine aspects:

(i)   *Total number of samples*: The tests differ in the number of samples presented to the panelists for evaluation, although only two samples are actually evaluated:
  (a) In the 2-AFC and the paired comparison tests, panelists evaluate two samples which are different from each other.
  (b) In the duo-trio, triangle, and 3-AFC tests, they evaluate three samples, of which two are identical and one is different.
  (c) In the A-not A test, although samples are presented to the panelist one at a time, the actual total number of samples evaluated by each panelist may be one or five samples in a series of 10 samples presented, depending on the protocol used.
(ii) *Coding of samples*: Except for duo-trio test where the reference is coded as reference or "R," each of the samples in all other tests are coded randomly with 3-digit numbers.
(iii) *Number and order of sample presentation*: The number and orders of presentation of samples given to the panelists vary depending on the test method used.
  (a) The 2-AFC and paired comparison tests having the least number of samples presented have also the least number of orders of presentation, A-B and B-A.

Table 9.4 Comparison of the different discrimination tests.

Parameter	Duo-trio	Triangle	2-AFC	3-AFC	Same-different Paired comparison	A not-A
Total number of samples	3	3	2	3	2	2
Number and type of samples	1 reference (R) and 2 coded samples, where 1 coded sample is R	3 coded samples where 2 samples are identical and one is different	2 coded samples	3 coded samples	2 coded samples	2 coded samples
Number and orders of presentation	Constant reference = 2 R_A BA R_A AB Balanced reference = 4 R_A BA R_A AB R_B BA R_B AB	6 ABB BAB BBA AAB ABA BAA	2 AB BA	3 ABB BAB BBA or AAB ABA BAA	2 AB BA	Variable
Method of sample presentation	Simultaneous	Simultaneous or sequential	Simultaneous	Simultaneous	Simultaneous	1–10 samples of either sample A or sample B in a series
Expected results	Determines if two samples are different but not the direction of difference	Determines if two samples are different but not the direction of difference	Determines which sample has the higher intensity of the specified sensory attribute	Determines which sample has the higher intensity of the specified sensory attribute	Determines if two samples are different but not the direction of difference	Determines if two samples are different but not the direction of difference
Probability of chance of guessing	1 in 2 or 50%	1 in 3 or 33%	1 in 2 or 50%	1 in 3 or 33%	1 in 2 or 50%	
Proportion of correct responses (%) for given d' values[a]						
0.0	50.00	33.33	50.00	33.33		
0.5	52.23	35.58	63.82	48.26		
1.0	58.25	41.80	76.02	63.37		
1.5	66.35	50.65	85.56	76.58		
2.0	74.68	60.48	92.14	86.58		
2.5	81.96	69.93	96.15	93.14		
3.0	87.65	78.14	98.31	96.88		

[a]*Source*: Ennis (1993). Tables 1–4 on probability of a correct response ($\times 10^4$) as a function of δ for the 2-AFC, 3-AFC, triangle, and duo-trio tests.

(b) The 3-AFC can be presented with three orders of presentation based on the two formats—ABB, BAB, and BBA or AAB, ABA, and BAA depending on the different sample presented.

(c) The duo-trio test requires two orders of presentation when constant reference (R) format is used—R_A-BA and R_A-BA; and 4 orders when balance reference format is used—R_A-BA, R_A-BA, R_B-BA, and R_B-BA.

(d) The triangle test requires 6 orders of presentation: ABB, BAB, BBA, AAB, ABA, and BAA.

(e) The A-not A test has variable number of presenting equal number of A and not-A in 10 sample-series tests.

(iv) *Method of sample presentation*: The samples may be presented to the panelists simultaneously or sequentially:

(a) In the duo-trio, 2-AFC, 3-AFC, and paired comparison tests, all samples are presented simultaneously.

(b) In the triangle test, the samples can be presented simultaneously or sequentially.

(c) The A-not A test presents sample sequentially one after the other in a series of 10 samples.

(v) *Expected results of the test*: Panelists may or may not be informed of the identity of the samples:

(a) In the duo-trio test, the panelists are informed that that one of the coded samples is identical to "R" which they should identify.

(b) Similarly, in triangle tests, panelists are also informed that one of samples is different while the other two are identical and they should identify the different sample.

(c) In the paired comparison test, the panelists are asked to indicate whether the two samples are the same or different.

(d) In the A-not A test, the panelist is asked whether the sample is A or not A.

(vi) *Results obtained from the tests*:

(a) The results of duo-trio, triangle tests, paired comparison and A-not A tests will inform the analyst whether there is an overall difference between two samples but not direction of the difference and the specific sensory attribute on which they differ.

The direction of the difference of specific sensory attribute is determined using 2-AFC and 3-AFC tests. The panelist is asked to identify which specific sensory attribute, preidentified prior to teating along with the direction of difference, has the higher intensity. In the 2-AFC test, the panelist is asked which of the two samples has the stronger intensity of the specified sensory attribute.

(b) In the 3-AFC test, the panelist must identify which sample has the stronger intensity over the other samples.

(vii) *Probability of chance guessing*:

(a) Duo-trio, 2-AFC, and paired comparison tests have 50% chance of guessing by the panelists.

(b) The triangle test has only 33% chance of guessing, which may have resulted in triangle test being considered as the best method in the old school of thought. The chance probability should not be used as basis for choosing the best discrimination test method.

(viii) *Magnitude of sensory differences between samples measured by d'*: The proportions of correct responses for various tests for given d' values indicate the varying power of the test. We will focus our discussion on the differences of duo-trio, triangle, 2-AFC, and 3-AFC tests. For a d' value of 0 indicating that no difference exists between samples, the number of correct responses is highest when using duo-trio and 2-AFC tests followed by triangle and 3-AFC tests. However, when the degree of difference increases to more than 0, say $d' = 2$, 2-AFC will result in the highest number of correct responses of 92% followed by 3-AFC with 87%, then duo-trio with 75%, and triangle test with 61%.

(ix) *Power of the test*: Based on these comparisons, it is clear that 2-AFC test is the most powerful discrimination test followed by 3-AFC and duo-trio tests. The least powerful among the four tests is the triangle test.

SELECTING THE BEST CONSUMER SENSORY TEST METHOD

Consumer acceptance of the food or food product can be measured using a variety of tests, different scales, and performed in various settings from the company's sensory laboratory to consumers' homes. The selection of the test is for the most part, dependent on the consumer test objectives. Particular attention needs to be given to steps that provide assurances of quality in the testing, data collection and analysis of the data. Of utmost importance, is to use a quantitative method and test protocols that will provide reliable and valid information, and produce actionable results. A properly planned and executed consumer sensory test can provide the key answers needed in product design and development.

REFERENCES

Amerine, M.A., R.M. Pangborn, & E.B. Roessler. 1965. *Principles of Sensory Evaluation of Food*. New York: Academic Press.

ASTM, Committee E18. 1979. In: *Manual on Consumer Sensory Evaluation*, ed. E.E. Schaefer. ASTM Special Technical Publication, 682. West Conshohocken, PA: ASTM Standard Method dire.

ASTM. 1996. Sensory Testing Methods. In: *ASTM Manual Series: MNL 26*, ed. E. Chambers IV & M.B. Wolf, pp. 38–53. West Conshohocken, PA: ASTM.

ASTM, Designation: E1885-04. 2011. *Standard Test Method for Sensory Analysis—Triangle Method*, 2011 Annual book of ASTM Standards V 15.08. ASTM International, West Conshohocken, PA: ASTM.

ASTM. 2010. E2164-08. *Standard Test Method for Directional Difference Test*, 2010 Annual book of ASTM standards V 15.08. ASTM International, West Conshohocken, PA: ASTM.

Bi, J. 1995. Nonparametric models for discrimaination methods and sensitivity analysis for triads. *Journal of Sensory Studies* 10:325–340.

Cardello, A.V., H.G. Schutz, C. Snow, & L. Lesher. 2000. Predictors of food acceptance, consumption and satisfaction in specific eating situations. *Food Quality and Preference* 11:201–216.

Chambers, E. & M.B. Wolf. 1996. *Sensory Testing Methods*, 2nd edn. West Conshohocken, PA: ASTM.

Chen, A.W., A.V.A. Resurreccion, & L.P. Paguio. 1996. Age appropriate hedonic scales to measure food preferences of young children. *Journal of Sensory Studies* 11:141–163.

Ennis, D.M. 1990. Relative power of difference testing methods in sensory evaluation. *Food Technology* 44(4):114–117.

Ennis, D.M. 1993. The power of sensory discrimination methods. *Journal of Sensory Studies* 8:353–370.

Epler, S., E. Chambers IV, & K.E. Kemp. 1998. Hedonic scales are better predictor than just-about-right scales of optimal sweetness in lemonade. *Journal of Sensory Studies* 13:191–197.

Gacula, M.C., S. Rutenbeck, L. Pollack, A.V.A. Resurreccion, & H.R. Moskowitz. 2007. The just-about-right intensity scale: functional analysis and relation to hedonics. *Journal of Sensory Studies* 22:194–197.

Hashim, I.B., A.V.A. Resurreccion, & K.H. McWatters. 1995. Consumer acceptance of irradiated poultry. *Poultry Science* 74:1287–1294.

Hautus, M.J., D. van Hout, & H-S. Lee. 2009. Variants of A-Not A and 2-AFC tests: Signal detection theory models. *Food Quality and Preferences* 20(3):222–229.

Hersleth, M., B. Mevik, T. Naes, & J. Guinard. 2003. Effect of contextual factors on liking for wine-use of robust design methodology. *Food Quality and Preference* 14:615–622.

Hersleth, M., O. Ueland, H. Allain, & T. Naes. 2005. Consumer acceptance of cheese, influence of different testing conditions. *Food Quality and Preference* 16:103–110.

Hough, G., I. Wakeling, A. Mucci, E. Chambers IV, I.M. Gallardo, & L.R. Alves. 2006. Number of consumers necessary for sensory acceptability tests. *Food Quality and Preference* 17:522–526.

IFT/SED. 1981. Sensory evaluation guideline for testing food and beverage products. *Food Technology* 35(11):50–59.

Jellinek, G. 1964. Introduction to and critical review of modern methods of sensory analysis (odour, taste, and flavour evaluation) with special emphasis on descriptive sensory analysis (flavour profile method). *Journal of Nutrition Dietetics* 1:219–260.

Kamen, J.M., D.R. Peryam, D.B. Peryam, & B.J. Kroll. 1969. Hewdonic differences as a function of number of samples evaluated. *Journal of Food Science* 34:475–479.

Kimmel, S.A., M. Sigman-Grant, & J.X. Guinard. 1994. Sensory testing with young children. *Food Technology* 48(3):92–99.

King, S.C., A.J. Weber, H.L. Meiselman, & N. Lv. 2004. The effect of meal situation, social interaction, physical environment and choice on food acceptability. *Food Quality and Preference* 15:645–653.

Kozlowska, K., M. Jeruszka, I. Matuszewska, W. Roszkowski, N. Barylko-Pikielna, & A. Brzozowska. 2003. Hedonic tests in different locations as predictors of apple juice consumption at home in elderly and young subjects. *Food Quality and Preference* 14:653–661.

Kroll, B.J. 1990. Evaluating rating scales for sensory testing with children. *Food Technology* 44(11):78–86.

Lawless, H.T. & H. Heymann. 1999a. *Sensory Evaluation of Food: Principles and Practices.* New York: Chapman & Hall.

Lawless, H.T., & H. Heymann. 1999b. Physiological and psychological foundations of sensory function. In: *Sensory Evaluation of Food,* 1st edn, pp. 28–74. New York: Chapman and Hall.

Meilgaard, M., G.V. Civille, & B.T. Carr. 2007. *Sensory Evaluation Techniques,* 4th edn. Boca Raton, FL: CRC Press.

Moskowitz, H.R. 1974. Sensory evaluation by magnitude estimation. *Food Technology* 28(11):16, 18, 20–21.

Moskowitz, H.R., A.M. Muñoz, & M.C. Gacula, Jr. 2003. *Viewpoints and Controversies in Sensory Science and Consumer Product Testing.* Trumbull, CT: Food & Nutrition Press.

Moskowitz, H. R., A.M. Muñoz, & M.C. Gacula. 2008. Hedonics, just-about-right, purchase and other scales. In consumer tests, in viewpoints and controversies. In: *Sensory Science and Consumer Product Testing.* Trumbull, Connecticut: Food & Nutrition Press, Inc.

Odesky, S.H. 1967. Handling the neutral vote in paired comparison product testing. *Journal of Marketing Research* 4:149–167.

O'Mahony, M. 1986. *Sensory Evaluation of Food.* New York: Marcel Dekker.

Peryam, D.R. & F.J. Pilgrim. 1957. Hedonic scale method of measuring food preference. *Food Technology* 11(9):9–14.

Pound, C. & L. Duizer. 2000. Improved consumer product development. Part one. Is a laboratory necessary to access consumer opinion? *British Food Journal* 102(11):810–820.

Resurreccion, A.V.A. & E.K. Heaton. 1987. Sensory and objective measures of quality of early harvested and traditionally harvested pecans. *Journal of Food Science* 52:1038–1040, 1058.

Resurreccion, A.V.A. 1998. *Consumer Sensory Testing for Product Development.* Gaithersburg, MD: Aspen Publishers.

Roessler, E.B., R.M. Pangborn, J.L. Sidel, & H. Stone. 1978. Expanded statistical tables for estimating significance in paired-preference, paired difference, duo-trio and triangle tests. *Journal of Food Science* 43:940–943.

Rothman, L. & M.J. Parker. 2009 Structure and use of just-about-right scales. In: *ASTM Manual Series MNL 63, eds. L. Rothman & M.J. Parker, pp. 1–13. West Conshohocken, PA: ASTM.*

Rousseau, B. & M. O'Mahony. 1997. Sensory difference tests: Thurstonian and SSA predictions for vanilla flavored yoghurt. *Journal of Sensory Studies* 12:127–146.

Schutz, H. 1965. A food action scale for measuring food acceptance. *Journal of Food Science* 30:365–374.

Stone, H. & J. Sidel. 2004. *Sensory Evaluation Practices*, 3rd edn. San Diego, CA: Academic Press, Inc.

Stone, H. & J. Sidel. 1985. *Sensory Evaluation Practices.* San Diego, CA: Academic Press.

10 What can sensory researchers do to characterize products? and...how does one select the best method?

Consumer acceptance of the food or food product can be quantified by using consumer affective tests. To translate consumer acceptance to product attributes, one must characterize this food, preferably using instrumental and physicochemical measurements, if there are methods that are known to validly characterize the sensory properties of the food. These tests are used to define the limits or range of product properties or attributes that correspond to acceptable products. Unfortunately, few of these instrumental and physicochemical methods exist, so in their absence, the product developer has to again depend on using sensory methods such as descriptive analysis ratings. This chapter describes how to go about establishing the relation between acceptance ratings against the descriptive analysis ratings or the physicochemical measurements. It describes how to develop mathematical models that can be used to predict consumer acceptance scores from descriptive analysis ratings or physicochemical measurements. Finally, the chapter discusses how to select the best method.

"OBJECTIVE" (i.e., INSTRUMENTAL AND PHYSICOCHEMICAL) MEASUREMENTS OF PRODUCT QUALITY

Product characterization may be conducted using various instrumental or physicochemical methods. When no instrumental or physicochemical test can be used to validly and reliably characterize an attribute, then sensory descriptive analysis methods are often used. When an instrumental test is available that accurately characterizes the sensory attributes of a product, it makes little sense to use a descriptive sensory panel to quantify the attribute. Instrumental and physicochemical measurements have been used to characterize attributes such as color, flavor, tastes, texture, and viscosity. There are literally hundreds of different instruments in today's market that can make these tests, so an exhaustive discussion of such objective measures cannot be done in a book of this type. Rather, we will just look at a few of the more popular methods. It is important to keep in mind that the methods are analogs to sensory assessment of product quality and product character. Indeed, many of the tests achieve their validation when they are shown to correlate with what the human judge perceives.

Sensory and Consumer Research in Food Product Design and Development, Second Edition.
Howard R. Moskowitz, Jacqueline H. Beckley, and Anna V.A. Resurreccion.
© 2012 Blackwell Publishing Ltd. Published 2012 by Blackwell Publishing Ltd.

To give a sense of today's technologies, instrumental color measurements of lightness, chroma, and hue angle can be obtained using a colorimeter, whereas texture can be quantified using measurements of shear, cutting force, work to cut, or parameters calculated from instrumental texture profile analysis using an Instron® universal testing machine/texture analyzer. A number of flavor attributes can be calculated from prediction equations using peak areas from gas chromatograms of headspace of known volatile compounds. Tastes can be quantified by analysis of chloride, sucrose, or acid. However, there are few sensory attributes that can be accurately predicted by instrumental and physicochemical measurements. In such cases, sensory descriptive analysis techniques are used to characterize critical attributes.

Objective measurements: Color

Appearance may be the single most important sensory attribute of food at the start of the purchase and consumption cycles. Decisions regarding whether or not to purchase and/or ingest a food are determined in great part by appearance. Color has also been shown to influence perception of sweetness (Clydesdale, 1991, 1993) and flavor intensity (Christensen, 1983). Tristimulus colorimetric methods have been developed that correlate highly with human perception of color (Clydesdale, 1978).

Objective measurements: Instrumental Texture Profile Analysis

Texture analysis has occupied a significant amount of research efforts, as engineers struggled with methods to record the physical properties that give rise to the complex texture percepts.

The Instron® Universal Testing Machine

Among the fundamental instrumental measures of texture, the place of honor goes to the Instron®. The Instron® Universal Testing Machine can be used to characterize textural properties of foods. The Instron® machine is sufficiently flexible to accommodate measurement cells that imitate the chewing process, and thus provide an instrumental analog to chewing and to the texture experience. Instrumental-based texture profile analysis, or TPA, can be made using the Instron® fitted with a compression cell with two cycles of deformation. The two cycles simulate the chewing action of teeth (Bourne, 1982). The instrumental properties of fracture, hardness, springiness, and cohesiveness, analogs to the human percept, are calculated from the force-deformation curves resulting from TPA. TPA curves have been generated on several products including pretzels and bread sticks (Bourne, 1982).

Objective measurements—Kramer Shear-Compression Test

The Kramer® machine provides a single integrated measure, rather than a profile of basic physical characteristics such as that provided by the Instron®. Thus, the Kramer system performs a single integrated test rather than providing a set of basic measures. The Kramer shear-compression test has been used for texture investigation of crisp foods such as potato chips, crunch twists, and saltine crackers (Seymour & Hamann, 1988). One cycle of deformation is used and the parameters of maximum force at failure, work done at failure, and force to shear are used to describe the food's texture attributes (Bhattacharya, Hanna, & Kaufman, 1986; Seymour & Hamann, 1988).

Objective measurements—snapping test

We can see another adaptation of fundamental texture measures by looking at snapping behavior. Sensory quality attributes such as crispness have been related to instrumental texture measurements. Kramer shear-compression test information is difficult to interpret and does not always mimic the action of biting into a sample for crisp foods. A 1-mm blunt blade attached to the Instron—with the sample placed on two bars sufficiently apart to prohibit friction during the deformation cycle—can be used to simulate the process. This test is called the snapping test. The snapping test has been used on crisp foods such as crackers (Katz & Labuza, 1981), crisp bread, potato crisps, ginger biscuits (Sherman & Deghaidy, 1978), snap cookies (Bruns & Bourne, 1975; Curley & Hoseney, 1984), and cream cracker biscuits (Gormley, 1987). The slope of the deformation curve is related to crispiness and fracturability.

DESCRIPTIVE ANALYSIS AND THE HUMAN JUDGE

It has long been recognized in the food and beverage worlds that machines can go just so far and no further in describing the sensory characteristics of a product. In lieu of these, the "subjective" human sensory response becomes critical for descriptive analysis. We concentrate in this section on how the human judge describes his or her sensory experience. The topics range from the use of the judge as an objective instrument to describe the nature of the perception to methods for quantifying perceived intensity.

Describing perceptions—panelists as descriptive instruments

We begin here with methods called descriptive analysis. The methods are formalized procedures whereby the panelists can record what they perceive and share these perceptions in a meaningful way with the panel leader and with other panelists who have been similarly trained. The information that the panelists provide when doing descriptive analysis often becomes a "fingerprint" of the product, used by product development to ensure ongoing quality and to determine whether the development is "on track."

Panelists are selected based on their demonstrated ability to perceive differences between test products, and their ability to communicate, that is, to verbalize perceptions (IFT/SED, 1981). This panel of judges is used as an analytical tool in the laboratory and is expected to perform like one (O'Mahony, 1991; Munoz & Civille, 1992). Descriptive analysis is a widely used technique in the sensory analysis of food materials. There are several standard techniques available such as: the flavor profile analysis (Cairncross & Sjostrom, 1950; Caul, 1957), the quantitative descriptive analysis (QDA) (Stone et al., 1974), the spectrum descriptive analysis (Meilgaard, Civille, & Carr, 2007), and the texture profile analysis (Brandt, Skinner, & Coleman, 1963; Szczesniak, Brandt, & Friedman, 1963).

Descriptive analysis testing is appropriate in several phases in the product development process. Descriptive analysis is needed to quantify the sensory attributes of a product. Panelists who have been trained in descriptive analysis are able to detect, identify, and quantify attributes of a food or product (Hootman, 1992). Thus, panelists who are qualified for discrimination and descriptive tests should not be used for acceptance testing regardless of their willingness to participate. The rationale is simple. An individual trained for descriptive analysis tends to have an analytical approach to product evaluation and, if used in the consumer panel, will bias the overall response required for the acceptance-preference task (Stone & Sidel, 2004).

Chambers and Wolf (1996) discuss descriptive analysis as the most common form of sensory testing, whereby the trained respondent is required to describe the product in terms of its characteristics and subjectively estimate the intensity of those characteristics by scaling procedures. Meilgaard, Civille, and Carr (2007) state that all descriptive analysis methods involve the detection and the description of both quantitative and qualitative sensory aspects of a product by trained panelists. In most cases, descriptive analysis is a sensitive method of testing that provides information regarding the sensory properties of the product that cannot be acquired through analytical means (Hootman, 1992).

Descriptive methods are based on: (1) panelist selection, (2) panelist training, (3) ability of the panelist to develop a descriptive language for the products being evaluated, (4) ability of panelists to provide reliable quantitative judgments similar to an instrument (calibration), and (5) analysis of data (Stone & Sidel, 2004).

Descriptive analyses in more detail

The descriptive tests can generate detailed information on aroma, flavor, and oral texture of the food or beverage being analyzed. Descriptive analysis can be classified into two aspects, qualitative and quantitative. In the qualitative aspect, the sensory parameters of the product are defined by terms such as attributes, characteristics, character notes, descriptive terms, descriptors, or terminology. These terms define the sensory profile or thumbprint of the sample. The quantitative aspect measures the degree or intensity of the identified characteristics from the qualitative analysis. This measurement is expressed through some value along a measurement scale (Meilgaard, Civille, & Carr, 2007).

Scales

Some common scales that are used in descriptive analysis are: category scales, linear scales, and magnitude estimation (ME) scales. Category scales are limited sets of words or numbers, where equal intervals are set between categories and the panelist "rates" the intensity of the product by assigning it a value (category) on the given scale. Recommended linear scales are 15 cm long with end points that are located at the ends or 1.25 cm away from each end; the panelist marks on the line his or her intensity rating of the product. The linear scale is also a popular scale because the product can be evaluated more accurately without the presence of "favorite numbers," but a disadvantage of linear scales is the difficulty to reproduce the results. Studies of the different scales focus on sensitivity (ability to pick up differences) and reliability (ability to provide reproducible measures). Galvez, Resurreccion, and Koehler (1990) compared unstructured line scales (ULS), semistructured line scales (SLS), and category scales on two samples of mung bean starch noodles, reporting that panelists discriminated better between noodles when they used the ULS method. They concluded that ULS was most reliable and most sensitive to product differences.

Scales—the special place of magnitude estimation

ME, also referred to as free number matching, is based on the approaches currently used by many experimental psychologists who look for the relation between physical intensity and sensory response. The literature of experimental psychology is filled with both methods and substantive results using this method. The research objective of ME is to create a scale that has ratio properties for subjective perception, on with ratio properties of physical measures. By doing so it is possible to understand at a scientific level how ratios of physical measures (e.g., forces in texture; concentrations of tastants in taste perception) co-vary with ratios of sensory perception.

The ME method allows the panelist to assign numbers so that the perceived ratio of the numbers "matches" the perceived ratio of the magnitudes of the stimuli. Panelists are free to choose any number that they wish. The first number is assigned freely, and the numbers that follow are assigned proportionally, to reflect perceived intensity. Moskowitz (1983) states four advantages of using ME, which include the following:

 (i) All fixed-point category scales contain arbitrary limits and ME eliminates these biases, which reduce scale sensitivity.
 (ii) ME has greater sensitivity than fixed-point scales in terms of revealing differences.
(iii) ME allows for a more meaningful correlation between the objective physical measure and the subjective magnitude. ME is a less biased measuring technique to reveal the function relating the two domains.
(iv) ME has captured the attention of many psychophysicists who use it in investigating sensory functioning. Meilgaard, Civille, and Carr (2007) state that ME is used in academic studies where a single attribute varies over a wide range of sensory intensities.

The panel

The descriptive analysis method requires that a panel be carefully selected, trained, and maintained under the supervision of a sensory analysis professional with extensive experience in using descriptive analysis (Einstein, 1991). The panel members must demonstrate motivation. The desire to participate is the first and most important prerequisite for a successful panel participant. Panel size ranges from 5 to 100 judges, where smaller panels are used for products on the grocery shelf, and larger panels are used for products that are produced in mass quantities (Meilgaard, Civille, & Carr, 2007). According to Stone and Sidel (2004), a panel has at least 10, but no more than 20 test subjects. Rutledge and Hudson (1990) and Zook and Wessman (1977) have published methods for selecting panelists.

Panel—selection criteria and approach
The first step in descriptive analysis, including preliminary screening, is the selection of the panel (Einstein, 1991). The selection of a panelist is based on at least four qualifications:

 (i) Availability of the panelist to attend several 2-hour sessions of training.
 (ii) Interest in the basic topic (flavor and odors).
(iii) Demonstrated taste (gustatory) and olfactory sensitivity.
(iv) Intellect.

In addition to the four foregoing qualifications, the test subjects are asked to participate in an interview, where they are asked about their interests, education, experience, and personality traits. Individuals who are too passive or dominant should be eliminated from the panel (Amerine, Pangborn, & Roessler, 1965).

The screening process is conducted in order to identify a group of panelists who will complete the training phase. A series of exercises should be designed, where candidates are shown samples that may be references or product (Einstein, 1991). The purpose of screening is to select candidates with basic qualifications such as normal sensory acuity, interest in sensory evaluation, ability to discriminate and reproduce results, and appropriate panelist behavior, which includes cooperation, motivation, and promptness (ASTM, 1981a). The results from the screening tests can provide the sensory practitioner with a sufficient amount of information about the prospective candidate (Einstein, 1991).

Panel—training
Einstein (1991) states that training comprises a carefully designed series of exercises that teach, practice, and evaluate the panelists' performance. The purpose of training panel members in sensory analysis is to familiarize an individual with test procedures, improve recognition and identification of sensory attributes in complex food systems, and improve sensitivity and memory so that precise, consistent, and standardized sensory measurements can be reproduced (ASTM, 1981b). According to Einstein, training is the key to successful employment of descriptive analysis. One of the main objectives in descriptive analysis training is to develop descriptive language, to be used as a basis for rating product attributes (Stone & Sidel, 2004). The training session is an opportunity for the panel leader to measure both individual and total panel performances. Communication within the group is critical in order to ensure that all attributes are understood and utilized in the same manner (Einstein, 1991).

Training makes a difference in performance. When comparing the performance of consumers and panelists who received 20 hours of training, Roberts and Vickers (1994) reported that consumers rated most attributes higher in intensity than did trained panelists. They found that trained panelists found fewer product differences than consumers did. Chambers, Allison, and Chambers (2004) compared the performance of descriptive panelists after short-term (4 hours), moderate (60 hours), and extensive (120 hours) training, and reported that panelists' performance increased with increased training. Panelists reported sample differences in all texture and some flavor attributes after only 4 hours of training. More differences were found after 60 hours, but panelists were able to ascertain differences in all texture and flavor attributes after extensive training for 120 hours. They concluded that extensive training may be required to reduce variation among panelists and increase their abilities to discriminate. Wolters and Allchurch (1994) likewise found that 60 hours of training increased the number of attributes that could be discriminated. Trained panelists can discriminate inconspicuous attributes better than untrained panelists and use a broader range of terms when describing a product's texture or Flavor Profile (Cardello *et al.*, 1982; Papadopoulos *et al.*, 1991).

Panel—reference stimuli
Rainey (1986) states that a reference standard is "any chemical, spice, ingredient, or product which can be used to characterize or identify an attribute or attribute intensity found in whatever class of products (whether food or nonfood, such as hot dogs or floor wax) is being evaluated by the trained panel." Generally, a reference standard should be simple, not complex.

References can be used as a tool in training a sensory evaluation panel because of at least seven reasons (Rainey, 1986). Specifically, references:

(i)   Help the panelist develop terminology to properly describe products.
(ii) Help in the determination of intensities and the identification of anchor end points.
(iii) Show action of an ingredient and the interaction of ingredients.
(iv) Reduce training time.

(v) Document terminology.

(vi) Provide identification of important product characteristics for the plant quality assurance program.

(vii) Assist the project team by providing important discussion tools to be used in new product development, product maintenance, product improvement, and cost reduction programs.

Test stimuli—the warm-up sample

A warm-up sample is a food sample given to a panelist before evaluation of test samples. In descriptive analysis, the warm-up sample is usually the control sample or one of the products being tested. Researchers use the warm-up sample in descriptive analysis in order to achieve more reliable results from the panel (Plemmons & Resurreccion, 1998). They found that a warm-up sample prior to test samples maximized reliability of descriptive analysis ratings, provided that the warm-up sample was used together with panel consensus attribute ratings. O'Mahony, Thieme, and Goldstein (1988) refer to the action of the "warm-up" as a short-term effect noted at the beginning of experimental sessions whereby performance rapidly improves, yet the effect rapidly dissipates after cessation of the activity.

Several methods for presenting warm-up samples in sensory analysis are cited in literature. Three methods for presenting the warm-up sample are:

(i) Assistance in panelist self-calibration, wherein panelists compare their responses to the consensus rating of intensity, for the warm-up sample (Malundo & Resurreccion, 1994; Hashim, Resurreccion, & McWatters, 1995; Chen, Resurreccion, & Paguio, 1996).

(ii) Present a reference sample first as a warm-up sample (Bett *et al.*, 1994).

(iii) Present a warm-up sample 5 minutes prior to the test to provide similar testing conditions for the first sample (Harper & McDaniel, 1993).

DESCRIPTIVE ANALYSIS TESTING METHODS

There are several testing methods for descriptive analysis. These include the Flavor Profile method, Texture Profile method, QDA method, and Spectrum Descriptive Analysis (Stone & Sidel, 2004; Meilgaard, Civille, & Carr, 2007). All of the methods have been developed in light of the fact that there are no basic or "primary" flavors or textures as there are primaries in color. As a consequence, researchers working in flavor and texture have had to create their own systems.

Flavor Profile® Method

The Flavor Profile® method was developed by Arthur D. Little, Inc., in Cambridge, Massachusetts. Cairncross and Sjostrom (1950) described the Flavor Profile® as a method of flavor analysis that makes it possible to indicate degrees of difference between samples on the basis of intensity of individual character notes, degree of blending, and overall amplitude, respectively.

The Flavor Profile® method is based on the organizing principle that "flavor consists of identifiable taste, odor, and chemical feelings factors plus an underlying complex of sensory impressions not separately identifiable" (Keane, 1992). According to Powers (1984a), the Flavor Profile® method is the most stringent sensory method in terms of length of time and effort required to train panelists to function as *human analytical instruments*. The Flavor Profile® involves a minimum of four assessors trained over a period of 6 months. The assessors evaluate the food or food product in a quiet, well-lit, odor-free room. It is suggested that a round table be used to facilitate discussion. Assessments take approximately 15 minutes per sample. One to three sessions are held. Assessors make independent evaluations and rate character note intensities using a 7-point scale from threshold to strong.

Texture Profile Method

The texture profile method was originally developed by the Product Evaluation and Texture Technology groups at General Foods Corp. in Tarrytown, New York (Munoz & Civille, 1992), to define textural parameters of foods. The method was developed to focus on aspects overlooked in the flavor profile method (ASTM, 1996). The method is described by Brandt, Skinner, and Coleman (1963) as the sensory analysis of the texture complex of a food in terms of its mechanical, geometrical, fat, and moisture characteristics; the degree of each present; and the order in which they appear from first bite through complete mastication.

The objective of the texture profile method is to: (1) eliminate problems dealing with subject variability, (2) compare results with known materials, and (3) establish a relationship with instrument measures (Szczesniak, Brandt, & Friedman, 1963). A moderator trained in texture profile analysis leads the analysis. A round table is used to facilitate discussion and evaluation. Assessors are trained on texture definitions, evaluation procedures, and standard reference scales. The assessors develop terminology, definitions, and evaluation procedures. Evaluation takes 5–15 minutes per product; a minimum of three replications is recommended. References are provided as needed and graphical rating scales are used.

Quantitative Descriptive Analysis (QDA) Method

The Tragon Corporation in Redwood City, California, developed the QDA method because of dissatisfaction among sensory analysts due to absence of statistical treatments using flavor profile and related methods (Meilgaard, Civille, & Carr, 2007). Statistical analysis is used heavily by QDA practitioners in order to select the appropriate terms, procedures, and panelists to be used for product analysis. Stone and Sidel (2004) described QDA as a method where "trained individuals identify and quantify, in order of occurrence, the sensory properties of a product or an ingredient. These data enable development of appropriate product multidimensional models in a quantitative form that is readily understood in both marketing and R&D environments." QDA is led by a descriptive analysis moderator. QDA requires 10–12 panelists, although in some tests 8–15 panelists may be involved. Training is best conducted in a conference-style room. During training, subjects develop terminology, definitions, and evaluation procedures. The training requires 2 weeks or approximately 8–10 hours. Products are evaluated in partitioned booths. Assessors take 3–20 minutes per product and a minimum of three replications are recommended. References are provided as needed. Fifteen-centimeter line scales are used in rating samples.

Spectrum Descriptive Analysis Method™

The Spectrum Descriptive method™ was designed by Civille. Modifications to the basic Spectrum™ procedure have been made over several years of collaboration with a number of companies (Meilgaard, Civille, & Carr, 2007). Spectrum™ was designed to provide a complete, detailed, and accurate descriptive characterization of a product's sensory attributes. The information received from this method of testing provides both qualitative and quantitative data. The method can be applied to many product categories including foods, beverages, personal care, home care, paper, and other products (Munoz & Civille, 1992). The Spectrum™ method is led by a moderator trained in the Spectrum Descriptive Analysis method™. Twelve to fifteen trained assessors are required. Assessors are trained in a room around a round table. Assessors develop the terminology, definitions, and evaluation techniques and agree on references to be used during the evaluations. The trained and calibrated panelists evaluate food in partitioned booths. Evaluation is approximately 15 minutes per product. Fifteen-point or 150-mm line scales are used.

Hybrid descriptive analysis methods

Descriptive analysis methods that employ a combination of some aspects of both QDA and the Spectrum Analysis methods are used by a large proportion of sensory practitioners (Einstein, 1991). These are called hybrid methods and are widely used in sensory descriptive analysis (Resurreccion, 1998).

MODELING RESPONSES TO THE PRODUCT, ESPECIALLY ACCEPTANCE

In the development of new food products, consumer acceptance of the food or food product is measured by affective tests. The characteristics of the product can quantified by sensory descriptive analysis ratings or instrumental and physicochemical measurements. The next question is whether there is a relation between what the product is sensed as being (from descriptive analysis), what the product comprises (in terms of ingredients, physical measures), and how well the consumer likes the product (from affective tests).

Mathematical modeling of relations between product characteristics and consumer acceptance are useful for at least three reasons:

  (i) *Learning*: What drives acceptance.
 (ii) *Optimization*: What combinations of factors generate a highly acceptable product.
(iii) *Quality control*: What combinations of features, either ingredients or sensory characteristics, define the product limits where consumer acceptance is within boundaries appropriate for maintaining product appeal in the marketplace.

The model or quantitative relation can be determined by plotting acceptance ratings against the descriptive analysis ratings or the physicochemical measurements. Mathematical models may be developed that can be used to predict consumer acceptance scores from descriptive analysis ratings or physicochemical measurements. These models comprise equations that can be used for predictive and control purposes

Modeling—Graphical Methods

Graphical methods are often used. In these approaches the data points are plotted in the appropriate coordinates (e.g., linear or logarithmically spaced). The overall acceptance ratings or attribute acceptance ratings are plotted on the ordinate against intensities (abscissa).

Typically each descriptor generates its own plot. The plot becomes the basis for setting quality specifications.

Munoz and Civille (1992) described a method to determine the nature of the relation between consumer acceptance and descriptive panel responses. Overall consumer acceptance of sensory attributes of color, flavor, and texture can be plotted against color, surface appearance, flavor, or texture attribute intensities obtained from descriptive analysis measurements or instrumental and physicochemical measurements. The relationships found, if any, may be linear or curvilinear.

An example of this type of analysis appears in Figure 10.1, which shows a curvilinear relation between overall acceptance and attribute intensity. When a decision on the lower limit of product acceptance is made by management, the product attribute specifications can be determined from plots of consumer acceptance versus attribute intensities similar to Figure 10.1.

This type of plot may be applicable to a wide variety of products and product attributes. Let us explicate the approach using the attribute of hardness. In this example, the product liked best was rated a 7.5 or like moderately on a 9-point scale, and the corresponding product was rated approximately 7.8 in texture profile analysis, which uses a 15-point scale for hardness. The graph indicates that the product will be likely used by consumers (minimum overall acceptance rating of 6) as long as the hardness intensity rating is between 4.95 and 10.9. Overall acceptance decreases with an increase or decrease in hardness. When mean hardness

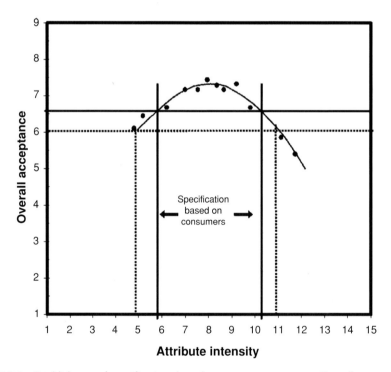

Figure 10.1   Establishment of specifications based on consumer response. Overall acceptance was measured using a consumer panel that rated product on a 9-point hedonic scale (1 = "dislike extremely"; 5 = "neither like nor dislike"; 9 = "like extremely"). Attribute intensity was determined from mean ratings by a descriptive panel using a 15-point (150-mm) line scale.

intensity ratings were 4.95 the mean overall acceptance of the product decreases to 6 or like slightly on a 9-point scale. When mean hardness intensity was rated 10.9, the overall acceptance of the product likewise decreased to 6.

There are important practical aspects emerging from this analysis, not just data-analytic methods. The company specification for product quality is now based on management's decision regarding the minimal level of overall acceptance of the product that can be tolerated. A number of companies may decide that they will only accept an overall acceptance greater than 6.5. In this case, the range of product hardness should be between 5.7 and 9.9 on a 15-point hardness scale. Should management decide to relax its standards and allow products with an overall acceptance of 6 into the marketplace, then hardness would range from 4.95 to 10.9. On the other hand, some companies may pride themselves on a high-quality product and allow only products with a mean rating of 7 or higher on the 9-point scale. In this second, more stringent case, the hardness intensity specification will narrow to the shorter range of 6.6–9 on the 15-point scale. This procedure, used to assess overall acceptance versus sensory attribute, requires the graphing of overall acceptance against a number of descriptive attribute intensities or instrumental and physicochemical measurements one variable at a time. Plotting can be tedious and time-consuming (Resurreccion, 1998), but today's computer programs may allow the plotting to be automated by a macro or a script.

Modeling—Descriptive/Predictive Equations

Mathematical models relating variables to each other may be developed using regression analysis. An example of a linear model is the following equation, relating acceptance to an independent variable X_1:

$$\text{Acceptance rating} = B_0 + B_1(X_1)$$

The acceptance score is the dependent variable, the independent variable is X_1. The intercept is B_0, which is the expected value of acceptance when $X_1 = 0$. Finally, B_1 is the regression coefficient (slope, rate of change of acceptance with unit changes in X_1). The independent variable may be the descriptive analysis ratings or instrumental and physicochemical measurements.

The relation between two variables may not necessarily be linear, which requires other forms of equations besides the simple linear equation. For example, Figure 10.2 shows a negative sloping, curvilinear relationship found between hedonic ratings for texture of snack chips and the energy/peak force needed to break a chip as determined by snapping test using the Instron universal testing machine (Ward, Resurreccion, & McWatters, 1995). The higher the energy/peak force to shear the snack chips, the lower the consumer texture acceptance rating was found to be. Conversely, the lower the energy/peak force the higher was the overall texture rating. The acceptance of the texture of snack chips can be related to energy/peak force to shear the snack chips using the nonlinear equation:

$$y = (0.38) \ln(x) + 0.97$$

Here, y is acceptance of texture and x is energy/peak force (Ward, 1995). Despite the nonlinear equation, however, the relation is monotonic. That is, there is no intermediate optimum or minimum; increases in the energy/peak force co-vary with decreases in texture liking, at least within the range tested.

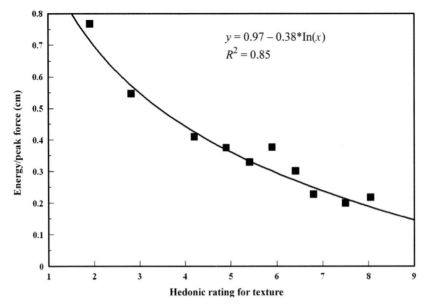

$$y = 0.97 - 0.38*\ln(x)$$
$$R^2 = 0.85$$

Figure 10.2 The relation of energy to peak force as the dependent variable versus the hedonic rating as the independent variable. The relation is shown with the hedonic rating as the independent variable, since in this case the goal is to identify a region of acceptable hedonic ratings, and in turn identify the Kramer values corresponding to that acceptance range. (E/PkF) as measured by Kramer shear. The overall texture acceptability of commercial and experimental snack chips was measured on a hedonic scale (1 = "dislike extremely"; 5 = "neither like nor dislike"; 9 = "like extremely"). Reproduced from Resurreccion, 1998. Figure 13.3, p. 215. Copyright 1998 by Aspen Publications, with kind permission of Springer Science and Business Media.

The foregoing nonlinear equation has use beyond simply describing the data. The equation summarizes how variables relate to each other. The equation allows the researcher to set boundaries on variables based on business criteria. The equation can be used to calculate the effect of energy/peak force on texture ratings; a high energy/peak force results in unacceptable texture in snack chip products and a low energy to peak force results in acceptable texture in snack chips. From this equation, an acceptance score of 7 (like moderately) or higher may be obtained in samples with an energy/peak force of 0.25–0.15 cm. If a product texture acceptance score of at least 6.0 (like slightly) is deemed minimally acceptable for the chip by management, then the equation tells the developer and the quality control inspector that the energy/peak force specification can be set accordingly at a maximum of 0.29 cm.

MODELING RESPONSE SURFACE METHODS

In the product development process, product optimization is an activity that is recommended to understand the relations between variables, and in turn to create the best product in its class. Optimization is a technique that can improve the efficiency of the research and development activity for both new as well as existing products (Schutz, 1983). According to Stone and Sidel (2004), optimization is a procedure used to develop the best possible product in its class. A widespread method used for optimization is response surface method (RSM), which is a useful statistical tool for investigation of complex processes (Chu & Resurreccion, 2004).

When the focus is on effect of two or more variables as they drive product acceptance and its interrelationship with sensory attribute intensities, RSM simplifies the process. RSM is combined experimental design and regression analysis. The objective is to predict the value of a response variable, or dependent variable, based on the controlled values of the experimental factors, or independent variables (Meilgaard, Civille, & Carr, 2007). The experimental design allows for the study of two or more variables simultaneously. The samples are evaluated by a consumer and descriptive panel and regression analysis generates the descriptive equations that will be used for prediction. From the parameter estimates, it can be determined which variable contributes the most to the prediction model, thereby allowing the product researcher to focus on the variables that are most important to the product acceptance (Schutz, 1983).

In RSM, the dependent variable is the acceptance rating; indeed, acceptance is the only subjective rating absolutely necessary to obtain optimal formulations. Contour plots of the prediction models allow the researcher to determine the predicted value of the response at any point inside the experimental region without requiring that a sample be prepared at that point (Meilgaard, Civille, & Carr, 2007).

RSM applications have enjoyed a long history in sensory research. The applications of RSM in a wide variety of areas including food research were reported as far back as 40 years ago by Hill and Hunter (1966). Optimization studies involving only food ingredients or formulations include those by Henselman, Donatoni, and Henika (1974); Johnson and Zabik (1981); Vaisey-Genser, Ylimaki, and Johnston (1987); Chow et al. (1988); and Shelke et al. (1990). In addition, RSM was used to optimize formulations of a peanut beverage (Galvez, Resurreccion, & Koehler, 1990), tortillas (Holt, Resurreccion, & McWatters, 1992), coffee whitener (Malundo & Resurreccion, 1993), peanut butter (Hinds, Chinnan, & Beuchat, 1994), extruded snack food (Thakur & Saxena, 2000), and reduced calorie fruit jam (Abdullah & Cheng, 2001). Studies used to optimize process variables and processing conditions were conducted on noodles (Oh, Seib, & Chung, 1985); lye peeling of pimiento peppers (Floros & Chinnan, 1988); mixing time, fermentation time, and temperature of Chinese steamed bread (Huang, Quail, & Moss, 1998); and roasting of hazelnuts (Ozedemir & Devres, 2000). Published literature involving both ingredient and processing variables include moisture content, cooking time, and holding temperature of mungbean noodles (Galvez, Resurreccion, & Ware, 1995); flour extraction, water content, frying time, and frying temperature of Puri, an Indian baked product (Vatsala, Saxena, & Rao, 2001); and ingredients and the roasting of peanuts in a chocolate peanut spread (Chu & Resurreccion, 2004); the potato dehydration process (Mudahar, Toledo, & Jen, 1990); and extrusion cooking (Bastos, Domenech, & Areas, 1991; Vainionpaa, 1991). In food formulation studies involving more than one ingredient, mixture experiments are adapted as opposed to factorial designs (Hare, 1974).

Examples of RSM technology abound. Let us look at two product development examples, for roasted low-fat peanuts and for chocolate spread. We will present the examples in detail, so that the reader can get a sense of just what is entailed in thinking about the RSM problem, how the study is designed and executed, and how the data are analyzed. It's not important to understand the specifics, but rather to get a sense of the process here, and what RSM can provide to the researcher and developer.

Using RSM to develop a product—example 1: roasted low-fat peanuts

The objective here was to develop specifications for an acceptable roasted, salted peanut with decreased oil content.

The experimental design

Three factors, percent salt, oil content, and degree of roast (color lightness, L, value), were varied to create an array of products of different physical, sensory, and presumably acceptance characteristics:

(i) Average degrees of roast = Hunter color lightness (L) values of 45, 50, and 55.
(ii) Salt levels = 1.5%, 2.0%, and 2.5% by weight.
(iii) Oil levels = 24%, 34%, and 47% by weight. The oil levels were not equidistant; the lowest level was 10% rather than 13% below the midpoint level due to the difficulty in further removal of oil in the peanuts by mechanical pressing.
(iv) The full-factorial design resulted in 27 treatments (3 oil levels × 3 degrees of roast × 3 salt levels = 27).
(v) It is important to note here that there are many different strategies for dealing with these types of design issues—one might use only a partial set of the 27 combinations (so-called fractional factorial designs), and one might use either two levels, three levels, or even more levels. Three levels are the simplest, when the researcher expects the relation to be nonlinear, with an optimum at a middle range, and when one suspects an interaction among the variables as drivers of the ratings.

Descriptive tests

Descriptive analysis tells the researcher about the panelist's perception of the product. As noted previously, the descriptive tests are not necessary to optimize liking. However, descriptive analysis provides a sensory signature of the different products in the RSM test. That sensory signature can be incorporated, leading to greater understanding of the effects of the independent variables on the sensory attributes of the prototypes:

(i) Panelists evaluated 27 roasted, salted, defatted peanut samples, prepared in three processing replications.
(ii) One processing replication was evaluated each day, over 3 days in a complete, balanced block design. Attribute intensities were rated using continuous line scales (Stone, 1992) on a computer score sheet.
(iii) Cluster analysis (CA) was used to evaluate performance of trained panelists and determine outliers (Malundo & Resurreccion, 1992), whose results were deleted from the data prior to analysis.

Consumer test

The consumer test provides the key data on which to do the optimization:

(i) A CLT (central location test) was conducted using 27 treatments and a replication with a block size of 54 (SAS Institute, 1980). Design parameters incorporated 13 consumer responses per treatment per block, for a total of 26 responses per treatment between two processing replications.
(ii) In a balanced, incomplete block design each consumer evaluated 5 of 27 samples from one replication. A total of 142 consumers were required for the study, to accommodate the different samples, given the requirement that each consumer could evaluate only 5 of the 27 samples.

(iii) The consumers assigned 9-point hedonic ratings for overall acceptance and then for the individual acceptability level of color, appearance, flavor, and texture. This type of data generates an "acceptance signature" of each product.

(iv) Participating consumers were females, as women were the primary household shoppers and there was no quota set for gender. The median age range was 35–44, and median household income was in the $40,000–$49,000 range. The race distribution was 73% white and 27% other races. About 50% of panelists consumed snacks one to three times per day. Typically in such acceptance tests it is important to record the information about the consumers in detail because in future corporate development work one wishes to repeat these types of studies to see how far the development has progressed. Unlike sensory attributes, which are presumed to remain more or less constant across people (so the variation across people is really due to unexplainable error), people of different ages, genders, shopping habits, income, and race might vary in what they like.

Looking for quantitative relations by regression modeling

After responses of one panelist found to be an outlier were deleted, scores of the remaining 10 panelists were analyzed by regression modeling to create a set of equations describing how the independent variables "drive" the individual acceptance rating. Overall liking, liking of appearance, color, taste, and texture generate separate equations.

A quadratic response surface model was fitted to the data using response surface regression analysis. The model or equation summarizes the relation between the ratings and the independent variables (fat, degree of roast, and salt studied) (Freund & Littell, 1986).

The regression program fits what is known as a simple polynomial model. The model included all linear and quadratic terms in the individual independent variables and all pairwise cross products of linear terms. Not all of the variables in the full models from the previously mentioned model were statistically significant. That is, one can incorporate terms in the model (e.g., squares or cross terms) that are actually "there" but really have no effect as predictors. In such cases, the appropriate action is to remove these extraneous terms. One can use them, but they do nothing and removing them makes the model or equation more parsimonious, stronger, and thus better. A subset of variables determined to contribute significantly to the models were selected. Response variables that resulted in significant regression models at $p \leq 0.05$ and a coefficient of determination (R^2) greater than 0.20 were selected as contributing to, and included, in the predictive model.

When a significant interaction term involved an insignificant linear term, the linear term was retained in the predictive model. Multiple regression analysis was used to finalize the models after significant variables were selected according to the procedure just described. Parameter estimates and coefficients of determination for prediction of overall liking and liking of color, appearance, flavor, and color of the defatted roasted peanuts are shown in Table 10.1 on the next page.

Graphically representing equal-intensity contours

Researchers who work with RSM have become accustomed to seeing two-dimensional plots showing how two of the independent variables combine to generate the dependent variable. When the study deals with more than two independent variables, as this study does, all but two of the independent variables are held constant. The two variables being studied are then used in conjunction with the equation. The dependent variable is set at a

Table 10.1   Parameter estimates and coefficients of determination (R^2) for independent variables used in prediction models for consumer acceptance of roasted, salted, defatted peanuts ($n = 142$)[a].

Overall liking	Liking attribute			
	Color	Appearance	Flavor	Texture
Independent variables in model				
Intercept	−21.23	−17.72	−15.49	−60.32	−17.14
Oil	0.63	0.54	0.49	0.71	0.52
Roast	0.45	0.40	0.33	1.93	0.35
Oil × Roast	−0.01	−0.0089	−0.0074	−0.0014	−0.0077
Roast × Roast	−	−	−	−0.014	−
R^2	0.2837	0.2297	0.2988	0.3249	0.3067

Source: Resurreccion, 1998. Table 13.1, p. 220. Copyright 1998 by Aspen Publications, reproduced with kind permission of Springer Science and Business Media.
[a]Factors are: Oil = oil content of full fat or defatted peanut; Roast = degree of roast (color lightness, L). Percent salt was not a significant factor in predicting acceptance. Panelists evaluated samples using a 9-point hedonic scale where 1 = dislike extremely and 9 = like extremely.

great number, and the computer program generates thousands of combinations of the two independent variables that, according to the equation, combine to generate the fixed level of the dependent variable. These thousands of points, in turn, generate a contour where the dependent variable is constant—thus giving rise to the so-called equal intensity (i.e., equal response-level) contour.

Contour for equal levels of consumer acceptance

The contour plots presented in Figure 10.3 (page 299) were generated using the prediction models for liking of texture (Figure 10.3a), flavor (Figure 10.3b), color (Figure 10.3c), appearance (Figure 10.3d), and overall liking (Figure 10.3e) using the Surfer Access System Program® (version 4.15, Golden Software, Inc., Golden, CO). In this study, the limit for product acceptance was set at a rating of at least 6 (like slightly) or higher on a 9-point hedonic scale for all attributes.

Interpreting the results of the RSM study

RSM in business applications searches for combinations of ingredients or process conditions that generate a desired (i.e., meaningful) acceptance level. Following this organizing principle, let's interpret the results shown by Figure 10.5. We have shaded all combinations of oil and degree of roast on the plots that would result in products with acceptance ratings equal to or greater than 6. Recall that the rating of "6" is just "liked slightly."

When ratings corresponding to 6 or higher for overall liking are used to *define a new product*, the data instruct us to consider as possible products those combinations of oil level from 48% to 39.1% oil at a degree of roast (color lightness, L) from 44 to 58, as well as all samples in the shaded area between 39.1% oil and L of 44, to 29.2% oil and L of 56. Of course, the prudent analysis would be to test a number of these combinations, simply to ensure that the model is correct. The model summarizes the data and is meant only to be a guide to development, rather than being the last step in development.

The researcher learns a great deal from modeling. The foregoing discussion dealt with liking. However, there is no reason to exclude the sensory attributes from modeling. When

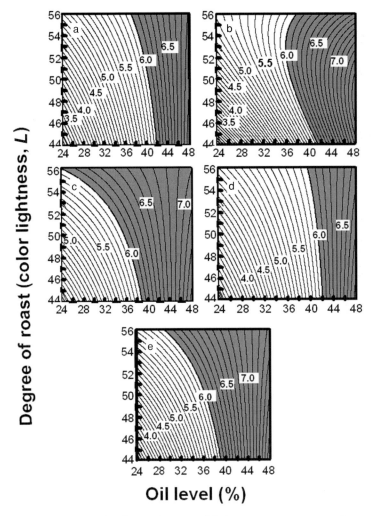

Figure 10.3 (a)–(e) Contour plots for prediction models for consumer acceptance for treatments with varying oil and roast levels. The five sets of contours are: a = texture, b = flavor, c = color, d = appearance, and e = overall liking. Ratings are based on a 9-point hedonic scale with 1 = "dislike extremely"; 5 = "neither like nor dislike"; and 9 = "like extremely." Shaded regions represent a hedonic rating of 6 ("like slightly") or greater for each attribute (Plemmons, 1997; Resurreccion, 1998, Figure 13.4, p. 221. Copyright 1998 by Aspen Publications, reproduced with kind permission of Springer Science and Business Media).

the researcher includes all of the attributes, sensory, and liking, the set of equations becomes a full product model that can estimate the likely acceptance and sensory profile of different attributes, telling the researcher what the consumer or expert panelist is likely to perceive for any particular combination. For this particular study, other sensory attributes (not shown) include brown color, roasted peanutty, raw/beany, burnt, astringent, oxidized, saltiness, and bitterness, respectively. It is worth noting that once panelists differentiate among products based on sensory characteristics, which can be very easily incorporated into the product model with today's PC-based regression techniques available in most statistical packages (see Table 10.2).

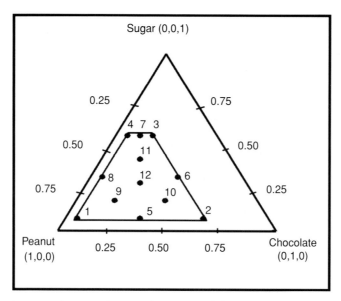

Figure 10.4 Constrained mixture design and experimental design points used at each level of roast. Ranges for each mixture component were as follows: peanut (x1) 0.25–0.90, chocolate (x2) 0.05–0.70, sugar (x3) 0.05–0.55. Experimental points were determined by the extreme vertices method (McLean & Anderson, 1966). Additional points selected for the design were the midpoints of each edge, the centroid of the constrained area, and three additional points in the center of the figure (Gacula, 1993). Points are labeled with numbers, indicating formulations studied. Values in parentheses indicate proportions of peanut, chocolate, sugar, respectively, where 1 equals 100% and 0 is equal to 0% (Chu & Resurreccion, 2004. Food & Nutrition Press, Inc., reprinted with permission from Blackwell Publishing Ltd.).

Table 10.2 Predicted attribute scores from descriptive analysis regression equations for characterization of roasted, salted, defatted peanuts of acceptable quality[a].

Experimentally varied factors	A	B	C
Oil (%)	39.1	29.2	32.0
Degree of roast (color lightness, L)	44.0	56.0	54.0
Salt (%)	2.0	2.0	2.0
Sensory ratings			
Brown color	81.0	56.8	59.9
Roasted peanutty	46.9	35.7	36.7
Raw/beany	19.4	25.6	22.4
Burnt	35.4	24.8	40.8

Source: Resurreccion, 1998. Table 13.3, p. 223. Copyright 1998 by Aspen Publications, reproduced with kind permission of Springer Science and Business Media.
A = minimum oil and dark degree of roast for peanuts with a hedonic rating of 6 (like slightly) on a 9-point scale for overall acceptance; B = minimum oil and light degree of roast for an acceptable ($\geq$6) product; C = oil level and degree of roast for an acceptable ($\geq$6) product with one-third less oil than full fat peanuts at 48% oil.
[a]Intensity scores are based on a 150-mm unstructured line scale. Prediction models are: brown color = $10.16 + 6.32X_1 + 0.33X_2 - 0.074X_1X_2 - 0.042X_1X_1$; Roasted peanutty = $30.18 + 0.11X_1 + 0.64X_2 - 0.037X_1X_2 + 0.031X_1X_1$; Raw/beany = $43.96 - 1.78X_1 - 0.22X_2 + 0.019X_1X_2 + 0.015 X_1X_1$; Burnt = $330.15 - 0.56X_1 - 9.80X_2 - 4.086X_3 + 0.08X_2X_3 + 0.083X_2X_2$.

Table 10.3   Parameter estimates and coefficients of determination (R^2) for factors used in prediction models for descriptive analysis of roasted, salted, defatted peanuts[a]

	Parameter estimates							
Factors	**Brown color**	**Roasted peanutty**	**Raw/ beany**	**Burnt**	**Astringent**	**Oxidized**	**Saltiness**	**Bitterness**
Intercept	10.16	30.18	43.96	330.15	57.98	48.16	36.20	42.02
Oil	6.32	0.11	−1.78	−0.56	−0.17	−0.25	−0.65	−0.19
Roast	0.33	0.64	−0.22	−9.80	−0.44	−0.20	−	−0.33
Salt	−	−	−	−4.09	−	−	−	−0.72
Oil × roast	−0.07	−0.04	0.02	−	−	−	−	−
Roast × salt	−	−	−	0.08	−	−	−	0.03
Oil × oil	−0.04	0.03	0.02	−	−	−	0.02	−
Roast × roast	−	−	−	0.08	−	−	−	−
R^2	0.5463	0.2412	0.2325	0.3247	0.0993	0.0976	0.1874	0.1920

Source: Resurreccion, 1998. Table 13.2, p. 222. Copyright 1998 by Aspen Publications, reproduced with kind permission of Springer Science and Business Media.
oil = oil content of full fat or defatted peanut; roast = degree of roast (color lightness, L); salt = percent salt.
[a]Regression models for each attribute were significant at $p < 0.05$.

Parameter estimates and coefficients of determination for factors and variables used in attribute intensity prediction models for the roasted, salted, defatted peanuts were determined from descriptive analysis data as presented in Table 10.3. Variables from the full regression models contributed significantly at $p \leq 0.05$ to eight predictive models, for brown color, roast peanutty, raw/beany, burnt, astringent, oxidized, saltiness, and bitterness attributes.

The strategy to use the model for other attributes is straightforward, following these steps:

(i) Ratings were set to a minimum of 6.0 for overall liking.
(ii) Maximum and minimum acceptable oil levels were 39.1% with a dark degree of roast ($L = 44$) and 29.2% with a light degree of roast ($L = 56$).
(iii) These levels were substituted into the regression models for brown color, roasted peanutty, raw/beany, and burnt attributes.
(iv) Their predicted attribute intensity ratings from descriptive analysis prediction equations for an acceptable product were calculated. They are presented in Table 10.3.
(v) Sample A represents a product with the minimum oil level at the darkest roast level ($L = 44$) for an acceptable quality roasted, defatted, salted peanut.
(vi) Sample B represents a product with the minimum oil level at the lightest roast level ($L = 56$) for an acceptable quality product.
(vii) Sample C represents a reduced oil level peanut with one-third less oil.
(viii) To generate an acceptable product (overall acceptance 6), peanuts with 32% oil must have a degree of roast resulting in a color lightness, L, no less than 54.
(ix) Factor levels (i.e., levels of the three independent variables) for products A and B were substituted into the prediction equations for the sensory attributes brown color, roasted peanutty, raw/beany, and burnt attributes.
(x) Results indicated that these defatted, roasted, salted peanuts had a brown color not too light or dark (descriptive score 56.8–81.0), a roasted peanutty intensity of 35.7–46.9, and a raw/beany and burnt intensity that is only slightly detectable (descriptive score 19.4–25.6, and 24.8–35.4, respectively.)

(xi) The attribute intensity scores for product C fall within the range of descriptive scores from samples A and B.

(xii) The bottom line: Attributes of acceptable defatted, roasted, salted peanuts were quantified through prediction equations. RSM allowed the product developers to quantify the effect of multiple factors important in formulating a quality roasted, defatted, salted peanut simultaneously and in considerably less time than required to graph acceptance scores versus descriptive attribute scores individually.

Using RSM to develop a product—example 2 chocolate peanut spread (a mixture)

In this second example, we deal with a different type of problem, albeit one that is just as common for optimization. We deal with the so-called mixture problem. All of the components are interrelated because the total must add up to a constant amount, typically expressed in percents.

In this second example, RSM was used to formulate process, analyze, and then optimize a chocolate peanut spread. The objective was to achieve high degree of overall acceptance to consumers for appearance, flavor, texture, and spreadability (Chu & Resurreccion, 2004). The key difference between the spread example here and the previous roasted low-fat peanuts is that now our components must add up to 100%. That is, the components can be separately varied, but not completely. Knowledge of the percent of any two of the three components automatically defines the level of the third component in the spread, since the sum is, by definition, 100%.

Experimental design

The mixture design that we used is known as a three-component constrained, simplex lattice design. Statisticians have developed the appropriate layouts; one doesn't have to know how to create them, but merely where to look. In our particular study, 36 formulations are shown in part schematically in Figure 10.4 (page 301).

The design comprised an additional process variable (degree of roast varying from light to medium to dark) along with varying levels of three ingredients. The ingredients do not vary from 0% to 100%, but rather vary within appropriate ranges. The design ensures that they add to 100%, and that each ingredient remains within its allowable range. The three variables were as following:

(i) Peanut 25–90%
(ii) Chocolate 5–70%
(iii) Sugar 5–55%

Consumer test

The consumer evaluation obtains judgments from 60 consumers, sufficient in these types of experimental designs to scale acceptance and provide sufficiently good data for modeling. Using the 9-point hedonic scale, the consumers evaluated overall acceptance, as well as the acceptance of the following sensory attributes: spreadability, overall, appearance, color, flavor, sweetness, and texture/mouthfeel. This type of protocol provides a rich data set for modeling because it deals with acceptance across all the relevant senses.

Statistical analysis and generation of contour plots

The key analysis here again is regression modeling. The model creates an equation for each rating attribute, using the statistically significant terms (percent each of peanut, chocolate, sugar). Here, however, we will go a step further, to create contour plots for each attribute, at each level of roast. These appear in Figure 10.5. Finding the areas of overlap, for high acceptance and any roast level is doable, by superimposing the plots (Figure 10.6).

Results

How do we interpret these results? The major question here is "how do we use the data to create a better product?" The peanut spread is just an example of what can be done.

The equal-acceptance contours for the mixture design are similar in nature to the equal-acceptance contour for the previous study on roasted peanuts. The key difference is that knowing two of the three formulation levels automatically define the level of the third variable, because the variables add up to 100%. Optimum formulations of chocolate peanut spread are represented by areas of overlap of regions of consumer acceptance of 6.0 for all attributes. These are all combinations of 29–65% peanut, 9–41% chocolate, and 17–36% sugar, adding up to 100%, at a medium roast.

In all of the work in RSM, whether regular combinations or mixture designs, it's important to remember that we are dealing with both empirical data and models. Our models come from the empirical data. The patterns in the data point to additional product formulations. Those formulations remain conjectures, until tested in what is known as a verification or validation phase. The happy new here for the peanut spread is that the verification phase reaffirmed the expected high performance of the optimum. More importantly, however, with RSM, strongly based in empirical data, there is generally good agreement between the predictions and the verification, as long as the predicted levels lie within the ranges tested, and reasonably near other products that were actually tested in the study. One runs into trouble when the verification products lie away from the set of products tested, or at the edges of the formulation levels.

What does the research learn from RSM testing?

It should become increasingly clear that RSM requires the researcher to do "homework," but in the end the researcher learns a great deal about the dynamics of the product. From the two RSM cases shown the researcher understands what variables make a difference to the consumer in terms of acceptance, how the product will score on attributes (either sensory attributes or liking of sensory attributes), and what levels correspond to the best product.

Yet, the analytic approaches may, in this case, seem like "overkill." Is there a better way? With the different products and with the extensive modeling, the natural question to ask is, "Why not select the best-performing product?" The objective of RSM is not to create the best product by the experiment, however, but rather to learn about the dynamics of the product as consumers perceive it. The different combinations are simply locations in the "ingredient space" or "processing space" that can be used to create the model. It is important to model the dynamics of the product, both to understand the product at the early stages of development but also to understand what is important. Such understanding becomes very useful when the time comes to change the product to accord with new business requirements, while at the same time having the task of maintaining product integrity.

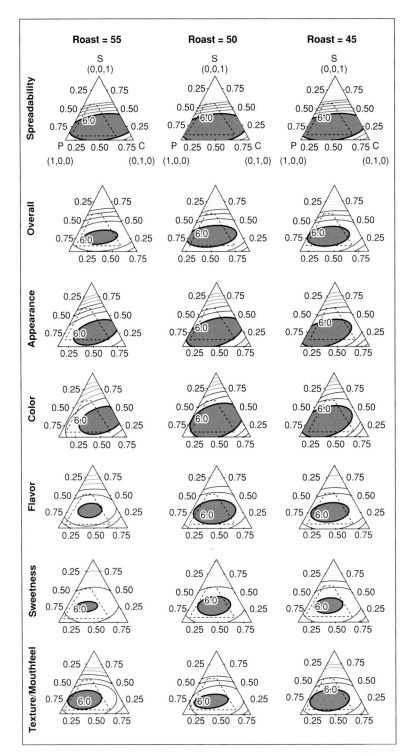

Figure 10.5 Contour plots for spreadability, overall acceptability, appearance, color, flavor, sweetness, and texture/mouthfeel at light, medium, and dark roasts. The shaded areas indicate maximized consumer acceptance (consumer rating = 6.0) rating for each sensory attribute. Values in parentheses indicate proportions of peanut (P), chocolate (C), sugar (S), respectively, where 1 equals 100% and 0 is equal to 0% (Chu & Resurreccion, 2004. Food & Nutrition Press, Inc., reprinted with permission from Blackwell Publishing Ltd.).

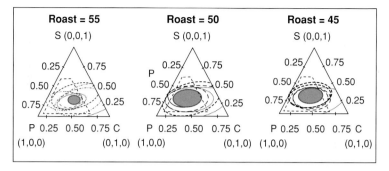

Figure 10.6 Regions of overlap on contour plots for all seven attributes at each level of roast, light, medium, and dark. Formulations obtained by superimposing contour plots, resulting in areas of overlap for prediction models for consumer acceptance ratings for chocolate peanut spread with treatments of varying levels of peanut (P), chocolate (C), and sugar (S) at a light (Hunter L-value 55 ± 1.0), medium (Hunter L-value 50 ± 1.0), and dark (Hunter L-value 45 ± 1.0) roast, for spreadability (- - - -), overall acceptability (●●●●), appearance (- - - ●), color (-●-●-), flavor (– – –), sweetness (—), and texture/mouthfeel (-●●-). Shaded regions represent a hedonic acceptance rating of at least 6.0 (= "like slightly") or greater for all attributes. Hedonic ratings are based on a 9-point scale with 1 = "dislike extremely"; 5 = "neither like nor dislike"; and 9 = "like extremely." Values in parentheses indicate proportions of peanut, chocolate, and sugar, at the vertices of a three component mixture design where 1 equals 100% and 0 is equal to 0% *Source:* Chu, C.A., and Resurreccion, A.V.A. 2004. Optimization of a Chocolate Peanut Spread Using Response Surface Methodology, *Journal of Sensory Studies,* Food & Nutrition Press, Inc., reprinted with permission from Blackwell Publishing.

DATA RELATIONS, CORRELATION, AND REGRESSION

We end this chapter with a short tutorial on statistical methods. Over the past decades statistical procedures have become of increasing import to the sensory world. The business use of statistics ranges from simple tests of inference (e.g., the T test for differences), onto studies of relations between variables (e.g., correlation and regression analysis), and on to mapping of the data and reducing the data into a form from which insights can be gathered.

Looking at the literature might lead the cynic to say that a lot of sensory analysis and consumer research is nothing more than esoteric statistics, brought to bear on relatively simplistic problems—such as what factors drive acceptance. In actuality, however, statistical analysis can be a boon to researchers who must synthesize patterns from the mass of subjective data and separate the patterns from the noise. Of course, it is important to be on one's guard not to let the statistical process overtake the rest of the effort, so that the researcher becomes a slave to what is *au courant*, what is esoteric, what is considered to be "powerful, state-of-the-art methods." As long as the statistical analysis is done to help thinking, not to replace thinking, the sensory researcher will be able to use these analyses for guidance. It is when the methods themselves run amok, with mapping and other high-level analyses spewed forth from all too easily used computer programs, with little understanding and even less effort, that statistics will become a problem.

Armed with those cautionary words, let's now look at some of today's statistical armory.

First . . . a note on probability

A lot of the following analyses talk about the "significance" of relations or decisions. Significance just means the proportion of times one makes the same qualitative decision when the study is repeated. The standard probability limit for sensory tests should be 0.05

or less, meaning that, if the test were to be repeated exactly the same, we would make the same decision in the proportion 1–0.05 or 0.95. That is, running exactly the same test would lead to the same answer 95% of the time. We're pretty confident that our measurements are robust and correct.

But, what about a probability of 0.33? This means that running the exactly same test would lead to the same answer only 67% of the time. We're not quite as confident. And so forth.

Type I and Type II errors in a layperson's terms. A standard probability value of 0.05 is used by most in hypotheses testing by industry practitioners. Many sensory professionals do this and support this practice blindly. The choice of the p-value (Type I error) is based on (1) acceptable risk and (2) repeatability of results. When risk and repeatability are not an issue, a higher p-value such as 0.10 or 0.15 can be used to provide "directional" information. Conversely, a higher p-value, such as 0.01, is used when the situation has high risk and repeatability of results is needed. The sensory professional should be able to explain and justify his or her choice of probability value for different test objectives (Moskowitz, Munoz, & Gacula, 2003).

Analytical methods: uncovering specific relations between variables—correlation analysis

Correlation analysis is a commonly used statistical technique to determine whether two variables are related. The nature of the relation is generally, but not always linear. A typical example is using correlation analysis to determine which individual descriptive attributes co-vary with overall liking. Correlation analysis may likewise be useful in searching for correspondence between consumer terms and descriptive analysis terms (Popper, Heymann, & Rossi, 1997).

The correlation coefficient, r, is a summary statistic, a single value that is used to determine the degree and the significance of the relation between two variables. Any time a correlation coefficient is calculated, it is an additional "good practice" to plot the data points in a scatterplot. The plot might show that the relation is not linear, but rather the relation is curvilinear. The linearity might have been assumed because there is a general linear trend, despite the curvilinearity (Jones, 1997).

Often, a correlation analysis is erroneously interpreted as a cause-and-effect relationship. A high correlation coefficient and a low probability that the correlation occurred by chance does not establish a causal relationship (Jones, 1997). To summarize, the correlation statistic conventionally shows the existence of, but not the quantitative nature of, a linear relation between two variables. There are methods to deal with nonlinear relations, but in such cases regression analysis rather than correlation analysis is appropriate.

The typical correlation statistic is the Pearson product moment correlation. For most people this is known as the correlation coefficient and is usually represented by the symbol r. Correlation coefficients (r) lie between the values of 1.00 and 1.00. A high absolute value indicates a high degree of relatedness, whereas the sign denotes whether the relation is positive ($+$) or negative ($-$).

Simple regression analysis

Regression analysis relates one or several independent variables to a dependent variable. Regression analysis is often used both for descriptive purposes (show the nature of the relation) and for predictive purposes (estimate the likely dependent variable corresponding to a specific level of the independent variable). When graphical representation of the observed data suggest a nonlinear response, regression using a polynomial or quadratic model may be appropriate.

The out from regression analysis is frequently reported in analysis of variance tables (Jones, 1997). Regression programs calculate the F statistic. From the value of the F statistic, and using appropriate statistical tables, it is straightforward to determine the statistical significance of the regression, which is the "equation" or "model" relating the independent and dependent variables. Regression analysis also generates the coefficient of determination (R^2). This statistic shows the percent of the variation in the dependent variable that can be accounted for by knowing the levels of the independent variable and using the regression equation to estimate the expected value of the dependent variable.

Residuals

The residuals are defined as the difference between the estimated or predicted value and the actual or obtained value, when one uses the regression equation to estimate. When one plots the residuals, they should look like a random distribution of points. A nonrandom distribution means that errors associated with poor fit of the regression may be due to a systematic effect, such as lack of a higher order term or a need to transform the data prior to generating an equation (Rothman, 1997).

Validating the regression model

Validation is an important step to establishing whether or not the equation emerging from the regression model is truly predictive, or whether the equation is merely a summary of the data. Validation can be done in several ways. For large data sets, validation can be accomplished by using a subset of the data. This is called cross-validation; the ability to estimate the value of hold-out samples. For small data sets, typically based on experiments, validation can be carried out using a new data set (Rothman, 1997).

MULTIVARIATE METHODS—REDUCING MANY VARIABLES INTO SIMPLER STRUCTURES

The methods distill many variables or many stimuli into simple structures that can be quickly understood and that allow relations to emerge. Multivariate methods have played a major role where assessments of quality are made against the background of all the other qualities of the food (Powers, 1981).

Multivariate analysis can help us understand the underlying properties that are measured in evaluating the quality of food and help establish which variables are determinants of food quality (Resurreccion, 1998). Multivariate methods are also suited for studying the data relationships between consumer evaluations and descriptive analysis results or physicochemical measurements (Resurreccion, 1988, 1998).

Within the last 25 years, there has been a remarkable expansion in the use of multivariate statistical methods to examine sensory data (Powers, 1984b). A lot of the expansion can be traced to the power of the personal computer, which has put multivariate analysis in the hands of anyone with a statistical package. In addition, as a generation of students has grown up with the widespread availability of these packages, multivariate analysis has evolved from a set of esoteric procedures to methods that just about any investigator can use easily and comfortably.

Principal components analysis

Principal components analysis (PCA) is a statistical technique used to extract structure from the variance-covariance or correlation matrix (Federer, McCulloch, & Miles-McDermott, 1987). PCA is often used to help interpret relations in data (Popper, Heymann, & Rossi, 1997). Computationally, PCA constructs linear combinations of the original data to account for the maximal variance in the data. These linear combinations are "mathematical primaries," having statistical meaning. Each linear combination is set up so that the set of newly constructed principal components is statistically independent of the others.

The sensory analyst uses PCA to reduce the number of variables to a smaller number of components with little loss of information. Example of the variables analyzed by PCA include: flavor and/or texture attributes, consumer acceptance attributes, physicochemical measurements, and ingredient levels, respectively. A simple correlation table may show that several groups of variables are related to each other more closely than they are to any other variable or set of variables. The mutually correlated variables would be located in the same principal component.

One way to visualize this concept is to think of the data as a cloud of points scattered in a three-dimensional space (Resurreccion, 1998). The axis through which most of the points in the cloud lie is the first principal component. The equation of the axis will consist of those variables that do the most to reduce the variability of the cloud. The second principal component is the axis selected at right angles to the first axis that produces the maximum reduction in unexplained variation. The third axis is likewise at right angles to the preceding axis and produces the maximum reduction in unexplained variation (Jones, 1997). In other words, the first component will account for the greatest portion of the variance, the second for the second largest portion, and so on until all the variance has been accounted for. Each of the principal components (new variables) is a linear combination of some of the original variables. Of course, there can be more than three such axes. As many components will be derived as necessary to account for the linear structure in the original variables (Resurreccion, 1988, 1998). In practice, the sensory analyst will determine the limit on the number of components to consider. It is therefore possible to create two or three principal components that can represent in excess of 70% of the observed variation (Resurreccion, 1998).

Examination of the original variables that are grouped in each of the principal components usually gives meaningful insight into the type of variation being explained by each principal component. In addition, these groupings of variables can be graphically presented to show product separation in the smaller two- or three-dimensional spaces that can be visualized. In many cases, the graphical representation of the results can be much more revealing than the numbers alone (Jones, 1997).

Factor analysis

In many ways, factor analysis is similar to PCA. Both multivariate techniques process the raw data to create a smaller set of "new variables" that are statistically independent of each other.

Factor analysis, really a name given to a whole series of procedures, constitutes an approach that attempts to deal with many attributes that behave similarly to each other and are thus correlated with each other. Factor analysis reduces this set of somewhat correlated attributes to a limited set of "basic attributes" (i.e., mathematical primaries) that are statistically uncorrelated with each other. Rather than having the products profiled on dozens of attributes, factor analysis processes the data and profiles the products on a limited set of predictors (so-called factors or principal components), which we just saw for PCA.

Factor analysis reduces a large number of variables to a smaller set of new variables, called factors, that can be used to explain the variation in the data. The objective is to create a smaller number of factors that together can replace the original variables measured in the study (Resurreccion, 1998). Although the algorithms used in factor analysis differ somewhat from those used by PCA, they are applied in similar ways to sensory data. It is not uncommon to start with a PCA to obtain some insights that can be used to initiate factor analysis (Jones, 1997) and in some instances may lead to similar results. In a sense we can think of factor analysis as a more advanced level of PCA, where the researcher first identifies the principal components and then creates newer variables by further processing these newly identified variables (e.g., by rotation).

In factor analysis, the factors are obtained by algorithms that work with correlations of the variables as opposed to variances, which are commonly used in PCA. In many cases, the axes found by factor analyses are treated by a mathematical operation called "rotation." There are several methods used for rotation and the method used must be specified in describing the results. The rotated axis yields a better alignment compared with the original axes. It is therefore possible to make a clearer interpretation of the resulting pattern of data points (Resurreccion, 1998).

Cluster analysis

CA is a general term for procedures that group variables or cases according to some measure of similarity (Ennis *et al.*, 1982). CA uses a variety of mathematical and graphical tools to locate and define groupings of data. The variables within a cluster are highly associated with one another, whereas those in different clusters are relatively distinct from one another (Cardello & Maller, 1987).

CA is primarily used for multivariate data and can be used to examine relationships either among variables or individuals (Jones, 1997). Although some clustering methods permit the use of nominal data, most methods require the data to at least be ordinal (Jones, 1997).

Several different methods can be used to create clusters. In general, clustering works by developing a measure of "distance" between observations or groups of observations. The most common "distance" is a geometric distance computed on pairs of observations or stimuli, using standard Euclidean formula for distance (Jones, 1997). This type of distance "metric" shows the absolute separation between two observations. Another approach uses a measure of dissimilarity in the patterns of two stimuli, based on their attribute scores. In this case the measure is no longer the Euclidean distance, but rather the statistic $(1-r)$, where r *is* the Pearson product moment correlation between two stimuli calculated on the different attributes. Whatever measure is used, the computations assign individuals to a cluster so as to minimize the distances among the points in the cluster.

Discriminant function analysis

Discriminant function analysis (DFA) determines which set of variables best discriminates one group of objects from another—*after* these two or more groups have been in other ways defined as falling into different classes (Resurreccion, 1988). DFA allows the sensory analyst to predict whether a specific food product sample will fall into one of several mutually exclusive, predetermined groups, on the basis of several variables measured for the food product (Frank, Gillett, & Ware, 1990; Ward, Resurreccion, & McWatters, 1995). These groups can be defined by criteria such as acceptance (consumer accepted or rejected) or simple classification (type of product). The sensory researcher is interested in understanding group differences or in predicting correct classification in a group based on the information on a set of variables or when probabilities of group membership must be determined.

In food quality measurements, the predictor variables are either instrumental measurements or sensory attribute ratings from descriptive analysis tests, or both. In turn, the food products are grouped into acceptable (hedonic score of 6 and above) or not acceptable (below 6) as determined through a consumer acceptance test (Resurreccion, 1998).

In some ways, DFA is similar to regression analysis. First, a "training set" of data is used to create the model. The training set generates a mathematical function that will give each observation the highest probability of being assigned to the known proper population. At the same time, the training set and modeling is used to minimize the probability that the same observation will be misclassified. Using instrumental or sensory descriptive analysis intensity ratings, the discriminant functions can predict, with probability value, the likelihood of membership in either the acceptable group or unacceptable group, respectively (SAS, 1985; Frank, Gillett, & Ware, 1990). Such approaches can turn into scoring machines, processing different stimuli, taking their measures, and then using the discriminant function to classify the product. In the most situations, only a subset of the original set of variables may be needed to create a useful discriminant function. The remaining measures do not add any more predictability.

DFA has additional facets beyond simple classification. One of these additional facets is basic learning. DFA helps to show how seemingly unrelated variables work together to describe and categorize not only a few new products, but existing products as well. It may be of interest to determine which of the sensory and instrumental variables, when used together, do the best job of distinguishing among several different products.

From such information, combinations of data can be obtained that will define the relations among various products, such as the likelihood that a new product will be perceived as being similar to a current, in-market product. This type of knowledge would allow tailoring a product to better compete in a specific market. Similarly, when a new product is developed, one could discover whether this new product matches one or more of the in-market products from which the discriminant function was generated. The sensory and instrumental data can be used to determine the closeness of the match by entering them into the function and finding the probabilities associated with the new product having come from each of the known populations (Jones, 1997; Resurreccion, 1998).

A related procedure, stepwise discriminant analysis (SDA), is used as a technique to combine variables that have significant discriminating power in predicting class membership (Powers, 1984b). In other words, SDA creates the best function with the fewest number of variables possible. One important problem for discriminant analysis must be pointed out. This is the fact that the mathematical techniques are indifferent to the theoretical relevance of one variable over another (Cardello & Maller, 1987). The SDA simply tries to do the best job in predicting membership of products into preassigned classes.

There are a number of issues with SDA. The first is the lack of theory underlying the analysis. Since there is no theory in SDA, the analysis is purely mechanical. The set of discriminating variables may not be the best (maximal combination). To identify the best solution, one would have to test all possible combinations (Klecka, 1980). When two variables are highly correlated and effect equal discrimination, only one will be selected. An important discriminator may not be selected at all (Resurreccion, 1988). Problems of colinearity among variables (so that variables are really restatements of each other) can be minimized by using common sense in interpreting the data from discriminant analysis and "forcing" in variables when necessary for theoretical or practical reasons (Cardello & Maller, 1987). When a variable is forced on the data by the project leader, it may result in an overlap between groups to an extent that the separation of groups is not that powerful (Sheth, 1970).

The second problem with SDA concerns validation of the analysis. When the same sample is used, both to create the classification function and to estimate how well that function performs, there will be an overestimation of the predictive power of the discriminant function. One suggestion splits the sample in half, using half for analysis and the other half for validation (Sheth, 1970). Snee (1977) split the collected data into two sets where one set of data is called estimation data and is used to estimate the model coefficients, whereas the complementary, remaining set of data is called prediction data and is used to measure the prediction accuracy of the model.

Canonical analysis

This analysis involves two sets of measurements that have to be correlated. The two sets are called the criterion set and the predictor set, respectively. The underlying strategy of the analysis maximizes the correlation between the two sets by creating a linear combination of each set. Canonical analysis is appropriate when the researcher focuses on the overall relationship between the predictor and criterion sets. The researcher may then isolate those predictor variables that contribute most to this overall relation. In addition, the analysis may reveal which of the criterion variables are most highly correlated (Sheth, 1970).

Multivariate multiple regression

This is a statistical technique that relates a set of independent variables such as instrumental measurements or sensory descriptive analysis ratings to a single dependent variable such as consumer affective responses. The regression selects a set of predictor variables, weighting them separately, in order to predict the dependent variable. In food quality evaluation, the most common use of multiple regression is to predict the magnitude of consumer affective responses on the basis of a series of measurements (instrumental or sensory) of the food (Cardello & Maller, 1987).

Partial least squares regression analysis

Partial least squares regression (PLSR) constructs predictive models when there are a number of predictor variables that are highly collinear (Tobias, 2002). There are two types of PLSR based on the number of dependent variables. The first is univariate PLSR, which works with one dependent variable. The second is multivariate PLSR, which works with several dependent variables simultaneously (Garthwaite, 1994).

PLSR has been used in many food research studies, generally those focusing on the relation between instrumental data and sensory data. Studies that use PLSR to interrelate sensory and instrumental measures encompass those dealing with appearance, taste/flavor, and texture, respectively:

(i) Several studies dealt with appearance versus physical factors—from studies involving processed diced tomatoes (Lee *et al.,* 1999) to cheddar cheeses (Truong *et al.,* 2002), and on to the dessert flan, *dulce de leche,* where instrumental color was related to corresponding descriptive appearance as shown by Hough, Bratchell, and MacDougall (1992).

(ii) Chemical and physicochemical measurements have also been related to sensory data as shown in studies by Pihlsgard *et al.* (1999) and De Belie *et al.* (2002) on liquid beet sugar and carrots, respectively.

(iii) Processing temperature and time have been related to sensory data sets to determine appropriate settings for equipment and have been applied to the processing of meat (Toscas, Shaw, & Beilken, 1999), dry-cured ham (Ruiz *et al.,* 2002), and bread (Sahlstrom, Baevre, & Brathen, 2003).

(iv) PLSR was also used to determine the effect of panel training and profile reduction on the development of vocabulary for the evaluation of meat quality (O'Sullivan *et al.,* 2002).

(v) The approach of relating multiple sets of data has also been done and is shown on studies involving cocoa drinks (Folkenberg, Bredie, & Martens, 1999) and apples (Karlsen *et al.,* 1999). In the study by Folkenberg *et al.* (1999) PLSR was used to relate sensory data with (1) physical, chemical, and rheological data, (2) main ingredients, and (3) consumer preference.

Preference mapping

Preference mapping (PREFMAP) is a member of the class of methods known as perceptual mapping. These mapping methods produce graphical displays from data. PREFMAP represents preference patterns by projecting external information into a configuration of points. With PREFMAP, integration between consumer reactions and descriptive is achieved. Preference information for each consumer taking part in a study is presented within a multidimensional space, representing the products evaluated (McEwan, Earthy, & Ducher, 1998).

PREFMAP is informative both for marketers and product developers. For the marketing research group, questions that can be addressed with PREFMAP analyses include the following:

(i) Where is my product positioned relative to my competitor's products?
(ii) Why is my product positioned there?
(iii) How can I reposition my existing products?
(iv) What new products should I create?
(v) Does the preference fall within the correct segment of the market? (Helgesen, Solheim, & Naes, 1997)

Three types of PREFMAP are used in the sensory research community: internal, extended, and external, as shown in Figure 10.7 (Cruz, Martinez, & Hough, 2002):

(i) Internal PREFMAP uses hedonic scores of the consumer as the active variables and sensory characteristics as supplementary variables (Rousset & Martin, 2001). Internal PREFMAPS creates a *consensus configuration* of the stimuli based solely on consumer

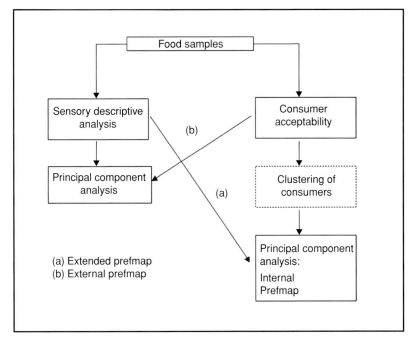

Figure 10.7 Schematic diagram of different preference mapping methods (adapted from Cruz, Martinez, & G. Hough, 2002. Food & Nutrition Press, Inc., reprinted with permission from Blackwell Publishing Ltd.).

preference data. A sample map and a consumer map are obtained corresponding to the scores and loadings of the PCA. The joint plot shows the consumers as directions of increasing preference on the sample map, where moving in the direction of preference means samples become "more liked," and moving in the opposite direction means samples become "less liked" (Helgesen, Solheim, & Naes, 1997).

(ii) External PREFMAP is obtained from PCA-analyzed QDA data and then correlates each of the consumers to this PCA space by regression analysis (Helgesen, Solheim, & Naes, 1997).

(iii) Extended PREFMAP, on the other hand, is created using from the correlation coefficients from the internal PREFMAP versus sensory analysis descriptors (Cruz, Martinez, & Hough, 2002).

PREFMAP delivers a visual representation of the stimuli. This visualization may suggest what intrinsic product characteristics drive consumer acceptability (McEwan, Earthy, & Ducher, 1998). For example, in the study conducted by Cruz, Martinez, and Hough (2002) on commercial mayonnaise, descriptive analysis of texture, flavor, and appearance were related to acceptability data. Extended PREFMAP showed that overall acceptance was better correlated with texture and flavor descriptors than with appearance descriptors.

In another study, this time dealing with rice and dry-cured ham samples (Rousset & Martin, 2001), hedonic scores were linked to objective assessment using internal PREFMAP. Results were able to show precise sensory differences between the two sets of samples giving the main reasons for acceptability or rejection.

Multidimensional scaling

Multidimensional scaling (MDS) is a multivariate statistical technique that captures the dissimilarities (or similarities) among stimuli by locating the stimuli in a geometrical space of smallest dimensionality. The panelists rate either the similarity or dissimilarity of various pairs of products. Computer techniques fit the stimuli in the space so that the rank order of the distances in the space matches, as closely as possible, the rank order of the dissimilarities. Dissimilarities, here, are simply the inverse of the similarities.

By evaluating the similarities of products, the researcher may create visual displays that represent the perceived dimensions of the respondents (Schiffman, Reynolds, & Young, 1981). Furthermore, there is no training of panelists. Thus, MDS may, in some ways, turn out to be less tedious and less expensive compared to other methods. MDS generates a large amount of information to be processed. MDS is considered a powerful tool for determining the attributes of a product category most relevant to consumers because it does not require the panelist to use descriptive language (Schiffman, Reynolds, & Young, 1981).

Sensory researchers have used MDS for the past five decades:

  (i) Heymann (1994) used MDS as a descriptive method to analyze vanilla samples employing untrained panelists. Her results demonstrated that MDS can be a useful *preliminary screening technique* for panelists but not a final descriptive system; the MDS space was unable to provide key descriptors or dimensions that could describe fine differences between samples.
 (ii) Chollet and Valentin (2001) used MDS to determine similarity relations among four matrices obtained on the effect of beer assessment training on verbal and nonverbal performance. Results show some similarities of vocabulary terms between trained and untrained subjects.
(iii) Tepper and Kuang (1996) utilized MDS procedures and a milk model system to understand the underlying dimensions of fat perception. The MDS models were able to show that addition of cream flavor affects perception of fat in dairy foods more than the addition of oil in the milk model system.
(iv) All in all, the usefulness of MDS is more as a heuristic, to represent stimuli for discussion purposes, rather than as a discovery system to learn really new facts that could help in product development.

Correspondence analysis

Correspondence analysis constitutes a descriptive and multidimensional approach using measures of association, such as the chi-square statistic. Correspondence analysis uses the frequency of consumers' terms (strong vs. weak points) generated during the tasting session, plotting the different stimuli and terms on a geometric map based on the frequency of association in actual use during the evaluation. For example, Rousset and Martin (2001) applied correspondence analysis to show correspondence between two sets of samples (rice and dry-cured ham) and attributes described by the consumers through a hedonic description of qualities and defects. The study was designed to directly collect the consumers' points of view about their perception, with the opportunity to simultaneously express positive and negative points about the samples. However, consumer terms generated more typically focused on differences involving popular and unpopular characteristics. Similar procedures were conducted by Rousset and Jolivet (2002) on acceptabilities of meat products, eggs, and fish by older consumers.

SELECTING THE BEST METHOD

Consumer acceptance of the food or food product can be quantified by using consumer affective tests and characterizing this food using sensory descriptive analysis ratings or instrumental and physicochemical measurements. The relation can be determined by plotting acceptance ratings against the descriptive analysis ratings or the physicochemical measurements. Mathematical models may be developed that can be used to predict consumer acceptance scores from descriptive analysis ratings or physicochemical measurements. These equations may be used to establish specifications for food that correspond to a predetermined degree of quality. We have seen the plethora of approaches, merely alluded to in the foregoing discussions. The armory of methods grows larger each year. Just keeping up with the methods can be a full-time job!

There is no single method that is superior to others in establishing correlations between physicochemical measurements and sensory assessment of quality (Powers, 1981). The ultimate choice must be made by the individual investigator (Cardello & Maller, 1987). Powers (1981) has suggested the following protocol to help select the best method:

(i) *Measure*: Acquire basic information regarding significant relationships between physicochemical measurements and sensory responses. The purpose of this first stage is to get to a point where one has enough objective facts to proceed with further exploration. It is always advisable, as a first step, to plot the results in order to better understand what is happening and to get a deeper feel for the data.

(ii) *Model*: Refine the data to permit fuller understanding of the instrumental, physicochemical, and sensory measurements. Researchers should watch out for outliers and variables or observations that do not belong in the data set.

(iii) *Validate*: In the final stage, carry out validation trials to determine whether the answers are "real" and the direction solid. Validation (or at least reliability) testing can be performed by full prediction testing on new data sets. If full prediction test with new data are not possible, then one ought to consider so-called split-sample validation, repeating the same data analysis method, and each time retaining part of the data for prediction testing.

REFERENCES

Abdullah, A. & T.C. Cheng. 2001. Optimization of reduced calorie tropical mixed fruit jam. *Food Quality and Preference* 12:63–68.

Amerine, M.A., R.M. Pangborn, & E.B. Roessler. 1965. *Principles of Sensory Evaluation of Food*. New York: Academic Press.

ASTM. 1981a. Selection of sensory panel members. In: *Guidelines for the Selection and Training of Sensory Panel Members*, pp. 5–17. West Conshohocken, PA: ASTM.

ASTM. 1981b. Training of sensory panel members. In: *Guidelines for the Selection and Training of Sensory Panel Members*, pp. 18–29. West Conshohocken, PA: ASTM.

ASTM. 1996. Sensory testing methods. In: *ASTM Manual Series: MNL 26*, eds E. Chambers IV & M.B. Wolf, pp. 38–53. West Conshohocken, PA: ASTM.

Bastos, D.H.M., C.H. Domenech, & J.A.G. Areas. 1990. Lung proteins: effect of defatting with several solvents and extrusion cooking on some functional properties. *Meat Science* 28(3):223–235

Bett, K.L., J.R. Vercellotti, N.V. Lovegren, T.H. Sanders, R.T. Hinsch, & G.K. Rasmussen. 1994. A comparison of the flavor and compositional quality of peanuts from several origins. *Food Chemistry* 51:21–27.

Bhattacharya, M., M.A. Hanna, & R.E. Kaufman. 1986. Textural properties of extruded plant protein blends. *Journal of Food Science* 51(4):988–993.

Bourne, M.C. 1982. *Food Texture and Viscosity*. New York: Academic Press.

Brandt, M.A., E.Z. Skinner, & J.A. Coleman. 1963 Texture profile method. *Journal of Food Science* 28:404–409.

Bruns, A.J. & M.C. Bourne. 1975. Effects of sample dimensions on the snapping force of crisp foods. *Journal of Texture Studies* 6:445–458.

Cairncross, S.E. & L.B. Sjostrom. 1950. Flavor profiles—a new approach to flavor problems. *Food Technology* 4:308–311.

Cardello, A.V. & O. Maller. 1987. Psychophysical bases for the assessment of food quality." In: *Objective Methods in Food Quality Assessment*, ed. J.G. Kapsalis, p. 61. Boca Raton, FL: CRC Press.

Cardello, A.V., O. Maller, J.G. Kapsalis, R.A. Segars, F.M. Sawyer, C. Murphy, & H.R. Moskowitz. 1982. Perception of texture by trained and consumer panelists. *Journal of Food Science* 47:1186–1197.

Caul, J.F. 1957. The profile method in flavor analysis. *Advanced Food Research* 7:1–6.

Chambers, D.H., A.M.A. Allison, & E. Chambers IV. 2004. Training effects on performance of descriptive panelists. *Journal of Sensory Studies* 19(6):486–499.

Chambers, E. & M.B. Wolf. 1996. *Sensory Testing Methods*, 2nd edn. West Conshohocken, PA: ASTM.

Chen, A.W., A.V.A. Resurreccion, & L.P. Paguio. 1996. Age appropriate hedonic scales to measure food preferences of young children. *Journal of Sensory Studies* 11:141–163.

Chollet, S. & D. Valentin. 2001. Impact of training on beer flavor perception and description: Are trained and untrained subjects really different? *Journal of Sensory Studies* 16:601–618.

Chow, E.T.S., L.S. Wei, R.E. Devor, & M.P. Steinberg. 1988. Performance of ingredients in a soybean whipped topping: A response surface analysis. *Journal of Food Science* 53(6):761–765.

Christensen, C.M. 1983. Effects of color on aroma, flavor and texture judgments of foods. *Journal of Food Science* 48:787–790.

Chu, C.A. & A.V.A. Resurreccion. 2004. Optimization of a chocolate peanut spread using response surface methodology. *Journal of Sensory Studies* 19:237–260.

Clydesdale, F.M. 1978. Colorimetry-methodology and applications. *Crit. Rev. Food Science Nutrition* 10:243–301.

Clydesdale, F.M. 1991. Color perception and food quality. *Journal of Food Quality* 14:61–74.

Clydesdale, F.M. 1993. Color as a factor in food choice. *Crit. Rev. Food Science Nutrition* 33(1):83–101.

Cruz, M.J.S, M.C. Martinez, & G. Hough. 2002. Descriptive analysis, consumer clusters and preference mapping of commercial mayonnaise in Argentina. *Journal of Sensory Studies* 17:309–325.

Curley, L.P. & R.C. Hoseney. 1984. Effects of corn sweeteners on cookie quality. *Cereal Chemists* 61(4):274–278.

De Belie, N., A.M. Laustsen, M. Martens, R. Bro, & J. De Baerdemaeker. 2002. Use of physico-chemical methods for assessment of sensory changes in carrot texture and sweetness during cooking. *Journal of Texture Studies* 33:367–388.

Einstein, M.A. 1991. Descriptive techniques and their hybridization. In: *Sensory Science Theory and Applications in Food*, eds H.T. Lawless & B.P. Klein, pp. 317–338. New York: Marcel Dekker.

Ennis, D.M., H. Boelens, H. Haring, & P. Bowman. 1982. Multivariate analysis in sensory evaluation. *Food Technology* 36(11):83–90.

Federer, W.T., C.E. McCulloch, & N.J. Miles-McDermott. 1987. Illustrative examples of principal component analysis. *Journal of Sensory Studies* 2(1):37.

Floros, J.D. & M.S. Chinnan. 1988. Seven factor response surface optimization of a double-stage lye (NaOH) peeling process for pimiento peppers. *Journal of Food Science* 53(2):631–638.

Folkenberg, D.M., W.L.P. Bredie, & M. Martens. 1999. What is mouthfeel? Sensory-rheological relationships in instant hot cocoa drinks. *Journal of Sensory Studies* 14(2):181–195.

Frank, J.S., R.A.N. Gillett, & G.O. Ware. 1990. Association of *Listeria* spp. Contamination in the dairy processing plant environment with the presence of Staphylococci. *Journal of Food Protection* 53:928–932.

Freund, R.J. & R.C. Littell. 1986. *SAS System for Regression*. Cary, NC: SAS Institute.

Gacula, M.C., Jr. 1993. *Design and Analysis of Sensory Optimization*. Trumbull, CT: Food & Nutrition Press.

Galvez, F.C.F., A.V.A. Resurreccion, & P.E. Koehler. 1990. Optimization of processing of peanut beverage. *Journal of Sensory Studies* 5:1–17.

Galvez, F.C.F., A.V.A. Resurreccion, & G.O. Ware. 1995. Formulation and process optimization of mungbean noodles. *Journal of Food Processing and Preservation* 19:191–205.

Garthwaite, P.H. 1994. An interpretation of partial least squares. *Journal of American Statistical Assoc* 89:122–127.

Gormley, T.R. 1987. Fracture testing of cream cracker biscuits. *Journal of Food Engineering* 6:325–332.

Hare, L.B. 1974. Mixture designs applied to food formulations. *Food Technology* 28(2):50–56, 62.

Harper, S.J. & M.R. McDaniel. 1993. Carbonated water lexicon: Temperature and CO_2 level influence on descriptive ratings. *Journal of Food Science* 58(4):893–898.

Hashim, I.B., A.V.A. Resurreccion, & K.H. McWatters. 1995. Consumer acceptance of irradiated poultry. *Poultry Science* 74:1287–1294.

Helgesen, H., R. Solheim, & T. Naes. 1997. Consumer preference mapping of dry fermented lamb sausages. *Food Quality and Preference* 8(2):97–109.

Henselman, M.R., S.M. Donatoni, & R.G. Henika. 1974. Use of response surface methodology in the development of acceptable high protein bread. *Journal of Food Science* 39:943–946.

Heymann, H. 1994. A comparison of free choice profiling and multidimensional scaling of vanilla samples. *Journal of Sensory Studies* 9(4):445–453.

Hill, W.J. & W.G. Hunter. 1966. A review of response surface methodology: A literature survey. *Technometrics* 8(4):571–590.

Hinds, M.J., M.S. Chinnan, & L.R. Beuchat. 1994. Unhydrogenated palm oil as a stabilizer for peanut butter. *Journal of Food Science* 59(4):816–820.

Holt, S.D., A.V.A. Resurreccion, & K.H. McWatters. 1992. Formulation evaluation and optimization of tortillas containing wheat, cowpea and peanut flours using mixture response surface methodology. *Journal of Food Science* 57:121–127.

Hootman R.C., ed. 1992. *Manual on Descriptive Analysis Testing for Sensory Evaluation*, pp. 1–4. West Conshohocken, PA: ASTM.

Hough, G., N. Bratchell, & D.B. MacDougall. 1992. Sensory profiling of dulce de leche, a dairy based confectionary product. *Journal of Sensory Studies* 7:157–178.

Huang, S., K. Quail, & R. Moss. 1998. The optimization of a laboratory processing procedure for southern-style Chinese steamed bread. *International Journal of Food Science Technology* 33:345–357.

IFT/SED. 1981. Sensory evaluation guideline for testing food and beverage products. *Food Technology* 35(11):50–59.

Johnson, T.M. & M.E. Zabik. 1981. Response surface methodology for analysis of protein interactions in angel food cakes. *Journal of Food Science* 46:1226–1230.

Jones, R.M. 1997. Statistical techniques for data relationships. In Consumer Responses and Analytical Measurements. In: *Relating Consumer, Descriptive, and Laboratory Data to Better Understand Consumer Responses*, ed. A. Munoz. West Conshohocken, PA: ASTM.

Karlsen, A.M., K. Aaby, H. Sivertsen, P. Baardseth, & M.R. Ellekjaer. 1999. Instrumental and sensory analysis of fresh Norwegian and imported apples. *Food Quality & Preference* 10:305–314.

Katz, E.E. & T.P. Labuza. 1981. Effect of water activity on the sensory crispness and mechanical deformation of snack food products. *Journal of Food Science* 46:403–409.

Keane, P. 1992. The Flavor Profile. In: *Manual on Descriptive Analysis Testing for Sensory Evaluation*, ed. R.C. Hootman. Philadelphia, PA: ASTM.

Klecka, W.R. 1980. Discriminant analysis. *Sage University Paper Series on Quantitative Applications in the Social Sciences*, pp. 07–019. Beverly Hills, CA and London: Sage Publications.

Lee, S.-Y., I. Luna-Guzman, S. Chang, D.M. Barrett, & J.X. Guinard. 1999. Relating descriptive analysis and instrumental texture data of processed diced tomatoes. *Food Quality & Preference* 10:447–455.

Malundo, T.M.M. & A.V.A. Resurreccion. 1992. A comparison of performance of panels selected using analysis of variance and cluster analysis. *Journal of Sensory Studies* 7:63–75.

Malundo, T.M.M. & A.V.A. Resurreccion. 1993. Optimization of liquid whitener from peanut extract. *Lebensm.-Wiss. u.-Technol* 26:552–557.

Malundo, T.M.M. & A.V.A. Resurreccion. 1994. Peanut extract and emulsifier concentrations affect sensory and physical properties of liquid whitener. *Journal of Food Science* 59:344–349.

McEwan, J.A., P.J. Earthy, & C. Ducher. 1998. Preference mapping: A review. Review no. 6, project no. 29742. Campden & Chorleywood Food Res. Assoc., Chipping Campden, UK.

McLean, R.A. & V.L. Anderson. 1966. Extreme vertices design of mixture experiments. *Technometrics* 8(3):447–454.

Meilgaard, M., G.V. Civille, & B.T. Carr. 2007. *Sensory Evaluation Techniques,* 4th edn. Boca Raton, FL: CRC Press.

Moskowitz, H. 1983. *Product Testing and Sensory Evaluation of Foods.* Westport, CT: Food & Nutrition Press.

Moskowitz, H.R., A.M. Munoz, & M.C. Gacula, Jr. 2003. *Viewpoints and Controversies in Sensory Science and Consumer Product Testing.* Trumbull, CT: Food & Nutrition Press.

Mudahar, G.S., R.T. Toledo, & J.J. Jen. 1990. A response surface methodology approach to optimize potato dehydration process. *Journal of Food Process and Preservation* 14:93–106.

Munoz, A.M. & G.V. Civille. 1992. The Spectrum Descriptive Analysis Method. In: *Manual on Descriptive Analysis Testing for Sensory Evaluation*, ed. R.C. Hootman, pp. 22–34. West Conshohocken, PA: ASTM.

O'Mahony, M. 1986. *Sensory Evaluation of Food*. New York: Marcel Dekker.

O'Mahony, M. 1991. Descriptive analysis and concept alignment. In: *IFT Basic Symposium Series: Sensory Science Theory and Applications in Foods*, ed. H.T. Lawless & B.P. Klein, pp. 223–267. New York, Basel, and Hong Kong: Marcel Dekker.

O'Mahony, M., U. Thieme, & L.R. Goldstein. 1988. The warm-up effect as a means of increasing the discriminability of sensory difference tests. *Journal of Food Science* 53(6):1848–1850.

O'Sullivan, M.G., D.V. Byrne, H. Martens, & M. Martens. 2002. Data analytical methodologies in the development of a vocabulary for evaluation of meat quality. *Journal of Sensory Studies* 17(6): 539–558.

Oh, N.H., P.A. Seib, & D.S. Chung. 1985. Noodles. III. Effects of processing variables on quality characteristics of dry nodles. *Cereal Chemistry* 62(6):437–440.

Ozedemir, M. & O. Devres. 2000. Analysis of color development during roasting of hazelnuts using response surface methodology. *Journal of Food Engineering* 45:17–24.

Pihlsgard, P., M. Larsson, A. Leufven, & H. Lingnert. 1999. Chemical and sensory properties of liquid beet sugar. *Journal of Agric. Food Chem* 47:4346–4352.

Plemmons, L.E. 1997. Sensory evaluation methods to improve validity, reliability, and interpretation of panelist responses. M.S. thesis, University of Georgia, Athens, GA.

Plemmons L.E. & A.V.A. Resurreccion. 1998. A warm-up sample improves reliability of responses in descriptive analysis. *Journal of Sensory Studies* 13:359–376.

Papadopoulos, L.S., R.K. Miller, G.R. Acuff, L.M. Lucia, C. Vanderzant, & H.R. Cross. 1991. Consumer and trained sensory comparisons of cooked beef top rounds treated with sodium lactate. *Journal of Food Science* 56(5):1141–1146.

Popper, R., H. Heymann, & F. Rossi. 1997. Three multivariate approaches to relating consumer to descriptive data. In: *Relating Consumer, Descriptive, and Laboratory Data to Better Understand Consumer Responses*, ed. A. Munoz, pp. 39–61. West Conshocken, PA: ASTM.

Powers, J.J. 1981. Multivariate procedures in sensory research: Scope and limitations. *Tech. Quart. Mast. Brewers' Assn* 18(1):11.

Powers, J.J. 1984a. Current practices and applications of descriptive methods. In: *Sensory Analysis of Foods*, ed. J.R. Piggott, p. 179. London: Applied Sci. Pub.

Powers, J.J. 1984b. Using general statistical programs to evaluate sensory data. *Food Technology* 38(6):74.

Rainey, B.A. 1986. Importance of reference standards in training panelists. *Journal of Sensory Studies* 1:149–154.

Resurreccion, A.V.A. 1988. Applications of multivariate methods in food quality evaluation. *Food Technology* 42(11):128, 130, 132–134, 136.

Resurreccion, A.V.A. 1998. *Consumer Sensory Testing for Product Development*. Gaithersburg, MD: Aspen Publishers.

Roberts, A.K. & Z.M. Vickers. 1994. A comparison of trained and untrained judges' evaluation of sensory attribute intensities and liking of cheddar cheese. *Journal of Sensory Studies* 9:1–20.

Rothman, L. 1997. Relationships between consumer responses and analytical measurements. In: *Relating Consumer, Descriptive, and Laboratory Data to Better Understand Consumer Responses*, ed. A.M. Munoz, pp. 62–77. West Conshockens, PA: ASTM.

Rousset, S. & P. Jolivet. 2002. Discrepancy between the expected and actual acceptability of meat products, eggs and fish: The case of older consumers. *Journal of Sensory Studies* 17:61–75.

Rousset, S. & J. Martin. 2001. An effective hedonic analysis tool: weak/strong points. *Journal of Sensory Studies* 16:643–661.

Ruiz, J., C. Garcia, E. Muriel, A.I. Andres, & J. Ventanas. 2002. Influence of sensory characteristics on the acceptability of dry-cured ham. *Meat Science* 61:347–354.

Rutledge, K.P. & J.M. Hudson. 1990. Sensory evaluation: Method for establishing and training a descriptive flavor analysis panel. *Food Technology* 44(12):78–84.

Sahlstrom, S., A.B. Baevre, & E. Brathen. 2003. Impact of starch properties on hearth bread characteristics. I. Starch in wheat flour. *Journal of Cereal Science* 37:275–284.

SAS Institute. 1980. *SAS Applications*. Cary, NC: SAS Institute.

SAS Institute. 1985. *SAS User's Guide: Statistics*. Cary, NC: SAS Institute.

Schiffman, S.S., M.L. Reynolds, & F.W. Young. 1981. *Introduction to Multidimensional Scaling.* New York: Academic Press.

Schutz, H. 1983. Multiple regression approach to optimization. *Food Technology* 37(11):46–48, 62.

Seymour, S.K. & D.D. Hamann. 1988. Crispness and crunchiness of selected low moisture foods. *Journal of Texture Studies* 19:79–95.

Shelke, K., J.W. Dick, Y.F. Holm, & K.S. Loo. 1990. Chinese wet noodle formulation: A response surface methodology study. *Cereal Chemists* 67(4):338–342.

Sherman, P. & D.S. Deghaidy. 1978. Force-deformation conditions associated with the evaluation of brittleness and crispness in selected foods. *Journal of Texture Studies* 9:437–459.

Sheth, J.N. 1970. Multivariate analysis in marketing. *Journal of Advertising Research* 10(1):29.

Snee, R.D. 1977. Validation of regression models: Methods and examples. *Technometrics* 19(4):415–428.

Stone, H. 1992. Quantitative descriptive analysis (QDA). In: *Manual on Descriptive Analysis Testing for Sensory Evaluation*, ed. R.C. Hootman. Philadelphia, PA: ASTM.

Stone, H. & J. Sidel. 2004. *Sensory Evaluation Practices*, 3rd edn. San Diego, CA: Academic Press, Inc.

Stone, H., J. Sidel, S. Oliver, A. Woolsey, & R.C. Singleton. 1974. Sensory evaluation by quantitative descriptive analysis. *Food Technology* 28(11):24–34.

Szczesniak, A.S., M.A. Brandt, & H.H. Friedman. 1963. Development of standard rating scales for mechanical parameters of texture and correlation between the objective and the sensory methods of texture evaluation. *Journal of Food Science* 28:397–403.

Tepper, B.J. & T. Kuang. 1996. Perception of fat in a milk model system using multidimensional scaling. *Journal of Sensory Studies* 11(3):175–190.

Thakur, S. & D.C. Saxena. 2000. Formulation of extruded snack food (gum based cereal-pulse blend): optimization of ingredient levels using response surface methodology. *Lebensmittel-Wissenschaft und-Technologie* 33:354–361.

Tobias, R.D. 2002. An introduction to partial least squares regression. Available at http://www.sas.com/rnd/ app/papers /pls.pdf. Cary, NC: SAS Institute Inc.

Toscas, P.J., F.D. Shaw, & S.L. Beilken. 1999. Partial least squares (PLS) regression for the analysis of instrument measurements and sensory meat quality data. *Meat Science* 52:173–178.

Truong, V.D., C.R. Daubert, M.A. Drake, & S.R. Baxter. 2002. Vane rheometry for textural characterization of cheddar cheeses: Correlation with other instrumental and sensory measurements. *Lebensm.-Wiss. u.-Technol.* 35:304–314.

Vainionpaa, J. 1991. Modeling of extrusion cooking of cereals using response surface methodology. *Journal of Food Engineering* 13:1–26.

Vaisey-Genser, M., G. Ylimaki, & B. Johnston. 1987. The selection of levels of canola oil, water, and an emulsifier system in cake formulations by response surface methodology. *Cereal Chem* 64(1):50–54.

Vatsala, C.N., C.D. Saxena, & P.H. Rao. 2001. Optimization of ingredients and process conditions for the preparation of puri using response surface methodology. *International Journal of Food Science Technology* 36:407–414.

Ward, C.D.W., A.V.A. Resurreccion, & K.H. McWatters. 1995. A systematic approach to prediction of snack chip acceptability utilizing discriminant functions based on instrumental measurements. *Journal of Sensory Studies* 10:181–201.

Wolters, C.J. & E.M. Allchurch. 1994. Effect of the training procedure on the performance of descriptive panels. *Food Quality Preference* 5:203–214.

Zook, K. & C. Wessman. 1977. The selection and use of judges for descriptive panels. *Food Technology* 11:56–61.

11 So what are the practical considerations in actually running a test? what do *I* need to know? what does the *rest of the company need to know?*

The principle for all sensory tests is that the test method should be selected on the basis of the objectives for the study. It is therefore necessary to have clear communication between the client or company and the sensory test manager.

T. Lawless and H. Heymann, *Sensory Evaluation of Food: Principles and Practices*

INTRODUCTION

Many often mutually conflicting considerations confront those who use a "sensory" or a "consumer" test, or request one, to support product development activities. Matters become even more complex when the sensory specialist outsources the execution of the test, as so often happens today. We write this chapter to guide the user. There is no one right answer about what to do. There are, however, considerations. It is to these considerations that we now turn. The chapter will become increasingly important in a virtual world, where skills may or may not exist, but where issues needing answers seem to multiply as the complexity of today's market overtakes company after company.

The key to a successful sensory test is managing the test so that, at the end of the process, the test provides the user with actionable information. How this can be ensured in a business environment where "virtual companies" are the rule, where professionals no longer work for the company but rather are retained on a project basis, where knowledge of the product and procedures, pitfalls, and shortcuts may no longer be resident in the mind of a long-term professional? What should be done—and how?

And then there are the other factors beyond people, such as context, for example. Understanding why consumers like one product and dislike another is not an easy task. We discuss those factors that influence consumer responses to acceptance tests. The degree of liking of a food may be affected by the context in which it was presented. These contextual factors may or may not be controllable by the sensory researcher but are useful in interpreting results of sensory testing.

We are going to look at a variety of different decisions to be made. Some of this material may repeat what has been said in other chapters. Nonetheless, by combining discussions of

Sensory and Consumer Research in Food Product Design and Development, Second Edition.
Howard R. Moskowitz, Jacqueline H. Beckley, and Anna V.A. Resurreccion.
© 2012 Blackwell Publishing Ltd. Published 2012 by Blackwell Publishing Ltd.

Table 11.1 Factors that the researcher should consider when creating the sampling plan.

The business that will be based on information from the sample.

Practicality of measuring a portion of the universe.

Costs of sampling, of the entire project, and the return on information investment.

The level of risk management is willing to take, such as type I and type II error (wherein type I error consists of rejecting the null hypothesis when it is true, and type II error consists of accepting the null hypothesis when the test hypothesis is true) and other risk factors.

Demographic characteristics of the population.

Source: Adapted from Resurreccion, 1998. Table 4.1, p. 73. Copyright 1998 by Aspen Publications, with kind permission of Springer Science and Business Media.

these considerations in a single chapter on "testing," and more specifically test execution, we hope to make the reader aware of the different types of decisions typically faced, and some of the criteria used to make them.

THE WHO: WHAT TYPE OF INDIVIDUAL SHOULD BE USED IN SENSORY TESTS FOR PRODUCT DEVELOPMENT?

Perhaps among the most important issues in planning a sensory test is the composition of the panel of individuals who will participate in the tests. The composition of the panel will depend on the type of sensory test being conducted and the risks/benefits involved. Table 11.1 lists some of these technical and business considerations; there are many more.

Expert panel for descriptive analysis

For descriptive analysis tests involving trained panelists, prospective participants are screened. Those individuals who pass the screening then move on to the training phase. The training phase presents opportunities for the sensory practitioner to track the performance of each panelist, and compare that performance to the group average and to expectations about performance based on previous experience. There are statistics to be computed from the panelist's performance, with those statistics providing insight into the panelist's capabilities.

Let us look at one example: reliability. During testing, one can compute a measure of reliability. When the reliability of rating is poor, the moderator can work with the panelist to "calibrate" him or her. The effort is worth it. One method used for calibration in descriptive analysis testing calls for a calibration session to be held one hour prior to the test period (Lee & Resurreccion, 2004; Chu & Resurreccion, 2005). During this pretest calibration session, trained panelists rate references, a warm-up sample and an "unknown" sample, and individual responses are tabulated and discussed. Panelists are then given results, usually in a computer printout, of their performance in rating identical samples and compared with the entire group. In addition, there is an opportunity for these respondents whose performance is not optimally reliable, or who differ from the rest of the panel, to taste the references or samples and adjust their rating according to group consensus.

Consumer panel—selecting the consumers

Whenever a consumer sensory test is conducted, a group of consumers are selected as a sample of some larger population, about which the sensory analyst hopes to draw some conclusions (Resurreccion, 1998). Therefore, the specific nature of a consumer panel is one of the most important considerations in planning the test. The consumer test should be conducted using panelists who match those of the intended users of the product, the target population, and who are naive in product testing (Lawless & Heymann, 1998; Stone & Sidel, 2004). With today's emphasis on testing, however, it may not be possible to find such inexperienced individuals with little history of panel participation, and so that requirement is on its way to disappearing. It is important not to work with individuals who have so much experience that we could call them professionalized consumers.

Knowing who the target population is going to be may seem simple, but it's not necessarily always known. This knowledge is essential to the success of the study. Proper identification enables the project leader to recruit the appropriate sample of test participants. The result is that one believes the data, because the results represent the reactions of the consumers to whom the product is marketed.

The target population may be defined as that segment of the population who regularly purchase, or is expected to use, the product or the product category. The panelists in a consumer sensory test should be selected on the basis of their frequency of use of a product, or product preferences, and their demographic characteristics. For example, children who drink milk should be used as panelists when the test concerns reactions to different flavors of flavored milk marketed to children. They do not need to drink a specific brand of milk, unless the objectives of the project demand that level of specificity. For a presweetened breakfast cereal, the target population may be children between the ages of four and twelve (Meilgaard, Civille, & Carr, 2007).

It is advisable to provide ways to verify one's information about the panelist. It is also good research procedure to obtain this updated, validating information during the study. Information needing update may involve key demographic characteristics and product usage and/or purchase frequency information of the sample to check on current usage patterns and changes in demographic characteristics. The validation check may be made by instructing panelists to fill out a demographic questionnaire during a break or after evaluation of samples. The questions can also provide a check on the recruitment procedure for the test. When there is some doubt about the sample obtained through the recruitment process, then alternative sampling procedures may be required (ASTM, 1979).

There is another benefit to acquiring additional information about the test participants. Responses to the screener and additional questions may be used to update existing databases. Although consumer attitudes toward food are not readily altered, food consumption patterns may change due to health, nutrition, or other reasons. Thus, for this reason periodic updating of the food use frequency information may be warranted.

Consumer panel—statistical sampling

Typically, panelists are screened on the basis of characteristics that match the target market. These characteristics may include age, gender, income, education, and product usage, among others. As a result, the sample is typically drawn from what one might consider a relatively homogeneous population. The assumption (which needs empirical verification each time) is that the differences in likes and dislikes will not differ substantially within this presumed homogeneous population. Thus, a simple random sampling will be sufficient.

Consumer panel—the issue of a random sample of panelists

Can a truly random sample or representative panel be recruited for a test? In reality, does every member of the defined population really enjoy an equal chance of being selected to represent the population? Moskowitz, Muñoz, and Gacula (2003) state that in practice, panelists who participate do so because they are interested for one reason or other, such as the reward, or want a free sample of the product. This precludes a truly random population. They likewise state that the issue of a truly random sample is probably not relevant at the early stages of product development, where most sensory researchers operate.

Consumer panel—employees versus nonemployees

Often, panelists who participate in consumer tests are recruited from within the company and participate in several such tests over a year, not just in one or two studies. These employees are laboratory, technical, or office workers who are asked to participate in the consumer panels. With no recruitment costs and no monetary panelist incentives, this appears to be the most economical approach to use. The low and fixed budget allows for a more accurate estimate of the cost of each test. Additional advantages of using employee panelists are their availability for a longer time and their familiarity with the testing procedure, therefore negating the need for orientation. The disadvantages are that employee panels are a source of bias and will not respond in the same manner as external or "real" consumer panelists. The most critical issue, often unaddressed, is the hidden costs of such an employee panel.

The importance of panelist selection for the test and frequency of test participation are critical issues in relation to the question of the objectivity and face validity of the results from employee acceptance tests. Gacula (Moskowitz, Muñoz, & Gacula, 2003) states categorically that it is a waste of time and resources to use employee panelists; therefore, they should never be used to gather preference, acceptance, or other forms of hedonic information on various product attributes. The practice of using employees in affective tests is not recommended for the reason that employees have definite biases that affect their responses (Resurreccion, 1998).

Although agreement exists that the use of employees in consumer affective tests is not recommended, just about all professionals in the field of testing agree that many of the sensory acceptance tests involve employees, primarily for reasons of convenience and budget. A similar situation may be met in university or government laboratories, where convenience samples consisting of students or staff are recruited to participate in acceptance tests (Resurreccion, 1998). Student and staff use in consumer panels should be avoided for the same reasons that employee panels should be. The accepted practice is to recruit consumer panelists from outside the organization. A trend has been to use local residents or contract with a local agency that recruits panelists for consumer acceptance tests.

Researchers faced with problems of cost and convenience are likely to be first to come up with workable solutions. There are those who recognize that when product maintenance is the objective, company employees and those residents who reside in close proximity to company offices, technical centers, or processing plants for the product samples being tested do not represent a particularly serious risk when asked to serve on the panel groups (Meilgaard, Civille, & Carr, 2007). However, they believe that for new product development, product improvement, or product optimization, these convenient and nonemployee individuals *should not be used to represent the consumer.*

EMPLOYEES AND BIAS

The bias associated with the use of employees (vs. naive consumers) in a consumer sensory test should be considered in assuring validity of the sensory test. Examples of biases when using employees are as listed in the succeeding sections.

Employee bias—employees are often unduly familiar with products

Employees are more familiar with the company's products than the average consumer. For the same reason that the product development team and all technical and production personnel involved in the development of the product need to be excluded from the evaluations, employees may have some knowledge of the product that may generate a biased response. Employees tend to favor or be more critical of products they manufacture. They either prefer products they manufacture or, if morale is bad, find reasons to reject products (Meilgaard, Civille, & Carr, 2007). When working with "blind" products, not identified with a brand, the risk remains that employees will recognize a product, despite efforts to disguise brand names or unique markings on those products. Such cases require consumer panels, not employee panels.

Employees do not rate characteristics of products the way consumers do

Employees may have additional knowledge about the product, such as product improvement efforts by the company. This extra knowledge may influence their ratings. For example, assume that the test product has a different appearance from the traditional product. Employees might rate their acceptance of the "improved" product higher than actual consumers would, even if the employees attempt to be objective in their ratings and consciously make the effort to disregard what they know about the product.

Furthermore, employees may respond in a manner that they think will please their supervisors. They have unusual patterns of product use because they may receive product for free, obtain the products from a company store at a greatly reduced price, or use the products due to brand loyalty. For these reasons, employee panels should be used with caution.

Overparticipation may lead to *professional* panelists

Another source of bias is a panelist's overparticipation (Moskowitz, Muñoz, & Gacula, 2003) in panels. When the panelist pool is small, and because recruitment is free, the same individuals may be used repeatedly, leading to their overexposure to the product. Employee panelists invariably evolve into professional panelists. There are employees who may volunteer to participate in consumer panels again and again because they enjoy the break, the rewards associated with participating, and the camaraderie with members of the panel and with the sensory practitioner who leads the panel or runs the study for consumer evaluation. Nonemployee consumers who start out as naive participants would likewise be transformed to professional panelists with frequent testing.

Demographic characteristics of employees do not represent those of the target market

Employees do not generally represent the target segment for the product and cannot be expected to give any indication of the response of the target market. The employee panels do not reflect the demographic characteristics of a larger nonemployee panel. Employee panels comprise adults, usually of a specific age range. Therefore, when the only available panels are employee panels, testing with children cannot be accomplished.

Due to the sources of bias just listed, the use of employee panelists in affective tests is not recommended. When there is no alternative but to ask employees to participate in a sensory acceptance test, effort should be made to use the same selection criteria that would be used for a nonemployee panel. At the very least, employees who volunteer to participate in consumer panels should be screened for product usage and their likes and dislikes for the product. Their food preferences should be compared with the stated food preferences of individuals from other populations obtained in the same fashion. Data on food preferences can be obtained from published surveys. Preferably, however, primary data from one's own surveys and information should be used as a basis for comparing one's own employees to bona fide consumers. Before making a decision on the use of results from acceptance tests involving employees, a series of comparative tests should be run to determine whether there is potential bias or not, when employee panels are to be used. These comparative tests should be identical for employee panels and consumer panels, respectively.

Stone and Sidel (2004) proposed that when employees are used as consumer panelists, these employees should be self-declared *likers* of the product to be tested or that their attitudinal scores to the product category being tested fall ±1 standard deviation of the grand mean of all survey participants. For the comparative analyses, both Stone and Sidel (2004) and Lawless and Heymann (1998) recommend that a split-plot analysis of variance be conducted. The analytical strategy to deal with the problem follows these six steps:

(i) The products are treated as the "within" variable. The panels (employee vs. nonemployee) are treated as the "between" variable.

(ii) When the *F* value for panels is significant, the employee and the outside consumer panels *are not similar.* The employee panel may not be substituted for the outside panel.

(iii) When the interaction between product and panel is significant, the panels are not rating at least one of the products similarly. The two panels *should not be substituted* for each other.

(iv) Neither the *F* ratio for the panels nor the *F* ratio for the interaction between the panels and product may reach significance. Only when both *F* ratios are statistically not significant can the panels be comparable.

(v) Lawless and Heymann (1998) remind the sensory practitioner that this approach relies on acceptance of the null hypothesis to justify panel equivalence. Therefore, the researcher must be concerned about the power and sensitivity of the test. The power of the test is ensured by a large sample size. In this test, small panel comparisons have too little sensitivity to detect any differences.

(vi) To ensure that a difference is not missed, the p value for significance should be tested at a higher level such as 0.10 or 0.20. The β-risk and chances of type II error become important in the comparison. If the employee panel is truly different from an outside panel, using the employee panel may result in serious management implications.

When employee panels will continue to be used, then statistical validation of their results must be established periodically. Satisfactory results from these statistical analyses allow substitution of preference behavior for demographic criteria in selecting subjects for acceptance tests. Similarly, when a pool of nonemployees participates regularly in acceptance tests in much the same way as employees would, then the qualifying procedures described for employees must be used and their responses should be monitored on a continuing basis. This practice should ensure reliable and valid results.

Organizations or "church panels" as sources of consumers

In recent years, some research guidance tests have started to use groups of individuals who are members of an organization, such as a church or organization raising money for scholarships (or other fund-raising objectives). When such groups are used, the panelists are usually not screened prior to their participation, although most researchers who use this method insist that panelists be nonrejecters of the product (Moskowitz, Muñoz, & Gacula, 2003). The advantage of using these organizations or church panels is mainly the cost, because in most cases the use of groups involves paying one amount for the group, rather than to each participant. Using this source of panelists minimizes recruitment and screening activities. The disadvantage of using church groups or organizations is that the participants *do not represent* the target market for the product. The frequency of testing may generate professional panelists.

Interest of the participants from the organization varies as panelists do not receive individual monetary rewards. In some cases, interest level for individual panelists can be high, for example, when the participant is interested in the fund-raising objective and views participation in the test as furthering that goal. However, interest often can be low, for example, when the panelist participates as a response to social pressure. Organization or church panels comprise individuals who volunteer their time, and thus in many other cases the members may not be willing to do more than the bare minimum to reward their group.

When working with church groups or organizations, the researcher may reduce the task demands on the panelist, often limiting the number of samples to 2–4, rather than expanding the sample set to deal with more strategic issues. Moskowitz, Muñoz, and Gacula (2003) caution against the use of church groups or organizations solely for their economy because using this low-cost approach is "growth blocking" and reduces the prospective benefits of testing. Relying on church panels and organizations becomes standardized, driving out anything that does not fit into its easy, neat purview of low cost, low-effort, simplistic research.

To minimize the overparticipation of panelists when using organizations as the source of consumers, the sensory practitioner should monitor the participation of volunteer groups and organizations and exclude frequent participants for a year. Effort should be expended to add new participants to the pool and replace dropouts on a regular basis.

HOW MANY PANELISTS WILL I NEED?

If any single question plagues the practitioner, follows around the sensory scientist like a shadow, and drives commercial consumer researchers (and others) to sheer distraction, it is the issue of base size. Unlike an instrument purchased off the shelf, the human judge is not there to be plugged in, turned on, and then to generate data. The human judge comes equipped with time pressures, excuses, and a host of irrelevancies that often interfere with the timely completion of experiments and almost never contribute positively.

Power of the statistical test as a driver of base size

The power of a test is the probability of finding a difference when one actually exists; that is, the probability of making a correct decision that the samples are different.

There are two types of errors that the sensory scientist can make when testing the null hypothesis (H_0) of any sensory method. These are the type I (α) and type II (β) errors, respectively. They are based on what happens when one proposes the so-called null hypothesis and then does the test:

(i) The *null hypothesis* typically proposes a general or default position. For example, the null hypothesis would state that there is *no relationship* between two measured phenomena or that a potential treatment has no effect.
(ii) The type I error (false positive) occurs when the sensory scientist rejects the null hypothesis (H_0) when it is true; that is, that the two samples actually do not differ, but the test says, incorrectly, that the two samples do differ. The type I error is driven by the value of α selected by the sensory scientist, usually 0.05 or 0.01. These levels correspond to one chance in 20 or one chance in 100, respectively, of making a type I error.
(iii) The type II error (false negative) occurs when the sensory scientist accepts the null hypothesis (H_0) when it is false; that is, that the two samples do, in fact, differ, but the test suggests that they are the same. The type II error is based on β and is the risk of not finding a difference when a difference exists.
(iv) The power of a test is the difference between 1 and β where: Power $= 1 - \beta$.
(v) Power depends on the degree of difference between the samples, the value of α selected, and *the size of the panel* (Lawless & Heymann, 1998).

Panel size

Panel size is a key issue in consumer sensory research. One of the most often cited problems with consumer sensory tests is the relatively small number of consumers participating in the evaluations, compared with the large numbers who participate in a market research test. Typical market research tests can involve 100–500 respondents, or even more, a base that is far greater than the number of participants in a sensory research test.

The use of large numbers of respondents in market research studies originates from the misconception that validity occurs when the researcher polls the response from a large number of consumers. Panel size (N) is a concern to management or sensory clients who want to maximize the assurance of making the right decisions but also want to save time and resources while conducting the test to learn about responses to the product(s). A larger panel of consumers in a test enhances the likelihood of finding product differences and thus discovering which of the products is the most acceptable (Stone & Sidel, 2004). Panel size influences cost and the quality of the research. The application of statistical sampling techniques permits the sensory scientist to use a subset of the population that will allow one to generalize the results (Resurreccion, 1998).

Researchers use a number of methods to estimate the required sample size. One method is simply guessing. When the panel size is selected without any basis except intuition, the process is inefficient and often leads to oversampling or undersampling. Oversampling originates from the misconception that validity is directly associated with the use of a large number of consumers and will result in inefficient use of time and resources. On the other hand, undersampling may lead to errors and the need to conduct another test. Usually a 95% confidence that a difference exists between two products, or that the product will be rated at a specified level by the entire population, is desired (ASTM, 1979). Since the panel costs money, cash-strapped researchers quickly learn to optimize the costs by cutting the panel size to the absolute minimum level necessary for them to make a statistically valid decision.

The published recommendation for the number of panelists needed for a descriptive test is 10–15, and 100 for consumer tests (ASTM, 1992; Meilgaard, Civille, & Carr, 2007). These recommendations are made with the caveat that the base size or number of panelists is not cast in concrete and should not be accepted without knowing the basis for these recommendations. Muñoz (Moskowitz, Muñoz, & Gacula, 2003) indicated that the sample size used in employee consumer tests ranges from 40 to 100. She adds that this small sample size should be used for screening purposes or in preliminary phases in the early product development process. Hough and coworkers (2006) presented estimates of sample size based on the average root mean square length (RMSL), the α value (type I error), and β value (type II error). They concluded that 112 consumers are needed considering the RMSL, an α value of 5%, and a β value of 10%.

For descriptive analysis tests, the sample size varies according to the predilections and resources of the research. Most analyses appear to work with 8–14 trained panelists, although the availability of the panelists may ultimately determine the number of participants in a test. When calculating panel size, panelists' variability is considered. Highly trained panelists will be less variable than consumers and therefore fewer panelists will be needed for a test. Many researchers rely on a small panel size when their panelists are highly trained. Chambers, Bowers, and Drayton (1981) reported that a panel of three highly trained panelists performed at least as reliably and discriminated among products as well as did a panel of eight less-trained panelists.

Calculations of sample size are based on statistical criteria. These criteria are the magnitude of difference in sample quality that seems to be important to users, the coefficient of variation or standard deviation from previous tests, and the risk that management is willing to take (ASTM, 1979; Gacula & Singh, 1984; Gacula, 1993). Gacula has specified three key factors (Gacula & Singh, 1984; Gacula, 1993; Moskowitz, Muñoz, & Gacula, 2003):

(i) *Relevant differences*: This is the magnitude of the difference to be detected (D). This magnitude is difficult to determine because of the several types of rating scales used in practice, the varying number of categories on a scale, and a variety of other factors that affect the results, such as the nature of the panelists.

(ii) *Variability*: Estimates of variance or standard deviation of the difference (S) in the case of paired comparison tests. This value is generally obtained from historical data.

(iii) *Criterion*: The selected significance level for the test for detection of the difference. The most common levels selected are $p = 0.01, 0.05, 0.10$. Levels of $p = 0.10, 0.15$, and 0.20 may be considered, depending on the type of test. The client sets the degree of precision that is desired. This tolerance level may be "within 5%" of the figure for the total population. At the same time, the client sets a confidence level at a desired value. For example, an error of 5% is set by the client (ASTM, 1979).

These three factors are considered to balance the cost and risk of obtaining the wrong conclusion. Different scenarios may generate the need for various numbers of panelists in the study (panel size, N). Detection of small differences for an attribute or response variable can end up requiring a large panel size, N. That increased panel size naturally will increase the cost to run the study. Conversely, using a small N creates risk of getting biased results. This is the reason Gacula (Moskowitz, Muñoz, & Gacula, 2003) considers N an approximation.

Relevance of the difference to be discovered in the test is of prime importance. One might spend a lot of money finding differences, being precise about the odds of finding those differences by chance alone, yet spend the money fruitlessly because the differences, no matter how clearly supportable statistically, are simply irrelevant to the consumer.

To avoid guesswork and to increase efficiency, it is recommended that sample size calculations must be made to determine the required number of responses for a test. Most statistical texts will give a formula for calculating sample size (Gacula & Singh, 1984; Gacula, 1993). The simplest well-known formula for calculating N is given by Gacula (1993) as:

$$N = [(Z^1 + Z^2)^2 S^2]/D^2$$

Here:

(i) $Z_1 = Z$ value corresponding to type I (significance level)
(ii) $Z_2 = Z$ value corresponding to type II errors
(iii) $S =$ standard deviation of the attribute in question
(iv) $D =$ specified difference to be detected at the prescribed type I and type II errors

Table 11.2 shows a practical application of these principles. The table shows two examples of calculations wherein all terms, except D, remain identical. Where $D = 0.75$, a sample size of 42 is required. In contrast, when D is decreased to 0.24, the sample size, N, increases to 378.

Panel size—a simulation study

Gacula (1993) conducted a simulation study in order to provide a realistic picture of the important role of S and D when calculating the appropriate sample size. Simulations show the statistical rules "in action":

(i) When testing two products on a 9-point hedonic scale with an assumed mean difference D equal to 0.0, and a standard deviation of 1.57 based on historical data, a panel size of $N = 100$ would be needed to test the null hypothesis.
(ii) However, the same conclusion would be reached whether $N = 100$ or $N > 100$ (such as a base size of 200, selected to be "sure" of the results).
(iii) The cost to work with a larger sample of panelists (far greater than 100) would have made a great difference in the cost compared to using a panel size of 100.

Table 11.2 Estimated sample size needed to test for two samples that differ only on the value of their difference, D.

Term	Description	Example 1	Example 2
Z_1	Tabular value for type I (α or significance level) error of $p = 0.05$	1.960	1.960
Z_2	Tabular value for type II (β) error of $p = 0.10$; Power $= 1 - \beta$ $= 1 - 10\%$ $= 90\%$	1.282	1.282
S	Standard deviation	1.500	1.500
D	Difference between samples A and B (from 0.0 to 1.0)	0.750	0.240
N	Sample size	42	378

(iv) Gacula's simulation demonstrates that factors such as the product, cost, and statistical considerations must be taken into account before one, by fiat, increases the number of participants in a test. These simulation programs can assist in the choice of an appropriate sample size, while minimizing cost.

(v) Guidelines regarding panel size have been published for the different types of test and are listed on Table 11.3. It should be noted that these are approximations.

Panel size for focus groups

Focus groups vary in size. Ideally, the base size comprises 8–12 participants who represent the target market for the product (ASTM, 1979; Sokolow, 1988; Chambers & Smith, 1991). The small panel size makes it difficult to draw a focus group sample that is truly representative of typical consumers or anticipated consumers of the product type involved:

(i) A group with 10 participants is probably close to ideal.

(ii) A focus group with fewer than eight participants does not generally provide adequate input.

(iii) A focus group having more than 10 participants will not likely provide sufficient time during a 1.5–2-hour session to cover the number of issues planned (Resurreccion, 1998).

Table 11.3 Recommended panel size for various consumer tests.

Consumer test	Number of panelists	Reference
Focus groups	8–12 (10 is ideal)	ASTM, 1979 Chambers and Smith, 1991; Sokolow, 1988
Laboratory test	50–100 25–50 (40 is recommended)	IFT/SED, 1981 Stone and Sidel, 2004
Central location test	100 50–300	Stone and Sidel, 2004 Meilgaard, Civille, and Carr, 2007
Home use test	50–100 75–300	Stone and Sidel, 2004 Meilgaard, Civille, and Carr, 2007

Panel size for laboratory acceptance and sensory tests

In the laboratory test, the panel comprises consumers who are recruited and screened for eligibility to participate in the tests from a consumer database comprising prerecruited consumers. Usually, the researcher aims for 25–50 ratings per product; at least 40 ratings per product are recommended by Stone and Sidel (2004), but 50–100 ratings per product are considered desirable (IFT/SED, 1981).

In smaller tests, comprising 20–30 ratings per product, it may be difficult to establish a statistically significant difference in a test with the small number of panelists. However, it is still possible to identify trends and to provide direction to the requestor. With 50–100 responses per product, statistical significance increases to a large extent.

Sidel and Stone (1976) provided a guide for selecting panel size based on the expected degree of difference in ratings, basing their guide on the 9-point hedonic scale and the size of the standard deviations. They warned their readers that different products have their own requirements, suggesting that smaller panels can provide statistical significance when the variability of the sample is small. Larger panels are needed as variability of the products increases (Resurreccion, 1998).

Panel size for central location tests

Tests that use "real" consumers have considerable face validity and credibility. Usually 100 more ratings per product are obtained, but the number may range from 50 to 300 (Meilgaard, Civille, & Carr, 2007), especially when segmentation is anticipated (Stone & Sidel, 2004). The increase in the number of consumers required in a central location test compared to the laboratory test is necessary in order to counterbalance the expected increase in variability due to the inexperience of the consumer participants and the "novelty of the situation." Several central locations may be used in the evaluation of a product. The increased number of respondents of 100 or more has advantages but also cost disadvantages compared to the laboratory test (Resurreccion, 1998).

Panel size for home use tests

Due to the lack of control over conditions of testing in a home use test, a larger sample than that required for a laboratory test is recommended. The minimum number of ratings per product is usually 50–100, but even more ratings per product might be desired. This number varies with the type of product being tested and with the experience of the respondent in participating in home use tests. With "test-wise" panelists, Stone and Sidel (2004) recommend that a reduction in panel size could be made because these panelists are familiar with how the test is conducted and feel more comfortable in the test situation. In home use tests that are multicity tests, 75–300 responses are obtained per city in three or four cities (Meilgaard, Civille, & Carr, 2007).

Estimating error through replications

Replication consists of running the panelists twice or more on the evaluation task. Replication in descriptive analysis testing is critical when panelist training is not intensive and when intensity references are not used. Limited training and absence of references result in highly variable data. In consumer tests, where N is large and replication is impractical, the residual error is compounded with panelist effect. Because of the large N, the compounded residual error would be an acceptable estimate of residual error.

Replication provides many functions such as increased statistical power and ability to test more than one processing run or batch or to estimate panelist variability from evaluation to evaluation. Replications are important to estimate residual or pure error. Without an estimate of residual error, tests of significance cannot be done. Furthermore, interaction between panelist and products cannot be estimated. Interaction is important in evaluating panelist performance and necessary in the small panels typically used in descriptive tests. Replication also provides an estimate of panel repeatability.

PRAGMATICS: HOW DO I GET THE RIGHT INDIVIDUALS TO PARTICIPATE?

Sensory and acceptance tests by their very definition require the human judge. With today's pressures on time, with the costs of research increasing, and with more and more people refusing to participate, a section on strategies for recruiting and getting the right people to participate is becoming increasingly relevant.

Recruitment

The recruitment of consumers is most frequently conducted by using a recruitment screener. Screener is short for "screening questionnaire." The screener contains a series of questions that assist the researcher in selecting an appropriate panel of consumers that match the target market. The screener may be an entirely separate questionnaire, different from the actual test questionnaire, administered during recruitment, whether by telephone or face-to-face. The screening questionnaire is generally created by attending to the needs of and soliciting the input from client management.

For example, in testing a dual-ovenable packaging for home-replacement meals, management may wish to include at least 70% of dual income households, or in testing a frozen shrimp product, management may wish to include 50% of households that have a family income above a specified level. The foregoing are straightforward requirements, which can be specified at the start of the study. Whether these requirements are actually necessary or not remains a much different question to answer and is left to those who request the study.

An example of a screening questionnaire used in recruitment appears in Table 11.4. A small number of questions, worded differently, may be added to the demographic questionnaire, which is filled out after evaluation of the samples to verify responses to food use and frequency of use questions. Some considerations that may be important to management or a client when screening prospective panelists are discussed later in this chapter.

Panel—appropriate users

One primary consideration in screening for consumer panelists is that the consumers should be actual or prospective consumers of the product. Muñoz (Moskowitz, Muñoz, & Gacula, 2003) feels this recruitment criterion is extremely important because the panelists need to be familiar with the product or the product category in order to have experience with the product, and also have the proper frame of reference.

Table 11.4   Sample of a screening questionnaire for panelists who are in a database.

Prescreen panelist using database information prior to calling		
Sample quotas for this study are		
Age	Male	Female
18–24	6	6
25–34	6	6
35–44	6	6
45–54	6	6
55–64	6	6

Panelist code #_____
Date called and status of call:

Hello, my name is_____and I am calling from the University of Georgia Experiment Station from the Department of Food Science. We are interested in consumers' opinions about peanut products. Do you have a few minutes to answer some questions about yourself and your use of peanut products? (YES—CONTINUE to 1; NO—TERMINATE and note the reason above under status of call.)

1. What year were you born? (write year____and place a check in the consumer's age group). Circle the consumer's gender.

18–24_____	(1981–1987)	M/F
25–34_____	(1971–1980)	M/F
35–44_____	(1961–1970)	M/F
45–54_____	(1951–1960)	M/F
55–65_____	(1940–1950)	M/F

(If under 18 or above 65, TERMINATE.) State, "We are only supposed to survey those between the ages of 18 and 65, but I thank you for your time and hope you will participate in other surveys that we may be calling for in the future. I hope you have a nice day."

2. When was the last time you participated in any test with us?
    (If less than 3 months, TERMINATE.)

    If answer is "I do not remember," prompt consumer with foods such as "Was it okra, honey spread, steak, hamburger?" (if answer is "yes" to any of these, terminate because consumer participated in a test within last 6 months). Otherwise, continue.

3. Do you have any food allergies?
    Yes _____
    No _____ Skip to Q#4

    If "yes," ask, "Which foods are you allergic to?" _____
    (If none of the foods mentioned include peanuts, CONTINUE. If allergic to peanuts, TERMINATE.)

4. How often do you eat peanuts?
    If once a month or more, _____ CONTINUE schedule for test.
    If less _____, TERMINATE.

5. We are conducting a taste test using some of the foods that I mentioned before. The test is scheduled for October 29 and 30. It will take approximately 1 hour and 20 minutes, which includes a 20-minute break, and you will be paid $10 for your time. Would you be interested in participating? (If "YES," CONTINUE.) What would be the most convenient time and day for you to come for the test? (Check available times and dates.)
    Which day would be best?

Tuesday, October 29	_____
Wednesday, October 30	_____

Table 11.4   (*Continued*)

The times for the test each day are:

10:00	_____
2:30	_____
5:30	_____

If YES, schedule for test. We can schedule up to 18 consumers per test but no more that a total of 60 consumers for the entire test.

If none of the times are acceptable, TERMINATE. State, "I am sorry none of the times fit into your schedule. May I put you on a waiting list for this test, and you can call us if you find out you can participate in one of the sessions?"

6. We need to verify your name, address, and phone number so that we can send you a reminder postcard through the mail about a week before the test that will have directions to the Sensory Evaluation Lab. If you find out you are unable to come to the test, please call us as soon as possible because it will be very important for us to schedule someone else to take your place.

Thank you very much for your time and participation. I hope you have a nice day.

Name of interviewer _____

Date _____

It should not be assumed that almost any group of people are potential consumers simply because of the broad distribution of consumption patterns that exist. Effort needs to be expended to recruit high proportions of consumers who are actual users of the product, product type, or product category. Recruiting the appropriate panelists can be done in various ways depending on the type of consumer test being conducted. Consumers can be screened at the time of testing as they are intercepted for a central location, mobile laboratory, or home use test and during the recruiting process for a laboratory test (Resurreccion, 1998).

Another approach recognizes the fact that product usage often correlates with demographic characteristics. One can select a population that has a greater than random chance of containing these individuals. An example is to work with homes near a school to find children who are candy eaters. Homes located far away from schools are probably less likely to have children of school age. Use of certain items is believed to be restricted to high-income households, or some products are believed to be more popular in certain regions of the country than are others. Another frequent practice in testing food products uses panels comprising only the major preparer and purchaser of food in the household, even when this criterion generates a larger proportion of female participants than in the user group. The ingoing belief is that the opinions of these individuals (those who shop or prepare the food) will be the primary determinants of food purchases for the entire household (Resurreccion, 1998).

Panel—how frequently should they use the product?

Classifications for frequency or use are heavy, medium, or light users of a product. Management or the specific client requesting the test should give direction regarding specified cutoff points for product frequency of use; the market research function would most likely have data leading to this decision. The sensory researcher assists in the decision regarding whether heavy users are needed, based upon the researcher's knowledge about the sensory impact of the product and how heavy users might find the sensory impact to be more important than do lighter or less frequent users. Certainly, heavy users are not needed for every test, but there are a few types of tests that may require a majority of heavy users of the product. These tests are for ingredient and process substitution, claim substantiation, comparisons with competitors and benchmarking, product matching, and establishment of quality control specifications (Moskowitz, Muñoz, & Gacula, 2003).

The desired frequency of use must be decided upon by the sensory practitioner and the client with input from the market research department. For example, a client and the sensory practitioner may choose to use a panel that has 60% heavy users and 40% regular users of a food product category. The client should keep in mind that the cost of recruiting will increase as the frequency of use requirement is made more stringent.

Panel—demographic characteristics

When running a consumer test, one requirement is to represent the target market for the product, or when there is no target market, then at least represent the range of potential consumers. One way to accomplish this objective recruits consumers from different markets, with the belief that such a broad sample represents the full range of consumers in the universe. The approach draws a sample for consumer tests that includes age, gender, residence in certain geographic regions of the country, income, education and employment, and race or ethnic background. The overall objective is to select a relatively homogeneous group from a fairly broad cross section, all individuals of which are expected to like and use the product or the product category, as well as exclude those individuals who exhibit extreme or unusual response patterns (Stone & Sidel, 2004). The cost of recruiting increases significantly, often quite considerably, when one wants to test a product with a narrow target market of consumers and recruit those hard-to-find individuals.

Let us look at the typical criteria—age, gender, and geographic location—as well as some other criteria:

(i) *Age*: Although age is very important when the product is targeted toward a specific age group, such as children, elderly consumers, etc., for other products that have broad appeal, a broad age group of consumers, ages of 18–64 are usually recruited for the test. The sample should be selected by age in proportion to their representation in the user population. It is important that current figures on the demographic characteristics of the target market be obtained before sampling (Resurreccion, 1998). Because available time is increased in retired elderly consumers, and young adults are often not interested in participating in such tests, special effort must be made in recruiting the proportion of each age category when age is not specified among recruitment criteria.

(ii) *Gender*: There are gender differences in the preference for certain foods. Gender is extremely important when the product is specifically developed to target one gender. For example, for a new restructured steak product, a company requested equal proportions of male and female consumers, when they projected that more men than women would compose the target market for their product. When the company does not specify gender, the panel would invariably comprise more females because females are generally more interested in participating in this type of activity. Otherwise, most clients would require both genders but make recommendations on their proportion to be used in the panel. The sample may be selected by gender in proportion to their representation in the user population. This specification to match the user population is especially important when gender effects are not known. In an example of tests on several types of coffee (Cristovam *et al.*, 2000), female consumers were more discriminating than were males in their preferences for light roast coffees. Males and females were likewise found to differ in their preferences for coffees to which milk was added. Similar findings were made in studies on starchy food (Monteleone *et al.*, 1998) and fat spreads (Bower & Saadat, 1998). When the practitioner uses a convenience sample of consumers that comprises a majority of consumers from one gender, for example, due to their availability or interest, the acceptance ratings for certain products may be biased due to the panel composition, and the results may lead to erroneous conclusions.

(iii) *Geographic location*: In some instances, differences in preference and usage of many products occur across geographic areas. It is therefore necessary to test products for national distribution in more than one geographic location. When cost prevents testing in different regions of the country, so that the study is executed in one market only, then it is advisable to recruit consumers who have lived in various regions of the country to represent these region differences and minimize bias. For example, consumers from different regions may have different sensory preferences for certain cuisines, such as the preference for hotter, spicier food in the US southwest region (Moskowitz, Muñoz, & Gacula, 2003). In other cases, consumers' expectations from one product may differ from one region to another; for example, in the United States consumers' preferences for the strength of coffee differ from one region to another.

(iv) *Other criteria*: There are other criteria that can be used, such as income and education, which are indicators of socioeconomic status; children in the household; dual income households; etc. These criteria may be added to the screening questionnaire if they are specifically required by the client. Additional criteria added to the screener will result in increased difficulty and time required for recruitment. The result is additional study cost. The client and the sensory practitioner should jointly decide on what additional criteria need to be added.

Experts, trained panels, and naive consumers

Here are "two major schools of thought on what types of individuals should be used to guide product development for early stage testing" (Moskowitz, 2000). There are those who believe that early stage product development should be guided by experts. These experts should have a sense of consumer tastes and be able to communicate these consumer desires to product developers. They believe that the expert possesses a superior ability to guide development. There exists no evidence to support this claim of superiority. The practice comes from the tradition of the company wine expert, the brewmaster, or the flavorist (Moskowitz, 2000). Although the majority of product developers today do not share this thinking that there is a true "golden tongue," the position that the "expert" is infallible continues in a number of companies (Moskowitz, 2000).

The other school of thought is that consumers, not experts, should be used for almost all tasks in early stage product development. Proponents of this school of thought argue that well-instructed consumers can describe products as well as experts (Moskowitz, 1996;

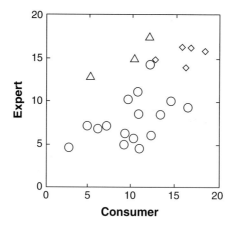

Figure 11.1 Standard errors across the 37 products tested in the study. The figure compares the standard errors for consumers versus for experts, after the scales used by experts have been adjusted to become a 0–100-point scale. The shapes refer to different sensory attributes: diamonds = visual attributes; triangles = texture attributes; circles = flavor attributes. The plot suggests that consumers are no more variable than experts in their rating of sensory attributes (Moskowitz, H.R. 1996. *Journal of Sensory Studies*. Food & Nutrition Press Inc., reprinted with permission).

Moskowitz, Muñoz, & Gacula, 2003). Whereas the majority of companies no longer rely on a company expert, many continue to count on a small group of trained panelists for product description. Moskowitz (1996) found high correlations between the trained panelists' and consumer ratings for a single attribute across products and observed that their ratings are interchangeable. He indicated that as a result of training, these individuals have a better descriptive vocabulary and sharper-tuned ability to describe products, but concludes that for many common sensory attributes, trained panelists and consumers appear to do equally well (Moskowitz *et al.*, 1979; Moskowitz, 1996).

Let us look at some data to address the problem (Figure 11.1). Scatter plots of consumer and trained panel ratings of 37 gravy products are shown in Figure 11.1. All correlations exceeded 0.90, indicating a high association between consumer and trained panel ratings for brownness of color, an appearance attribute; strength of tomato flavor, a flavor attribute; and oiliness of texture, a mouthfeel attribute.

However, the argument continues. Here is just a sampling of the different points of view:

(i) Lawless and Heymann (1998) state that consumer panelists should not be relied on for accurate description information.

(ii) Moskowitz is of the opinion that the practice of using trained individuals rather than consumers for early stage product development has retarded progress in the field (Moskowitz, Muñoz, & Gacula, 2003).

(iii) An opposing viewpoint is that consumers are biased when they are asked about product attributes, and ought only to be asked to rate overall liking.

(iv) Husson, Le Dien, and Pages (2001) stated that there has not been enough research to determine whether or not consumers are biased when they are asked about product attributes.

RUNNING THE STUDY: A POTPOURRI OF ISSUES

The samples and the actual evaluation

Test execution: how many samples can a panelist validly rate?

When consumers are used, a majority of sensory practitioners believe that the number of samples to be tested should be a minimum, preferably two samples. Moskowitz, Muñoz, and Gacula (2003) regard this belief as a myth, based on studies involving taste, smell, and foods. Moskowitz reported that across many studies it is shown clearly that panelists can track the magnitude of the stimulus through their rating of intensity (Moskowitz, Muñoz, & Gacula, 2003). On the other hand, Gacula (Moskowitz, Muñoz, & Gacula, 2003) states that one objective of providing only two samples per evaluation is to estimate *carry-over effects,* and not because one believes consumers to be incapable of rating more than two samples. In the case of highly flavored products, two samples are recommended to provide estimates of carry-over effects. When carry-over effects are large, then reanalysis of the data is warranted.

Further, Gacula (Moskowitz, Muñoz, & Gacula, 2003) advises that a "control or reference" should be used when more than two samples are tested to prevent so-called wandering scores. When the number of samples to be tested is large (e.g., 12 samples), panelists can be asked to return for further tests to complete the set of samples. When this is not possible, the samples are administered using a balanced incomplete block design presentation to allow for convenience to the panelists and logistical considerations in conducting the test.

Are booths needed for the evaluations?

In the evaluation of samples, good laboratory and environmental conditions, free from distraction, are needed. Environmental conditions should be of high quality, such as proper lighting, comfortable temperature, air free from odors, and supplies for expectoration of samples as needed.

In laboratory tests, isolated booths where panelists cannot see each other or talk to each other are used. Booths are always desirable (Lawless & Heymann, 1998; Stone & Sidel, 2004); however, it is possible to conduct valid tests without isolated booths. When space is limited and resources are limited, it is not possible to construct booths. Furthermore, in a facility with no booths, such as in a focus group room or a hall (church or hotel), evaluations cannot be conducted in booths. In such cases, it is important to have the necessary panelist-to-staff ratio to ensure that panelists are not distracted or cause distractions to other panelists. One technique to minimize distractions seats all panelists facing one direction, and has ample space between panelists during the evaluations.

Scales and subjective measurement

Can consumers use scales? Some sensory scientists believe that consumers have difficulty rating product attributes and working with scales. For these reasons, the skeptics recommend only fully anchored scales for consumer tests. On the other hand, studies reported successful use of scales, even those scales without anchors other than at the end points (Hough, Bratchell, & Wakeling, 1992; Rohm & Raaber, 1991. Muñoz's (Moskowitz, Muñoz, &

Gacula, 2003) experience has led her to believe that consumers can easily use scales that are not fully structured, as long as the scales are sound, with properly selected end anchors and possibly a midpoint. Muñoz further stated that consumers are capable of using any scale provided they can understand it. She recommends a short orientation prior to the consumer tests, wherein the scales are shown to the consumers and their main characteristics are explained. Furthermore, examples should be provided to demonstrate the scale properties and their use. Consumers' questions should be encouraged until the use of the scales is clear.

Fully anchored scales pose a problem

The use of fully anchored scales poses problems in regard to the selection of appropriate word anchors for each category. Muñoz (Moskowitz, Muñoz, & Gacula, 2003) states that finding word anchors that ensure equidistance between scale categories is difficult and, when unsuccessful, could generate a flawed scale, and consequently a biased measurement. This problem is exacerbated in global research when the scales are translated into other languages.

Scales should be easy to use

Choose scales that are easy for the panelists to use. When possible, use the same scales for all attributes in order make the respondent's task easier. When the evaluations are easy to perform, the researcher will obtain better data. Although the researcher may make every effort to remind the respondent to use each scale, individually, according to its scale points, panelists get tired of changing focus. Panelists adopt a strategy in order to simplify their task. Often, after panelists adopt one way of thinking about the scale, they apply that strategy to all scales. In light of the respondent's tendency to do this, the use of a common scale is recommended. One has to consider, however, the opposite tendency. In an effort to refrain from confusing the respondent, the researcher who uses a single scale format might end generating ratings that correlate across scales, with little "new information" from any scale other than the first scale to be used with a product. The resolution of the issue of different scales versus the same scale must be considered in light of the desire for good data, and the strategy of respondents to simplify the task and minimize their efforts.

What types of scales should be used in the evaluations?

Sensory evaluation involves the quantification of sensory perceptions through the use of scales. Quantification allows the collection of numerical data that can be analyzed statistically. The three types of methods used for scaling in sensory evaluation are:

(i) *Category scaling*, where panelists assign numerical values to perceived sensations based on specific and limited responses (Lawless & Heymann, 1998).
(ii) *Ratio scaling used in magnitude estimation*, where panelists assign any number they wish, to reflect the ratios between sensations.
(iii) *Line scaling*, where the panelists make a mark on a line that corresponds to the strength of sensation or degree of liking for a sample.

Table 11.5 Four levels of measurement used in sensory evaluation.

Nominal scales are used to denote membership in a category, group, or class.

Ordinal scales are used in ordering or ranking.

Interval scales are used to denote equal distances between points and are used in measuring magnitudes, with a zero point, which is usually arbitrary.

Ratio scales are used in measuring magnitudes, assuming equality of ratios between points and the zero point is a "real" zero.

Source: Resurreccion, 1998. Table 2.5, p. 35. Copyright 1998 by Aspen Publications, reproduced with kind permission of Springer Science and Business Media.

Levels of measurement

How are the numbers assigned to the different scaling methods? Four levels of measurement are used in sensory evaluation. These are the nominal, ordinal, interval, and ratio scales (see Table 11.5; Resurreccion, 1998):

(i) *Nominal scale*: The nominal scale uses numbers to label, code, or classify events, items, or responses (Lawless & Heymann, 1998; Stone & Sidel, 2004). The scale can be related to days on a wall calendar, or the baseball program where the players are numbered (ASTM, 1979). Examples of the use of nominal scales are the labels or code questions such as those on demographic characteristics of panelists, gender (male, female), or employment classification (full-time, part-time, retired, not employed). No assumption is made that the numbers reflect an order. Appropriate analysis of the responses may be to use frequency counts, distributions, or modes (the response given by the most panelists). Different frequencies of response for different samples can be compared by chi-square analysis (Stone & Sidel, 2004) or other nonparametric statistical method (Siegel, 1956).

(ii) *Ordinal (ranking) scale*: Ordinal scales are those that use numbers assigned to organize the rank order of samples from "highest" to "lowest" or "most" to "least." Ranking is one of the most commonly used sensory evaluation methods that use an ordinal scale. These tell whether the sample has a quality of being equal to, greater than, or less than another sample. For example, a number of peanut butter samples might be rank ordered for perceived roasted "peanutty" aroma from highest to lowest, or a number of breakfast bars might be rank ordered from most preferred to least preferred. Analysis of ranked data is straightforward. One uses tests such as the simple rank-sum statistics found in published tables (Basker, 1988) that updates and modifies previous approaches (e.g., Kramer's rank-sum test). In addition, different methods that can be used include those appropriate for nominal scales and in particular those methods classified as nonparametric methods. These nonparametric methods make no assumptions about the nature of the underlying distribution of responses, and require only that the data have ordinal-level properties. The methods include the Wilcoxon signed ranks test, Mann–Whitney, Kruskal–Wallis, Friedman's two-way analysis of variance, chi-square, and Kendall's coefficient of concordance (Resurreccion, 1998).

(iii) *Interval scale*: The interval scale assumes that the intervals or distances between adjacent points on the scale are all equal. Examples of an interval scale in the physical sciences are the Fahrenheit and centigrade scales for temperature, where each has equal divisions between values. In sensory evaluation, an example of an interval scale is the

9-point hedonic scale for liking. Interval scales help define the differences between samples. In addition, because interval scales are considered to be truly quantitative scales, a large number of statistical tests are used to analyze results. These statistical indices and procedures include means, standard deviations, T test, analysis of variance, correlation and regression, and factor analysis among the many statistical methods that are appropriate to analyze the data.

(iv) *Ratio scale*: In ratio scaling, panelists assign a number to reflect the ratios between sensations. Magnitude estimation is the most frequently used ratio scale in consumer testing. Magnitude estimation assumes that the ratios of the ratings reflect the ratios of the respondent's perceptions or feelings. Magnitude estimation further assumes that there is a fixed and meaningful zero point or origin of the scale (e.g., the Kelvin scale of temperature, which has a meaningful zero, unlike the Fahrenheit and centigrade scales).

What criteria should the researcher use to select the appropriate measurement scale?

Selection of a scale for a specific test is one of the most important steps in implementing a consumer test. Criteria that should be used in selecting or developing a scale for a sensory test appear in Table 11.6.

Hedonic scale

In the world of scales and test methods, the 9-point hedonic scale has gained special consideration because of its suitability in measurement of product acceptance and preference. It was developed by Jones, Peryam, and Thurstone (1955) and Peryam and Pilgrim (1957). Further research resulted in a 9-point bipolar scale with a neutral point in the middle and nine statements that describe each of the points or categories. The equally spaced categories are described by the statements as shown in Table 11.7.

The hedonic scale is easily understood by panelists and is easy to use. The reliability and validity of the 9-point hedonic scale in the assessment of several hundred food items has been established in the food science literature (Meiselman, Waterman, & Symington, 1974; Peryam *et al.*, 1960). A number of sensory scientists espouse the use of the 9-point

Table 11.6 Criteria for selecting or developing scales.

(i)   The scale should be valid. It should measure the attribute, property, or performance characteristic that needs to be measured as defined by the objectives of the study.
(ii) The scale should be unambiguous and easily understood by panelists. Questions as well as the responses should be easily understood by the panelists.
(iii) The scale should be easy to use. Consumer tests involve panelists who are not trained.
(iv) It should be unbiased (Stone & Sidel, 2004). Results should not be an artifact of the scale. Bias may result from the word or numbers used in a scale. Unbalanced scales result in biased results.
(v)   It should be sensitive to differences. The number of categories used and the scale length will influence the sensitivity of the scale in measuring differences.
(vi) The scale should consider end point effects.
(vii) The scale should allow for statistical analyses of responses.

Source: Resurreccion, 1998. Table 2.6, p. 37. Copyright 1998 by Aspen Publications, reproduced with kind permission of Springer Science and Business Media.

Table 11.7   The 9-point hedonic scale.

9	Like extremely
8	Like very much
7	Like moderately
6	Like slightly
5	Neither like nor dislike
4	Dislike slightly
3	Dislike moderately
2	Dislike very much
1	Dislike extremely

hedonic scale exclusively. The 9-point hedonic scale has been widely used, researched, and recommended for a number of years, and validated in the scientific literature (Stone & Sidel, 2004). Stone and Sidel (2004) state that the stability of responses and the extent to which the data can be used as a sensory benchmark for any particular product category is of particular value when using the 9-point hedonic scale.

Modifications to the 9-point scale have been suggested, such as eliminating the neutral point ("neither like nor dislike"), or simplifying the scale by eliminating options such as the "like moderately" and "dislike moderately" points on the scale or truncating the end points by eliminating the "like extremely" and "dislike extremely" points, but these suggested modifications were either unsuccessful or had no practical value. These modifications would introduce biases such as the problem of end-use avoidance (the hesitation of panelists to use the end categories). A major concern has been that the scale is a bipolar scale but the analysis of the results is unidirectional. However, there is no evidence that consumers have difficulty with the scale and that the statistical analysis presents a problem (Stone & Sidel, 2004).

Another major concern is the number of available scales versus the use of the scales by panelists. Truncating a 9-point scale to a 7-point scale may leave the consumer panelist with only a 5-point scale. It is, therefore, best to avoid the temptation to truncate scales (Lawless & Heymann, 1998). Too few scale points hinder discrimination. When using a short scale, such as a 3-point category scale, the researcher loses a great amount of information about the differences among products. Unstructured line scales with three word anchors, one at each end and the midpoint, are not subject to this problem because in theory the line scale comprises an infinite number of locations.

Acceptance of a product can be quantified by the mean and the standard deviation. When tested against a wide array of competitive products, the scale provides an order of acceptance of the different products. By modeling, using the 9-point scale as the dependent variable, the researcher can compare the scores of products to the highest possible scores that the product category may achieve. Furthermore, parametric statistical analysis provides useful information about product differences. These analyses include procedures such as analysis of variance (ANOVA) of acceptance. Stone and Sidel (2004) plotted responses of 222 consumers who used the 9-point hedonic scale in evaluating 12 products and observed that this resulted in a sigmoid plot, which indicates that the scores are normally distributed. Stone and Sidel (2004), contrary to O'Mahony (1982) and Vie, Gulli, and O'Mahony (1991), however, claim that data from this scale should not be assumed to violate the normality assumption.

Consumer responses from use of a hedonic scale can likewise be converted to ranks or paired preference data. In order to convert paired preference data, it is necessary to count the number of panelists who rated one product higher and analyze the result using $P = 1/2$, or binomial distribution. The 9-point hedonic scale has yielded results that are reliable and valid. Thus far, further efforts to improve the scale have proven to be unsuccessful. Therefore one can say that researchers should continue to use the 9-point hedonic scale with confidence (Stone & Sidel, 2004).

Hedonic scales for testing with children
In testing with children, the facial hedonic scale is popular as a method to determine consumer acceptance and preferences. These may be simple "smiley face" scales or may depict a popular cartoon character familiar to children (Figure 11.2) and those with limited reading or comprehension skills (Lawless & Heymann, 1998). Stone and Sidel (2004) cautioned that young children may not have the cognitive skills to infer that some of the face scales depicted in Figure 11.2, specifically scales (a) and (b), are supposed to indicate their internal responses to a test product (see facing page).

In studies using the facial scales, the 9-point "smiley face" scale (Figure 11.2e) did not perform well with children 3–5 years old (Chen, Resurreccion, & Paguio, 1996). The 3-point scale has been used in studies with children (Birch *et al.*, 1990; Birch & Sullivan, 1991). Scale lengths from the 5-point scale (Fallon, Rozin, & Pliner, 1984) and the 7-point scale (Kimmel, Sigman-Grant, & Guinard, 1994) have been used with children around 4 years of age. The ingoing assumption is that the scale has to be matched to the cognitive abilities of the child.

Children perform as well as adults in many cases, leading to the conjecture that perhaps worries about the child's ability to use scales may be less well grounded than one might have thought. For example:

(i) In preference tests with children ages 5–7 years old, Kroll (1990) used the 9-point scale with verbal descriptors (see Table 11.8) and found this age group to be able to use the scales effectively.
(ii) Chen, Resurreccion, and Paguio (1996) demonstrated that a 5-point facial hedonic scale with specific descriptors (Kroll, 1990) could be effectively used and understood by 47–59-month-old children (i.e., 4 year olds).
(iii) A 7-point facial hedonic scale with appropriate verbal descriptors (Kroll, 1990) could be used with children ages 60–71 months (i.e., 5-year-olds). The facial scales used by Chen, Resurreccion, and Paguio (1996) are shown in Figure 11.2 (scales (c) and (d)).
(iv) Kroll (1990) likewise found that children ages eight and above are able to complete self-administered questionnaires, thus eliminating the need for costly one-on-one interviewers.
(v) However, unstructured scales are not recommended for children who have been observed to place their responses on the extreme ends of the scale, and either "love" or "hate" a product rather than using the entire scale (Moskowitz, 1985).

What should I know about questionnaire design?

The questionnaire is the most critical aspect of the study. It is important because it is the means by which the research question is presented to the panelist. The questionnaire is the instrument for collecting panelists' responses to the question. Unfortunately, questionnaire design is an often overlooked and underestimated skill.

Preparation of the questionnaire is initiated once the decision has been reached on the test method, experimental design, and other information that was considered important to the test

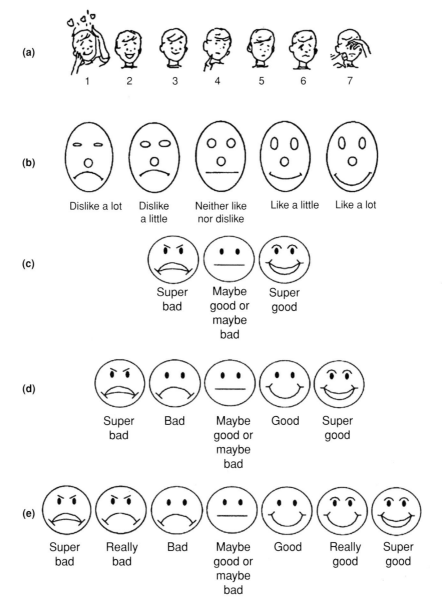

Figure 11.2 Examples of facial hedonic scales used in determining preschool children's preferences. (a) and (b), pictorial scales that are found in the literature that appear to have been used for measuring children's responses by products (reprinted from Stone, H., & J.L. Sidel. 1993. Face scales. In: *Sensory Evaluation Practices,* 2nd edn, ISBN 0-12-672482-2, p. 87, with permission from Elsevier; (c), (d), and (e), pictorial scale for hedonics suitable for young children; Chen, Resurreccion, & Paguio, 1996. Food and Nutrition Press, Inc., reprinted with permission from Blackwell Publishing Ltd.; Resurreccion, 1990. Fig. 11.1 (a), (b), (c), (d), and (e). p. 171. Copyright 1998 by Aspen Publications, reproduced with kind permission of Springer Science and Business Media).

Table 11.8   Rules on question structure and wording.

(i)	Keep the questions clear and similar in style. To avoid confusion, the direction of the scales should be uniform.
(ii)	Use directional questions to address differences that are detectable and can differentiate products.
(iii)	Questions should be interestingly worded and encourage consumers to respond.
(iv)	Consider the importance of including a personal question such as, "What is your family income?" Respondents may consider this question obtrusive and may not answer the question.
(v)	Overelaboration can produce contradictions.
(vi)	Do not overestimate the respondent's ability to answer specific questions such as those involving recall or estimation, for example, "How much salt do you use when you prepare food?"
(vii)	Avoid double negatives.
(viii)	Questions that talk down to respondents should be reworded.
(ix)	Questions should be simple, direct, and encourage consumers to respond.
(x)	Questions should be actionable, that is, lead to specific actions in terms of modifying the product.

Source: Adapted from Resurreccion, 1998. Table 2.4, p. 32. Copyright 1998 by Aspen Publications, with kind permission of Springer Science and Business Media.

objective. In most cases, the project leader develops the questionnaire (Resurreccion, 1988). In many cases the study has been run before, so that the same questionnaire can be used.

Questions to be answered
The questionnaire design begins by asking and answering the following simple questions:

(i)   What is the objective of the test?
(ii)   What questions should be asked?
(iii)   What is the sequence of questions to be asked?
(iv)   How should each question be worded?

The questionnaire should be designed so that consumers participating in these tests should not only understand the questions but be able to answer the questions in the frame of reference intended (ASTM, 1979). Questionnaire wording has been shown to influence the data collection; therefore, wording of the questionnaire is extremely important. Ambiguity should be prevented as much as possible. Lawless and Heymann (1998) recommend that the researcher do "homework" by preparing a flowchart to help design the questionnaire. Some *rules* (really rules of thumb) for question wording were proposed by various authors such as ASTM (1979) and Meilgaard, Civille, and Carr (2007) and have been adapted for use in this chapter. They are listed in Table 11.8.

Structuring the questionnaire—will questionnaire construction affect overall liking ratings?
Popper *et al.* (2004) investigated how the inclusion of attribute questions affects a respondent's rating of overall liking. Five groups of respondents evaluated four variations of a dairy dessert. Each group used a different questionnaire comprising of:

(i)   overall liking only;
(ii)   overall liking plus intensity scales, which asked respondents to rate the strength of 10 sensory characteristics;

(iii)   overall liking plus attribute liking scales, which asked respondents to rate their liking of appearance, flavor, and texture of the products;
(iv)   overall liking plus just-about-right scales, which asked respondents to indicate whether the level of 10 sensory characteristics was "too high," "too low," or "just about right;" and
(v)   overall liking plus attribute liking and just-about-right scales.

They reported that:

(i)   Including just-about-right questions on the questionnaire changed the average overall liking ratings of the products.
(ii)   Including attribute liking questions also tended to alter overall liking ratings but to a lesser extent than did including just-about-right questions.
(iii)   Including intensity scales had no effect.

Question sequence
Questions should be carefully and strategically positioned in the questionnaire. The placement of the questions on the questionnaire may affect the usefulness of the information obtained. Furthermore, bias may result from the positioning of questions in a certain order.

Two factors to consider when deciding on question sequence are the importance of the question and the order in which the attributes are perceived. The most important questions should be asked first. This order attempts to ensure that data are collected from the consumers before extraneous factors or boredom set in (Resurreccion, 1998). Next, the sequence of questions is based on the order attributes are perceived: visual appearance, aroma, flavor, and taste, then texture and aftertaste. After appearance, flavor, or texture are perceived next. Muñoz (Moskowitz, Muñoz, & Gacula, 2003) gives two cogent reasons for asking about flavor before texture, stating that there is no difference when flavor or texture is rated after appearance, but in some cases it is better to rate flavor before texture because of these reasons:

(i)   Flavor may require retasting and texture evaluations will not be influenced by retasting.
(ii)   In general, consumers are more familiar with flavor than texture characteristics. When flavor is evaluated first, consumers will feel more comfortable than when texture is evaluated first (Moskowitz, Muñoz, & Gacula, 2003).

When the panelist has to manipulate the product to evaluate texture attributes, texture by feel may follow flavor appearance (Moskowitz, Muñoz, & Gacula, 2003). This sequence is not mandatory but it is efficient because the panelist can evaluate attributes according to the sequence they are perceived after the sample is placed in the mouth. Within the general flavor and texture attributes, however, specific attributes should be positioned according to the order they are perceived (Moskowitz, 2001).

Position of the overall liking rating in the questionnaire
Should the overall liking question be located at the start or at the end in a questionnaire when questions about other attributes are asked? This may well be the most frequently asked question in questionnaire design. There is a lot of unsubstantiated information regarding this issue. The only thing certain about the placement of the overall question is that there is no

definitive answer. Position of the overall liking question depends on the product or type of food being evaluated and how it is consumed, and the objectives of the study.

(i) *Placement as the first question*: Placement of the overall liking question first on the questionnaire is preferred by many sensory professionals. Their reasons for this preference are:

 (a) *Importance*: The overall liking question is the most important one in the questionnaire. Important questions are asked first to ensure that responses are obtained before any source of distraction or influence occurs.

 (b) *True rating*: First and immediate response to the product will be obtained; this is thought to be the true and most realistic response.

 (c) *Minimize attribute effects*: To minimize bias and influence from responses to attribute questions.

 (d) The panelist will *not* be given the opportunity to justify his or her response to the overall liking question by first analyzing his or her responses to the attribute questions.

 (e) *Caveat*: In multiproduct tests, wherein a monadic sequential presentation order is used and attribute questions are included in addition to the overall liking question, the beneficial effect of putting the overall liking question first is realized only with the first sample. With subsequent samples, positioning of the overall liking question first is minimized and the response to overall liking will not be devoid of influence from the responses to remaining questions.

(ii) *Placement as the last question*: The overall liking question is positioned last:

 (a) When the sensory practitioner believes that overall liking should be decided by the panelist *only after* having evaluated and provided ratings on the specific attributes.

 (b) When evaluating certain foods that warrant positioning of the overall liking question last. When consumers must manipulate the product, or use the product first in order to "validly" evaluate the product on overall liking such as in chewing gum, which changes with the length of time the product is in the mouth; or products with lingering basic tastes or feeling factors like those in pungent, spicy, or mentholated products.

 (c) When product performance over a period of time is needed prior to rating overall liking. Examples include products used with skin care or other personal products.

 (d) In multiproduct tests wherein attribute questions are asked in addition to the overall liking question. ASTM (1998) recommends putting the overall acceptance question at the end of the questionnaire in these cases, so the influence of attribute ratings will affect all products equally.

The foregoing thinking (option #2) is shared by Gacula (Moskowitz, Muñoz, & Gacula, 2003) who states that overall liking or overall evaluation of consumer products is a process of sensory integration of various product properties. Hence, the overall liking question should be placed at the end of the questionnaire. Gacula expresses this schematically as:

Overall liking = f (diagnostic sensory attributes; f refers to the function relating sensory attribute ratings to overall liking ratings)

In addition to overall liking, many practitioners ask panelists to rate the degree of liking of specific attributes. These attributes are those that are presumed to be important as determiners of overall liking. Examples include the general attribute, such as liking of flavor, or even liking of a specific flavor question (Meilgaard, Civille, & Carr, 2007). Whereas attribute questions can be useful in identifying the reasons consumers like or dislike a product, there is concern that such questions can be a source of bias (Stone & Sidel, 2004).

Position of demographic questions
When demographic questions are asked in a questionnaire, they should not be asked at the start of the questionnaire, but rather located at the end, after the most important research issues have been answered. When using consumers recruited from a database and their demographic information is available, asking demographic questions often fails to add any additional information (Resurreccion, 1998).

How long should the questionnaire be?
How many questions should be asked to ensure that the panelist does not get tired or bored? There is no optimal length of a questionnaire. The questionnaire should be the shortest possible length that will answer project objectives (Resurreccion, 1998).

A short questionnaire may not provide all the answers needed for certain products or objectives, so a longer questionnaire will be needed. Questionnaires may range from 1–2 questions or, on the other extreme, 48 questions or even more. In a four-product test with 12 questions per product, a total of 48 questions are not unusual. The shorter questionnaires have been used when the sensory researcher needs to have the panelist focus on these questions; survey research professionals work with longer questionnaires. Moskowitz, Muñoz, and Gacula (2003) state that questionnaire length may be a surrogate for emotional assurance to the marketing brand manager that the study will generate valid results.

The number and choice of attributes on a questionnaire can be reduced by input from the product developer, who may have had a long history of working with the product type. For new products, focus groups are invaluable in narrowing the critical attributes (Resurreccion, 1998), as demonstrated by McNeil, Sanders, and Civille (2000) in developing a questionnaire for peanut butter. In such cases, questionnaire length and panelist fatigue are no longer problematic. The real issue is that consumers do not like being tied up for a long time. Knowing that boredom or apathy easily sets in, and may result in the panelist walking out or tuning out and providing random responses, is essential. Thus, Moskowitz, Muñoz, and Gacula (2003) offer the recommendation that the researcher should try to think about the task from the panelist's point of view. The questions that a researcher should ask include the following:

(i)   Are there too many questions?
(ii) Are the questions redundant?
(iii) Are the questions meaningful?
(iv) Are the questions self-explanatory or will an interviewer be needed?

Successful questionnaire design follows when the researcher lightens the panelist's load by thoughtful design of the questionnaire. When designing a long questionnaire, it is critical that the interest of the panelist is maintained to ensure that the responses are valid and will help to answer the objectives of the test. Muñoz (Moskowitz, Muñoz, & Gacula, 2003) stated that consumers are able to handle long questionnaires when they are given sufficient time to respond to them, when the questionnaire has been formatted properly, when attributes are clearly understood, and finally when the orientation process prior to the test shows/explains the questionnaire to panelists and gives panelists the opportunity to ask any questions.

There is little research on the way questionnaire length affects validity of responses and quality of the data. This gap in the research has led a number of sensory practitioners to be conservative and limit their questions to fewer than 10. Other practitioners have had success in using lengthy questionnaires. For instance, Vickers *et al.* (1993) reported that adding key diagnostic questions that included attribute liking and intensity questions, resulting in increasing the length of the questionnaire from a single overall liking question to a list of 19 questions, had no effect on the liking scores or on the subjects' ability to discriminate among cake samples presented to 450 consumers.

Respondent "fatigue"
In most ordinary circumstances panelists do not exhibit fatigue. Moskowitz, Muñoz, and Gacula (2003) state that sensory fatigue occurs when the sensory system is stressed far more than what occurs under normal conditions. To lose the sense of smell requires the individual to be exposed to a constant unwavering stimulus. An example of such a stimulus would be to hold the panelist in an enclosed chamber for an extended period with the odor stimulus made to fill the room. In cases claiming fatigue, panelist boredom is usually the culprit. In fact, a great deal of so-called panelist fatigue can be traced to such boredom. The *experimentum crucis* to demonstrate this psychological boredom can be done by offering the panelist a $50 bill. The sensory loss often clears up, rather immediately!

Do purchase intent questions have a place on a sensory test?
Should the questionnaire include questions on purchase intent, price determination, and brand preference? The answer depends on the objectives of the study. When the test focuses mainly on product properties such as attribute hedonics and diagnostics, then the practice questions are not included. When the decision is made to incorporate these questions, they should be located at the end of the questionnaire.

Should I use open-ended questions?
The use of open-ended questions is controversial and not recommended by Stone and Sidel (2004). However, others (Meilgaard, Civille, & Carr, 2007) consider space on a score sheet for open-ended questions desirable.

Is questionnaire format important?
Consumers have less difficulty with a questionnaire when it has been properly formatted. The format of the questionnaire should be an important aspect of questionnaire design but is often overlooked. Sensory practitioners need to pay attention to the appearance of the questionnaire. The appearance of the questionnaires should not be cluttered or have unclear print or tiny fonts and should be consistent from page to page. When panelists do not have to adapt mentally to a new format for each question or each new page, they can concentrate on the product evaluation.

How important is pretesting the questionnaire?
The questionnaire should be pretested to eliminate ambiguous terms, confusing items, and cues. In addition, the pretest will ensure that the consumers understand the questionnaire. Pretesting ensures that the respondents will have no difficulty following directions or completing the tasks within a reasonable length of time. Ideally, pretesting of the questionnaire

Table 11.9   Recommended practice for questionnaire design.

(i)	Keep the questionnaire as short as possible to carry out the objectives of the study.
(ii)	Placement of the overall liking question should be first when the overall liking is the most important question, no influence of the attribute ratings is desired, and when evaluating only one sample. Overall liking should be located last when the influence of responses to attribute questions is desired, in multiproduct tests, and tests on products that require time to evaluate or are based on perceptions of performance.
(iii)	A recommended sequence of attribute questions is: appearance, aroma, flavor, texture, and overall acceptance.
(iv)	Questions should be interestingly worded and encourage consumers to respond.
(v)	Tasks should be appropriately selected and eliminate panelist boredom and apathy.
(vi)	Consider the importance of including a personal question such as, "What is your family income?" Respondents may consider this question obtrusive and may not answer the question.

on untrained individuals who possess the characteristics of the target population should be used. The pretest will determine whether the questionnaire is too long or poorly designed (Resurreccion, 1998). Table 11.9 lists recommended practices for questionnaire design. Lastly, it is important to consider coding and tabulation of the data for manual or computer data processing, and the analysis of the data and the statistical package to be used. Taking these latter activities into account will result in considerable savings of time and effort (Resurreccion, 1998).

WHAT CONTEXTUAL FACTORS INFLUENCE ACCEPTANCE?

Do appearance, taste, and flavor/aroma influence preference or rejection of products?

Appearance, taste, and flavor/aroma of a food contribute greatly to the consumer's perception of the quality and liking of a food. One's first encounter with food products is often visual and will affect subsequent willingness to accept a product. Thus, the effect of visual sensation should not be underestimated. Color is often the first sensory attribute consumers use to judge the acceptability of a food or beverage, and they may react negatively to a product if its color does not meet expectations (Tepper, 1993). Color seems to have more effect on the flavor and acceptability and on the perception of sweet (Francis, 1995). For example, appropriate coloration increased acceptability of fruit punch flavored beverages (Clydesdale, Gover, & Fugardi, 1992), and that regular colored chips were preferred over dark chips in normal lighting but the dark chips were preferred over the lighter ones when the panelists could not see the color (Maga, 1973).

The effect of color on other variables becomes less significant when viewed in the context of other attributes of the food (Tepper, 1993). For example, color manipulation of orange juice reduced consumer acceptance of the color but had little influence on ratings of flavor, sweetness, or overall liking (Tepper, 1993). Similarly, the effect of color on the acceptance of two food products, chicken bouillon and chocolate pudding, from two groups (young and elderly adults) resulted in young adults rating significantly lower in terms of appearance the chicken bouillon and pudding with no added color compared to those with standard or with a higher intensity of color. Elderly adults did not find any difference in appearance acceptance between the products. Neither group reported any differences in terms of taste and flavor (Chan & Kane-Martinelli, 1997).

Taste and odor intensity grow as a power function of physical intensity, but the corresponding acceptability ratings may not follow the same order. Sweetness, for example, increases in acceptability with increasing physical intensity up to a certain point, whereupon acceptability declines with further increases in intensity (Cardello, 1994). Added aromatic flavor may increase acceptability of a product. In a study of milk with and without the addition of an aromatic flavor, the flavored milk was rated significantly higher in liking compared to the plain milk (Lavin & Lawless, 1998).

Food packaging—should commercial packaging be used in testing products for sensory acceptance?

Packaging acts both as a protective barrier between the contained food product and the environment and as a powerful communication tool with the consumers. Visual appearance of the food itself is a significant driver of its acceptability, as is the visual appearance of its packaging, particularly its shape, color, design, including associated logos, symbols, brands, and item names.

When commercial packaging is used in a product test, it may be better accepted than if presented in a generic packaging. Cardello (1995) found that respondents liked food better when the test samples were packaged and presented in a commercial type of package, as opposed to a typical military ration package.

Food temperature—does serving temperature affect acceptance?

Food products must be served at the appropriate temperature in order for the product to meet consumers' expectation upon the occasion of consumption. The taste of a beverage is expected to be most hedonically positive at temperatures at which the beverage is or most commonly consumed. Tasting a beverage at an unfamiliar temperature can decrease liking for the beverage, at least temporarily, indicating that the temporary rejection is, at least in part, the result of unmet expectations based on learned ideas of appropriate serving temperature for the beverage.

In the United States, foods are served at different temperatures. What exactly is the relation between temperature and acceptability for various foods and beverages? Acceptability of 10 foods commonly served hot, for example, entrée items, increases from 40°F to 140°F, with increasing serving temperature. Conversely, in the case of milk and lemonade, beverages that are normally served cold, acceptance decreases with increasing serving temperature (Cardello & Maller, 1982). Finally, coffee, which may be served either hot or cold, shows high acceptance at both temperature extremes and low acceptance at ambient (room) temperature (Cardello & Maller, 1982).

Temperature changes may induce other sensory changes, which in turn may alter hedonic responses to a food as discussed previously. We saw in the preceding paragraph that the acceptability of food or beverages served at different temperatures is significantly altered by changing the consumers' expectations concerning the temperature at which the food or beverage is typically served. Zellner *et al.* (1988) studied the acceptance of guanabana and tamarind juices served at cold and at room temperature to two groups. The data showed that information about temperature and normal serving can affect the ratings of acceptance. When informed that the juices are normally served at room temperature, the "test group" given this information assigned significantly higher acceptance ratings for juices served at room temperature versus to the same juice when served cold. The no-information "control group" gave lower ratings of acceptance. The differences in preferred temperatures implicated consumer expectations as a factor in food acceptance.

"Alliesthesia," a theoretical mechanism of the perceived pleasantness of a stimulus, is influenced by its usefulness to the body. Taste, olfactory, and thermal sensations are affected by alliesthesia, and the general conclusion to be reached is that bodily deprivation levels can modulate the perceived acceptability of food. Thus, a cold beverage on a hot day seems to taste better than the same beverage on a cold day.

How do attitudes, opinions, and beliefs influence overall liking? Food neophobia

Neophobia is a tendency of consumers to avoid unfamiliar food. An underlying reason for a novel food rejection from the evolutionary standpoint is anticipation of danger, or a fear that a new food will physically endanger you. It has been suggested that neophobia can be considered as a mechanism that protects individuals from consuming potentially toxic food (Rozin & Vollmecke, 1986). Subjects with high food neophobia are less likely to have tasted or eaten unfamiliar foods than those with low food neophobia.

Neophobia varies greatly in the human population and depends on many factors. In children or young people, neophobia decreases with age (Pelchat & Plinger, 1995), whereas the elderly population (66–80 years) was more neophobic than the other age groups (Tuorila *et al.*, 2001). Men were more neophobic than women, and food neophobia scores decreased with increasing education and with the degree of urbanization (Tuorila *et al.*, 2001). Cultural difference may also influence neophobia. For example, people from the United States and Finland are less willing to try novel foods as compared to people from Sweden (Ritchey *et al.*, 2003).

Food neophobia may be decreased by several means:

(i) Providing information about the new food. An example of it was demonstrated by Pliner (1994) on 5–11-year-old children. When they were asked why they refused to try new food, the typical response was that it would not taste good, that is, they expected a new food to be unpalatable than a familiar one. Pelchat and Pliner (1995) found that providing children with verbal information that a new food tasted good increased their willingness to taste.

(ii) Food exposure and simple exposure has an effect on subsequent consumption of novel foods (Cardello *et al.*, 1985). Tuorila *et al.* (2001) found that tasting experience was particularly beneficial for the highly neophobic subjects, as their willingness ratings were greatly enhanced by the earlier experience. One of the examples of mere exposure to a novel food to enhance its acceptability could be providing shoppers with a free sample. Offering information about a new product, such as product name, description of ingredients, and use context, could act like exposure and decrease neophobial behavior (Tuorila *et al.*, 1994).

(iii) Another way to enhance a new food acceptance is to associate or make resemblance of a novel food to a familiar one (Tuorila *et al.*, 1994).

Cultural and social influences

Cultural and social factors are powerful extrinsic factors that can control food acceptance and perceived quality. The powerful influence of culture in food acceptance is easily seen in the strict dietary avoidance of pork by Jews and Muslims, the avoidance of beef in India, or the avoidance of eggs by certain tribes in Africa. Individuals who avoid the food items commonly perceive them as "unclean" or "disgusting" and might therefore be acceptable and well liked by those individuals who are not part of the culture (Cardello, 1998).

Other social and cultural factors, such as income level and status in society, can also influence liking/disliking, as reflected in the gourmet and health food preferences of many "upscale" consumers. Some social effects on consumer liking of foods derive from the social influences of parents, peers, leaders, and heroes (Cardello & Schutz, 2006).

One factor, however, appears to be less relevant than previously thought. The impact of social influence on food preferences appears to be low. For example, although food preferences of children resemble those of their parents, the correlations are fairly low, accounting for only about 2–6% of the variance; the food choices induced by observing a heroic figure in the Duncker (1938) study dissipated within a few days (Pliner & Mann, 2004).

Cross-cultural effects

Understanding the factors that influence food choice in different cultures is essential for decision-making in the development of successful products especially for a company planning to expand their export markets.

Knowledge of cultural differences in testing can be quite important. For example, there are differences among consumers in terms of how they use the hedonic scale, and how one interprets the data. Yeh *et al.* (1998) compared the usage of the 9-point hedonic scale between American, Korean, Chinese, and Thai consumers. They reported that Chinese, Korean, and Thai respondents use a smaller range of the 9-point hedonic scale (referred to as Asian central tendency and possibly dislike category avoidance) than did the Americans. The Americans used more extreme scale categories and often avoided the middle categories. Direct translations of the 9-point category hedonic scale from non-Western cultures may result in less statistical power to detect sample hedonic differences.

Cross-cultural differences might play a role in affecting preferences for basic tastes:

 (i) Chinese subjects living in the United States assigned higher pleasantness ratings to the higher concentrations of sucrose in water and showed a tendency to rate sucrose on cookies as tasting more pleasant compared to the US subjects (Bertino & Chan, 1986).
 (ii) In a comparison of Australian and Japanese consumers, no cross-cultural differences in hedonic ratings of sweet, salty, and bitter were observed, though Japanese indicated preferences for stronger umami taste of MSG (monosodium glutamate, also known as sodium glutamate) and the two highest intensities of sourness (Prescott *et al.*, 1992).
 (iii) Prescott *et al.* (1998) found that there was no cross-cultural difference in the perception of manipulated taste intensity in orange juice (citric acid), grapefruit juice (caffeine), salad dressing (citric acid), and cornflakes (sodium chloride) between Japanese and Australian consumers. Huge variability of odors and flavors differs across cultures as well.

(iv) Common food odors such as banana, peppermint, lemon, vanilla, and strawberry were universally liked across 16 population groups in North and Central America, South Africa, Europe, and India, whereas the preference for majority of the other food odors was influenced by traditional food habits and availability of regional flavor sources (Pangborn, Guinard, & Davis, 1988).

(v) There are other cross-cultural differences. The relative importance of health, mood, convenience, sensory appeal, natural content, price, weight control, familiarity, and ethical concern as motives in food choice were assessed in Japan, Taiwan, Malaysia, and New Zealand by Prescott and Young (2002). Health, natural content, weight control, and convenience are the most important factors for food choice for the Taiwanese and Malaysian consumers; sensory appeal for the New Zealand consumers; and price and ethical concern for the Japanese. Their findings are important to understand differing motives for food choice cross-culturally and also provide insight of which food claims may be useful in promoting choice in the countries studied.

Product category—are apples judged as apples or as fruit?

Product categorization affects the degree of hedonic contrast. Hedonic contrast is rating one product better ("good") when the product is presented with a "less good" version. An example of hedonic contrast was demonstrated in a study on gourmet ("good") and ordinary canned ("less good") coffee where gourmet coffee was more liked than the canned coffee. The degree of contrast in this case was high since both products were categorized as "coffee." However, degree of contrast may be decreased when the two coffee products are placed in two categories, such as gourmet and ordinary. In the same study, hedonic contrast was smaller for those who thought of these two types of coffee as belonging to different beverage categories (Zellner, 2007). Therefore, a researcher or a company introducing a new product to the market might consider the context in which it is presented and its categorization. Introducing a product that is designed to be in "ordinary" category will be judged as less preferred if introduced together with a one from "supreme" category. To decrease hedonic contrast, a stimulus can be introduced as being unique without comparing it with better versions.

Product information—how will information affect consumer liking and behavior toward a specific food?

Product information includes brand and label information, ingredient information, nutrition and health claims, and information about processing and preservation techniques used, and are sometimes referred to as "information framing" effects. One hypothesis of how information may influence affective ratings of food is through their effect on consumers' expectations of liking (Cardello & Schutz, 2006).

Branding or food labeling—what's in a label?

From the consumer's perspective, because most processed foods are sold packaged and labeled with product identity (name), a food is often associated with that name. In restaurants, food choices are selected by name. How food products are named or labeled has a big influence even without the brand name identification (Meiselman, 1996).

In a sensory testing, food samples are traditionally presented in a "neutral" fashion with respect to brand such that comparisons between samples would not be biased by any information about their nature or origin. This is called testing the products "blind." Yet, even a generic label can affect the response; it doesn't have to be a brand. Although some cues are unavoidable, Cardello *et al.* (1985) found that when traditional soups and dental liquid diets (developed for jaw-fractured patients) were labeled as soups, the traditional soups were rated higher than the special diet preparations. However, when the same two items were both labeled as "dental liquid diets" and consumed with a straw rather than a spoon, relative hedonic ratings were reversed.

Consumer tests of several tomato purees showed that knowledge of brand name affected not only hedonic judgments but sensory perception as well. Overall liking scores were higher when products were labeled with brand names. When the consumers knew the brand name, importance of attributes such as fresh tomato taste and odor were minimized when the panelists were rating overall liking (Di Monaco *et al.*, 2003). Additionally, in pasta products presented blind or with information, panelists significantly preferred branded products (Di Monaco *et al.*, 2004).

The foregoing conclusions come from studies with adults (Di Monaco *et al.*, 2003, 2004). Yet, when we move to teens, we see a lesser influence of brand on responses to chocolate breakfast cereal and seasoned cheese crackers, despite their strong affinity for branding (Allison, Gualtieri, & Craig-Petsinger, 2004). Hedonic responses were measured for overall liking and for specific product characteristics, such as seasoning or sweetness (or "diagnostic" attributes). Contrary to research on adults, the branded concept descriptors for both products did not prove an advantage over the context flavor descriptors for overall liking by teens (Allison, Gualtieri, & Craig-Petsinger, 2004).

Specific aspects of the information on the label may exert either a positive or a negative effect on liking. The effect of some soybean oil label attributes of brand name (familiar and unfamiliar), price (high and low), nutritional information (with and without cholesterol, and rich in vitamin E), and soybean type (with and without the term "transgenic") on consumer intention to purchase was evaluated by conjoint analysis (Carneiro *et al.*, 2005). The term *transgenic* had a negative impact on the purchase intention by consumers who declared the intention was not to buy the transgenic soybean oil. However, economics can modify that negative impact; the lower price attached to some test in the conjoint study had a positive impact on purchase intention (Carneiro *et al.*, 2005).

Kihlberg *et al.* (2005) studied the effect of liking of four bread types, given the information on flour origin (conventional vs. organic farming system) and health effect (cholesterol-reducing effect vs. no information). They found a greater positive effect on liking resulted from information about organic production than health information.

Ingredient formulation, nutrition, and health claims—will information about health benefits of a food enhance likelihood of consumption of that food?

This effect was demonstrated by Tuorila and Cardello (2002) on juice samples containing 0, 0.3, and 0.6% KCl, respectively. The various groups of panelists tested one of the products, respectively. Each group was informed that the juices contained functional ingredients designed to improve one of the following health benefits: physical endurance and energy, mental alertness and memory, and mood and emotional well-being. Thus the groups differed in terms of what they were informed as to the benefit of the product. A control group received no information. The benefit of *improved physical and cognitive performance* was found more likely to motivate subjects to want to consume functional juice products than did the benefit of *improved emotional well-being*. The information about health benefits affected liking for a food, but the effect observed was not as pronounced as the effect on likelihood of consumption.

Stubenitsky *et al.* (1999) worked with consumers who were not habitual users of reduced-fat products. Full-fat or reduced-fat pork sausages and milk chocolate snack bars were tested in home use tests for 12 weeks under one of two conditions: product information provided or product information withheld, respectively. In the initial tests, all products received high scores for acceptability. The authors reported no consistent shifts in hedonic ratings of the reduced-fat compared with the full-fat products over the 12 weeks. The "reduced-fat" information had a small, negative effect on acceptance ratings for the chocolate snack bars, but not for the sausages, indicating generally high and sustained consumer acceptance over extended periods (Stubenitsky *et al.*, 1999). An increasing interest in nutritional and health information could provide opportunities for food industries to use the ingredient listing and health claims as a marketing tool.

Processing and preservation techniques

Technologies that engender consumer concern, for whatever reason, would be expected to produce lower levels of liking for these foods (Cardello, 2003). Food processed by novel and emerging food technologies, for example, biotechnology, ionizing radiation, pulsed electric fields, ultraviolet laser treatment, pose challenging problems for researchers interested in the factors responsible for consumer choice, purchase behavior, and acceptance of these foods.

The importance of "production method" such as genetic engineering (transfer of genetic material between species), protein engineering (altering the characteristics of microorganisms without transferring genetic material), and traditional selective breeding and "benefit" toward the health of the consumer, product quality, the environment, animal welfare, or the manufacturer on the purchase likelihood decisions of consumers for novel cheeses was assessed by Frewer *et al.* (1997). The happy news comes from those consumers who said that process was important to them. They were positive to new technologies. Positive correlations between perceived benefit and need, and purchase likelihood were observed for those 79% of respondents who considered process important (Frewer *et al.*, 1997).

There is the converse of that; concerns associated with the safety of a food processing/preservation technology can influence liking/disliking of tasted samples. In another study, the effect on product liking of chocolate pudding using 20 food processing/preservation technologies was examined by Cardello (2003). The greater the level of concern for any technology, the lower the expected liking for a product processed by that technology.

The meal context—should a reference food be served alone or accompanied with another food?

Evaluating a food as part of a meal and offering a choice of foods, the acceptability of the food is enhanced when evaluated as part of a meal rather than as a single item (King *et al.*,

2007). King *et al.* (2004, 2007) determined that meal context and the ability to choose food items had the strongest positive effects on acceptance ratings. For example, acceptability of salad and tea were significantly higher when the items were presented as part of a meal versus individual items (King *et al.*, 2004), and testing foods (cannelloni or lasagna, breadsticks, salad, tea) as separate items rather than part of a meal, selected from the restaurant menu, yielded lower scores for all items except for lasagna (King *et al.*, 2007).

Appropriateness of food for use situations

Appropriateness of judgments for food use situations as part of routine sensory evaluation can provide valuable information to guide product development. Knowing the appropriateness "profile" may allow the developer to maximize product utility in the intended use situation without jeopardizing the validity of preference/acceptability judgments (Cardello & Schutz, 1996). Cardello and Schutz (1996) conducted 29 laboratory taste tests with 27–38 consumers each on one or more food products for preference/acceptability, then rated for their appropriateness in 10 different use situations such as when hungry, for breakfast, on a cold day, when tired, and so forth. Results indicated that:

  (i) appropriateness ratings had very similar patterns for products that varied little in basic physical properties that might influence use;

 (ii) significant differences existed among products in their appropriateness ratings for certain use situations, enabling useful distinction on the products; and

(iii) products that did not differ in preference/acceptability had significant differences in their appropriateness for certain food use situations.

The environment: where and under what physical conditions the food is eaten

We end this chapter with ambience and its effect on the rating of acceptance. Factors other than the physical location of the test might also affect food acceptance. These are lighting, noise, and physical dining environment, but there could be others.

The lighting, sounds, colors, and design can have an impact on a meal and its acceptance (Edwards & Gustafsson, 2008). The influence of noise at four levels including low noise, loud noise, loud music, and a silence control on the gustatory affective ratings and preference for sweet or salty were conducted by Ferber and Cabanac (1987); and median affective rating for sucrose was significantly higher in loud noise and with loud music, while no change was observed for salt. Manipulating the ethnicity of the environment also influenced response to the same food where it was shown that when the environment reinforces the food theme, acceptability increases (Bell *et al.*, 1994).

Décor and venue make a difference, as we might expect. Meiselman, Hirsch, and Popper (1988) compared eating in a dining room setting and in the field (with soldiers). They found that situational facilitating factors (such as providing tables, chairs, dishes, and food heating facilities) significantly increased the caloric intake. The arrangement of the eating place and, in particular, the salience and accessibility of food influence food consumption. Results from field studies suggest that decreasing accessibility of food decreases consumption (Hirsch & Kramer, 1993).

Test location—will identical foods perform differently in different settings?

When the same food is served in different environments, acceptance of the food can be very different. The situation in which food is consumed is an important context factor (Meiselman, 1996). Thus the results from a test might be suspected when the test is conducted in an environment substantially different from a real eating environment.

Experiments bear out this potential that the test environment can make a substantial difference. In two independent demonstrations, preprepared food was served in different environments: first, identical prepared meals were served in both a training restaurant and in a student cafeteria; second, a prepared main dish was served in a food science laboratory class, and as part of an entire meal in two student cafeterias and in a training restaurant. In the training restaurants and in the student cafeterias, people selected and paid for their meals. The acceptability ratings of the food served varied across the three different environments in the following order: restaurant ratings were greater than those from laboratory settings, which were greater than from a cafeteria setting. Differences in acceptance were attributed to contextual effects and the expectations they produce, actual product differences, and a number of possible covariates. Ratings of sensory attributes tended to mirror the acceptability effects. In their first study, the meal was rated higher in the training restaurant than in the student refectory. In the second study, main dish ratings of overall acceptance and ratings of the attributes of flavor, texture, and color were all in the same order: training restaurant ratings were greater than those from the laboratory, which were greater than those from the dining halls. The main finding in these two demonstrations is the consistently higher rating of restaurant food over cafeteria food, when the food is identical (Meiselman, Johnson, & Crouch, 2000).

Similar results where obtained by Pound, McDowell, and Duizer (2000) on evaluation of consumer responses of attribute liking and intensity as well as overall liking of commercial chocolate in four types of testing situations (central location, in-home, teaching laboratory, and formal sensory laboratory). Perceptions of certain attributes in chocolate were found to differ in different testing situations, but liking scores of these attributes did not. Consumers were found to be more critical of attributes when tested in a formal sensory laboratory. All four locations tested gave similar results, meaning that conducting sensory panels at home is as valid a method of collecting consumer opinion as traditional locations (Pound, McDowell, & Duizer, 2000).

Finally, and in opposition to the foregoing significant effects of test environment, Hersleth *et al.* (2005) on the other hand found no significant effect when the physical environment, together with social interaction, was incorporated in the study. Overall liking of cheese samples in three different locations—at a laboratory, at a central location, and at home—were conducted. The main results showed that changing the environments and the degree of social interaction in the consumer tests had no significant effect on hedonic ratings for the products. They concluded that lack of a natural meal context during testing, similar expectations in the three testing situations, and high familiarity of the product category may explain the high degree of consistency in hedonic ratings.

FINALIZING TEST PROTOCOLS

A number of decisions have to be considered in running or requesting a sensory test. The key deciding factor in finalizing the test is whether test protocols will answer the goals and objectives of the test and whether the information gathered will lead to actionable information. Cost and practical considerations are likewise important but should not compromise the validity of the results.

REFERENCES

Allison, A.A., T. Gualtieri, & D. Craig-Petsinger. 2004. Are young teens influenced by increased product description detail and branding during consumer testing? *Food Quality and Preference* 15:819–829.

ASTM. 1998. Standard E 1958-98: standard guide for sensory claim substantiation. In: *ASTM Annual Book of Standards*. West Conshohocken, PA: ASTM.

ASTM, Committee E18-1979. 1979. *Manual on Consumer Sensory Evaluation*, ed. E.E. Schaefer, p. 52. ASTM Special Technical Publication 682.

ASTM, Committee E480-84. 1992. *Standard Practice for Establishing Conditions for Laboratory Sensory Evaluation of Foods and Beverages*. ASTM Special Technical Publication.

Basker, D. 1988. Critical values of differences among rank sums for multiple comparisons by small taste panels. *Food Technology* 42(2):79–84.

Bell, R., H.L. Meiselman, B.J. Pierson, & W.G. Reeve. 1994. Effects of adding an Italian theme to a restaurant on the perceived ethnicity, acceptability, and selection of foods. *Appetite* 22:11–24.

Bertino, M. & M.M. Chan. 1986. Taste perception and diet in individuals with Chinese and European ethnic backgrounds. *Chemical Senses* 11:229–241.

Birch, L.L., L. McPhee, L. Steinberg, & S. Sullivan. 1990. Conditioned flavor preferences in young children. *Physiological Behavior* 47:501–505.

Birch, L.L. & S. Sullivan. 1991. Measuring children's food preferences. *Journal of School Health* 61(5):212–213.

Bower, J.A. & M.A. Saadat. 1998. Consumer preference for retail fat spreads: an olive oil based product compared with market dominant brands. *Food Quality and Preference* 9:367–376.

Cardello, A.V. 1994. Consumer expectations and their role in food acceptance. In: *Measurement of Food Preferences*, ed. H.J.H. MacFie & D.M.H. Thomson, pp. 253–297. London: Blackie Academic and Professional.

Cardello, A.V. 1995. Food quality: relativity, context and consumer expectations. *Food Quality and Preference* 6:163–170.

Cardello, A.V. 1998. Perception of food quality. In: *Food Storage Stability*, ed. I.A. Taub & R.P. Singh, pp. 1–38. Boca Raton: CRC Press.

Cardello, A.V. 2003. Consumer concerns and expectations about novel food processing technologies: effects on product liking. *Appetite* 40:217–233.

Cardello, A.V. & O. Maller. 1982. Acceptability of water, selected beverages and foods as a function of serving temperature. *Journal of Food Science* 47:1549–1552.

Cardello, A.V., O. Maller, H.B. Masor, C. Dubose, & B. Edelman. 1985. Role of consumer expectancies in the acceptance of novel foods. *Journal of Food Science* 50:1707–1714, 1718.

Cardello, A.V. & H.G. Schutz. 1996. Food appropriateness measures as an adjunct to consumer preference/acceptability evaluation. *Food Quality and Preference* 7:239–249.

Cardello, A.V. & H.G. Schutz. 2006. Sensory science: measuring consumer acceptance. In: *Handbook of Food Science, Technology, and Engineering,* vol. 2, ed. Y.H. Hui, p. 56 (2–16). Boca Raton: CRC Press.

Carneiro, J.D.S., V.P.R. Minim, R. Deliza, C.H.O. Silva, J.C.S. Carneiro, & F.P. Leao. 2005. Labeling effects on consumer intention to purchase for soybean oil. *Food Quality and Preference* 16:275–282.

Chambers, E., IV, J.A. Bowers, & A.D. Dayton. 1981. Statistical designs and panel training/experience for sensory analysis. *Journal of Food Science* 46:1902.

Chambers, E., IV & E.A. Smith. 1991. The use of qualitative research in product research and development. In: *Sensory Science Theory and Applications in Foods*, ed. H.T. Lawless & B.P. Klein, pp. 395–412. New York, Basel, and Hong Kong: Marcel Dekker.

Chan, M.M. & C. Kane-Martinelli. 1997. The effect of color on perceived flavor intensity and acceptance of foods by young adults and elderly adults. *Journal of the American Dietetic Association* 97:657–659.

Chen, A.W., A.V.A. Resurreccion, & L.P. Paguio. 1996. Age appropriate hedonic scales to measure food preferences of young children. *Journal of Sensory Studies* 11:141–163.

Chu, C.A. & A.V.A. Resurreccion. 2005. Sensory profiling and characterization of chocolate peanut spread using response surface methodology (RSM). *Journal of Sensory Studies* 20(3):243–274.

Clydesdale, F.M., R. Gover, & C. Fugardi. 1992. The effect of color on thirst quenching, sweetness, acceptability and flavor intensity in fruit punch flavored beverages. *Journal of Food Quality* 15:19–38.

Cristovam, E., C. Russell, A. Paterson, & E. Reid. 2000. Gender preference in hedonic ratings for espresso and espresso-milk coffees. *Food Quality and Preference* 11:437–444.

Di Monaco, R., S. Cavella, T. Iaccarino, A. Mincione, & P. Masi. 2003. The role of the knowledge of color and brand name on the consumer's hedonic ratings of tomato purees. *Journal of Sensory Studies* 18:391–408.

Di Monaco, R., S. Cavella, S. Di Marzo, & P. Masi. 2004. The effect of expectations generated by brand name on the acceptability of dried semolina pasta. *Food Quality and Preference* 15:429–437.

Duncker, K. 1938. Experimental modification of children's food preferences through social suggestion. *Journal of Abnormal and Social Psychology* 33:489–507.

Edwards, J.S.A. & I.-B. Gustafsson. 2008. The five aspects meal model. *Journal of Foodservice* 19:4–12.

Fallon, A.E., P. Rozin, & P. Pliner. 1984. The child's conception of food: the development of food rejections with special reference to disgust and contamination sensitivity. *Child Development* 55:566–575.

Ferber, C. & M. Cabanac. 1987. Influence of noise on gustatory affective ratings and preference for sweet or salt. *Appetite* 8:229–235.

Francis, F.J. 1995. Quality as influenced by color. *Food Quality and Preference* 6:149–155.

Frewer, L.J., C. Howard, D. Hedderley, & R. Shepherd. 1997. Consumer attitudes towards different food processing technologies used in cheese production—the influence of consumer benefit. *Food Quality and Preference* 8:271–280.

Gacula, M.C., Jr. 1993. *Design and Analysis of Sensory Optimization*. Trumbull, CT: Food & Nutrition Press.

Gacula, M.C., Jr. & J. Singh. 1984. *Statistical Methods in Food and Consumer Research*. Orlando, FL: Academic Press.

Hersleth, M., O. Ueland, H. Allain, & T. Naes. 2005. Consumer acceptance of cheese, influence of different testing conditions. *Food Quality and Preference* 16:103–110.

Hirsch, E.S. & F.M. Kramer. 1993. Situational influences on food intake. In: *Nutritional Needs in Hot Environment*, ed. B.M. Marriott, pp. 219–239. Washington DC: National Academy Press.

Hough, G., N. Bratchell, & I. Wakeling. 1992. Consumer preference of dulce de leche among students in the United Kingdom. *Journal of Sensory Studies* 7:119–132.

Hough, G., I. Wakeling, A. Mucci, E. Chambers IV, I.M. Gallardo, & L.R. Alves. 2006. Number of consumers necessary for sensory acceptability tests. *Food Quality and Preference* 17:522–526.

Husson, F., S. Le Dien, & J. Pages. 2001. Which value can be granted to sensory profiles given by consumers? Methodology and results. *Food Quality and Preference* 12(5–7):291–296.

IFT/SED. 1981. Sensory evaluation guideline for testing food and beverage products. *Food Technology* 35(11):50–59.

Jones, L.V., D.R. Peryam, & L.L. Thurstone. 1955. Development of a scale for measuring soldiers' food preferences. *Journal of Food Science* 20:512–520.

Kihlberg, I., L. Johansson, O. Langsrud, & E. Risvik. 2005. Effects of information on liking of bread. *Food Quality and Preference* 16:25–35.

Kimmel, S.A., M. Sigman-Grant, & J.X. Guinard. 1994. Sensory testing with young children. *Food Technology* 48(3):92–99.

King, S.C., A.J. Weber, H.L. Meiselman, & N. Lv. 2004. The effect of meal situation, social interaction, physical environment and choice on food acceptability. *Food Quality and Preference* 15:645–653.

King, S.C., H.L. Meiselman, A.W. Hottenstein, T.M. Work, & V. Cronk. 2007. The effects of contextual variables on food acceptability: a confirmation study. *Food Quality and Preference* 18:58–65.

Kroll, B.J. 1990. Evaluating rating scales for sensory testing with children. *Food Technology* 44(11):78–86.

Lavin, J.G. & H.T. Lawless. 1998. Effects of color and odor on judgments of sweetness among children and adults. *Food Quality and Preference* 9:283–289.

Lawless, H.T. & H. Heymann. 1998. *Sensory Evaluation of Food: Principles and Practices*. New York: Chapman & Hall.

Lee, C.M. & A.V.A. Resurreccion. 2004. Descriptive profiles of roasted peanuts stored at varying temperatures and humidity conditions. *Journal of Sensory Studies* 19:433–456.

Maga, J.A. 1973. Influence of freshness and color on potato chip sensory preferences. *Journal of Food Science* 38:1251–1252.

McNeil, K.L., T.H. Sanders, & G.V. Civille. 2000. Using focus groups to develop a quantitative consumer questionnaire for peanut butter. *Journal of Sensory Studies* 15:163–178.

Meilgaard, M., G.V. Civille, & B.T. Carr. 2007. *Sensory Evaluation Techniques*, 4th edn. Boca Raton, FL: CRC Press.

Meiselman, H., E.S. Hirsch, & R.D. Popper. 1988. Sensory, hedonic and situational factors in food acceptance and consumption. In: *Food Acceptability*, ed. D.M.H. Thomson, pp. 77–87. London: Elsevier.

Meiselman, H.L. 1996. The contextual basis for food acceptance, food choice, and food intake: the food, the situation and the individual. In: *Food Choice, Acceptance and Consumption*, ed. H.L. Meiselman & H.H. MacFie, pp. 239–263. New York: Blackie Academic & Professional.

Meiselman, H.L., D. Waterman, & L.E. Symington. 1974. Armed Forces Food Preferences, Tech. Rep. 75-63-FSL. US Army Natick Development Center, Natick, MA.

Meiselman, H.L., J.L. Johnson, & J.E. Crouch. 2000. Demonstrations of the influence of the eating environment on food acceptance. *Appetite* 35:231–237.

Monteleone, E., L. Frewer, I. Wakeling, & D.J. Mela. 1998. Individual differences in starchy food consumption: the application of preference mapping. *Food Quality and Preference* 9:211–219.

Moskowitz, H.R. 1985. *New Directions in Product Testing and Sensory Evaluation of Foods.* Trumbull, CT: Food & Nutrition Press.

Moskowitz, H.R. 1996. Experts versus consumers: a comparison. In: *Descriptive Sensory Analysis in Practice*, ed. M.C. Gacula, Jr., pp. 127–145. Trumbull, CT: Food & Nutrition Press.

Moskowitz, H.R. 2000. R&D-driven product evaluation in the early stage of development. In: *Developing New Food Products for a Changing Marketplace*, ed. A.L. Brody & J.B. Lord, pp. 277–328. Lancaster, PA: Technomic Publishing Co.

Moskowitz, H.R. 2001. Interrelations among liking attributes for apple pie: research approaches and pragmatic viewpoints. *Journal of Sensory Studies* 16:373–391.

Moskowitz, H.R, J.G. Kapsalis, A. Cardello, D. Fishken, G. Maller, & R. Segars. 1979. Determining relationships among objective, expert and consumer measures of texture. *Food Technology* 23:74–88.

Moskowitz, H.R., A.M. Muñoz, & M.C. Gacula, Jr. 2003. *Viewpoints and Controversies in Sensory Science and Consumer Product Testing.* Trumbull, CT: Food & Nutrition Press.

O'Mahony, M. 1982. Some assumptions and difficulties with common statistics for sensory analysis. *Food Technology* 36(11):75–82.

Pangborn, R.M., J.-X. Guinard, & R.G. Davis. 1988. Regional aroma preferences. *Food Quality and Preferences* 1:11–19.

Pelchat, M. & P. Pliner. 1995. Try it; you'll like it: effects of information on willingness to try novel foods. *Appetite* 24:153–166.

Peryam, D.R. & F.J. Pilgrim. 1957. Hedonic scale method of measuring food preference. *Food Technology* 11(9):9–14.

Peryam, K.R., B.W. Plemis, J.M. Kamen, J. Eindhoven, & F.J. Pilgrim. 1960. Food Preferences of Men in the Armed Forces. Chicago: Quartermaster Food and Container Institutor of the Armed Forces.

Phillips, B.K. & K.K. Kolasa. 1980. Vegetable preferences of preschoolers in day care. *Journal of Nutrition Education* 12(4):192–195.

Pliner, P. 1994. Development of measures of food neophobia in children. *Appetite* 23:147–163.

Pliner, P. & N. Mann. 2004. Influence of social norms and palatability on amount consumed and food choice. *Appetite* 42:227–237.

Popper, R., W. Rosenstock, M. Schraidt, & B.J. Kroll. 2004. The effect of attribute questions on overall liking ratings. *Food Quality and Preference* 15:853–858.

Pound, C., K. McDowell, & L. Duizer. 2000. Improved consumer product development. Part 1: Is a laboratory necessary to assess consumer opinion? *British Food Journal* 102:810–820.

Prescott, J., D.D. Laing, G. Bell, M. Yoshida, R. Gillmore, S. Allen, K. Yamazaki, & R. Ishii. 1992. Hedonic responses to taste solutions: a cross-cultural study of Japanese and Australians. *Chemical Senses* 17:801–809.

Prescott, J., G. Bell, R. Gillmore, M. Yoshida, M. O'Sullivan, S. Korac, S. Allen, & K. Yamazaki. 1998. Cross-cultural comparisons of Japanese and Australian responses to manipulations of sourness, saltiness and bitterness in foods. *Food Quality and Preference* 9:53–66.

Prescott, J. & O. Young. 2002. Motives for food choice: a comparison of consumers from Japan, Taiwan, Malaysia and New Zealand. *Food Quality and Preference* 13:489–495.

Resurreccion, A.V.A. 1988. Applications of multivariate methods in food quality evaluation. *Food Technology* 42(11):128, 130, 132–134, 136.

Resurreccion, A.V.A. 1998. *Consumer Sensory Testing for Product Development*. Gaithersburg, MD: Aspen Publishers.

Ritchey, P.N., R.A. Frank, U.-K. Hursti, & H. Tuorila. 2003. Validation and cross-national comparison of the food neophobia scale (FNS) using confirmatory factor analysis. *Appetite* 40:163–173.

Rohm, H. & S. Raaber. 1991. Hedonic spreadability optima of selected edible fats. *Journal of Sensory Studies* 6:81–88.

Rozin, P. & T.A. Vollmecke. 1986. Food likes and dislikes. *Annual Review of Nutrition* 6:433–456.

Sidel, J.L. & H. Stone. 1976. Experimental design and analysis of sensory tests. *Food Technology* 30(11):32–38.

Siegel, S. 1956. *Nonparametric Statistics for the Behavioral Sciences*. New York: McGraw-Hill.

Sokolow, H. 1988. Qualitative methods for language development. In: *Applied Sensory Analysis of Foods*, vol. 1, ed. H.R. Moskowitz, pp. 4–20. Boca Raton, FL: CRC Press.

Stone, H. & J.L. Sidel. 2004. *Sensory Evaluation Practices*, 3rd edn. San Diego, CA: Academic Press.

Stubenitsky, K., J.I. Aaron, S.L. Catt, & D.J. Mela. 1999. Effect of information and extended use on the acceptance of reduced-fat products. *Food Quality and Preference* 10:367–376.

Tepper, B.J. 1993. Effects of a slight color variation on consumer acceptance of orange juice. *Journal of Sensory Studies* 8:145–154.

Tuorila, H., H.L. Meiselman, R. Bell, A.V. Cardello, & W. Johnson. 1994. Role of sensory and cognitive information in the enhancement of certainty and liking for novel and familiar foods. *Appetite* 23:231–246.

Tuorila, H., L. Lähteenmäki, L. Pohjalainen, & L. Lott. 2001. Food neophobia among the Finns and related responses to familiar and unfamiliar foods. *Food Quality and Preference* 12:29–37.

Tuorila, H. & A.V. Cardello. 2002. Consumer responses to an off-flavor in juice in the presence of specific health claims. *Food Quality and Preference* 13:561–569.

Vickers, Z.A., C.M. Christensen, S.K. Fahrenholt, & I.M. Gengler. 1993. Effect of questionnaire design and the number of samples tasted on hedonic ratings. *Journal of Sensory Studies* 8:189–200.

Vie, A., D. Gulli, & M. O'Mahony. 1991. Alternate hedonic measures. *Journal of Food Science* 56:1–5.

Yeh, L.L., K.O. Kim, P. Chompreeda, H. Rimkeeree, N.J.N. Yau, & D.S. Lundahl. 1998. Comparison in use of the 9-point hedonic scale between Americans, Chinese, Koreans, and Thai. *Food Quality and Preference* 9:413–419.

Zellner, D.A. 2007. Contextual influences on liking and preference. *Appetite* 49:679–382.

Zellner, D.A., W.F. Steward, P. Rozin, & J.M. Brown. 1988. Effect of temperature and expectations on liking for beverages. *Physiology & Behavior* 44:61–68.

12 Evolving sensory research

As sensory professionals continue to integrate themselves in both the scientific and the business communities, their focus widens and by necessity their roles evolves. Today we see research run by sensory professionals in areas as diverse as studies in hunger/satiety/ consumption, plant quality control, international product screening/optimization, and creation of machines to automatically assess product batches coming out of production and storage. The path forward is an integration of the different facets of sensory science. These facets include, on the one hand, adoption of techniques it has already developed and "battle-tested," and on the other, assimilation of methods and world views from allied fields of science such as food intake.

APPLIED VERSUS ACADEMIC RESEARCH IN PRODUCTS AND CONCEPTS

Sensory analysts straddle business and academia. The majority of sensory analysts spend their days doing studies on corporate products. A minority of sensory analysts do the same type of work, but in universities and publicly funded research institutions. A still smaller minority of sensory analysts, work within the domain of small private enterprises, either as single owner practitioners or market researchers.

The differences between the academic and the industry researchers are interesting and significant. Most academics are rewarded on the basis of being able to bring new methods into their field, present them at conferences, publish the results, and in general add to scientific knowledge. One can see the contributions of these individuals at conferences such as the Pangborn Conference held every 2 years. The contributions of the academics tend to be smaller scale studies using advanced methods, with new types of conclusions. Sometimes the results are clear and useful, other times the results or presentations are hazy and less obvious. In contrast, the industry-based researcher in sensory analysis is grounded in the requirement to produce useful data that can be easily communicated. Their papers in the Pangborn Conference and other meetings tend to be, not surprising, less theoretical, more basic, more practical, as their jobs would dictate.

Market researchers constitute a third party to this dynamic between academic and industry sensory researchers. Market researchers come from a different intellectual heritage than do

Sensory and Consumer Research in Food Product Design and Development, Second Edition.
Howard R. Moskowitz, Jacqueline H. Beckley, and Anna V.A. Resurreccion.
© 2012 Blackwell Publishing Ltd. Published 2012 by Blackwell Publishing Ltd.

the more conventional sensory researchers. Market researchers usually have less scientific training, and more often than not find themselves by accident in the marketing research job, often without the proper knowledge. Furthermore, academic and business sensory researchers carry in them the desire to do things themselves, and to establish their own credibility within the field. In contrast, most market researchers see themselves as doing a specific job, providing insights, but do not feel that they are part of a coherent, dynamic field. Sensory professionals do have that feeling of belonging to the group.

Marketing researchers have a difficult time dealing with sensory researchers. Turf wars prove out that generality. Both groups of researchers aim to be the *primary* information gatherers in the field. Sensory researchers prove their mettle by presenting new techniques, becoming the experts in their companies, and serving as the low-cost suppliers. Often sensory researchers end up as in-house market researchers, doing the low-cost, standard projects. This is an inevitable consequence of positioning oneself as the low-cost supplier. Often the less prepared, but more business-facile market researcher gets to work on the large-scale, high-visibility product projects. Market researchers are generally more tuned into the business, because they report through marketing, whereas sensory analysts report through R&D, and deal with the service functions in the company.

The foregoing describes the past. One of the key questions for the next decade, and indeed the next generation of researchers, is how the different professionals will evolve. We are already seeing many academics in universities becoming contract testers in the sensory industry. A number of departments of food science receive money on a regular basis from clients to perform sensory tests. The costs are lower than going to a conventional sensory specialist in a small, private company, or to a market researcher. It is not clear whether this type of involvement will produce better sensory research in the long run because the academics are motivated by science, or whether the academics will do service work of a routine type to fund their laboratories.

A second trend is the emergence of the sensory researcher as another species of consumer researcher. Whereas in the previous decades sensory analysts prided themselves on working with expert panelists, on profiling the product in terms of sensory aspects, and on quality control using difference tests, more recently sensory analysts have gotten into the field of consumer research. Sensory researchers run small-scale guidance panels to measure product acceptance. Sensory researchers are becoming involved in ideation and early stage "fuzzy front end" research, to identify new ideas. Sensory researchers have become amateur ethnographers, in a quest to understand person-product interaction. The emerging question is how this evolution of the sensory researcher will affect the roles of the market researcher and the sensory researcher in corporations. Already there are fights for budgets, for responsibility, for recognition, as the two professions knock heads in their attempt to survive in an environment of intense competition and limited resources. Whether the two fields will finally merge in the corporate structure but keep their functional identities, whether the sensory researcher will simply become an adjunct to market research, or whether there will evolve from this competition a new species of consumer researcher that combines the best of both remains to be seen.

SERVICE VERSUS SCIENCE: THE RESEARCHER AS CONTRACTOR VERSUS THE RESEARCHER AS PROFESSIONAL

Sensory researchers have struggled over the years to win acceptance for their field, and for themselves. Whereas the beginnings of sensory research comprised "taste test kitchens," and often less respect than one might wish, the past decade has seen sensory research accepted by university and industry alike. The methods that sensory researchers use have become standard, accepted in companies for product development, and accepted and even demanded of flavor suppliers by customers to demonstrate the adequacy of their submissions.

Acceptance comes at a price. The practicality and applicability of sensory research has generated an industry—academic sensory researcher as contractor to industry. Many sensory researchers in academia now position themselves as able to provide the necessary data for food companies. As a consequence, in the sensory world there is a growing subcommunity of academic contractors to industry, who then use the proceeds from their business to pay for new research. It is not clear as yet whether this demand for service work by industry on the sensory professional will lead to better sensory methods, or whether the money will corrupt the sensory professional, who will turn the university job into a low-cost supplier. It is vital, however, for the sensory professional to keep in mind the real goal of the academic—to improve methods and science, rather than simply to do the same thing at a low cost. Only time can tell whether the corporate funding of academic sensory research will raise the corporate standards, or lower the academic ones.

THE SEARCH FOR ACTIONABILITY—BEYOND PURITY TO PRACTICALITY

In today's environment, the search is on for research that can answer problems and move the business ahead. The traditional approach to research, to keep it pure and scientifically unassailable, is slowly, sometimes painfully giving way to a more pragmatic consideration. Researchers can no longer afford to position themselves as pure scientists, working within the corporation, to create the necessary data. Management will not stand for that.

This spotlight on actionability affects the sensory professional. During the years when the focus of the sensory professional was on building a spotless laboratory, one of the major refrains heard was that the data had to be precise. Vice presidents of R&D had enough time and little enough pressure to support the sensory researcher's effort to obtain solid data. Management accepted the need to train experts so that these experts would profile products with great precision and reliability. Management acknowledged that it might be expensive to do many replicate measurements in order to reduce the noise. No one questioned how the data would be analyzed, and what particular actions could be made in an operational sense to improve the product. It was tacitly agreed that somehow R&D product developers would be able to understand the numerous "spider plots" generated for sensory profiles, and in some way translate this information to product modifications. The corporate data files in R&D bear silent witness to this halcyon period where sensory researchers considered themselves scientists first and businesspeople second.

Today's situation differs considerably. Sensory researchers have evolved and continue to evolve. The first part of the evolution came from basic, simple "model building" using equations. The focus was on simple product systems, comprising one to three different independent variables. These models could reveal changes needed either in sensory attributes or in physical formulations to achieve increased acceptance. This process has been going on for about a decade and a half, as the sensory researcher has learned about, internalized, and then promoted the use of experimental design. Sensory researchers began to realize that it was necessary to do something with the data and that data for its own sake, even pure data well collected, did not do the job.

The second part of the evolution continues today, with increasingly complex models, having 4–15 variables modeled in products, rather than the simple 1–3 variables covered in the early stages. This more complex situation involves other measures as well, such as yield, cost of goods, instrumental measures and expert panel measures. The conventional statistical programs, so easily able to deal with one to three independent variables, do not do so well when called on to do high-level modeling, constrained optimization, and reverse engineering. New software is being used, and the sensory professional is becoming increasingly sophisticated in both thinking through the problem, and using high-level technology to answer it.

This second stage of evolution, requiring more complex thinking and modeling, results from the pressures of the market, rather than from any innate development in sensory research. That market pressure to maintain share, cut product cost, accelerate development, create niche products, requires the developer and thus the sensory researcher to expand their capabilities. The problem cannot be easily addressed with the simplistic study focusing on the one to three variables only. The developer and the sensory researcher are increasingly forced to consider far more variables. In such situations, the sensory researcher not only expands his or her scope of development techniques but also, through the effort to solve the complex problem, becomes a more integral part of the innovation team.

INTERNET BASED RESEARCH—FIRST CONCEPTS, THEN PACKAGES, THEN PRODUCTS

We hear today a great deal about the Internet. Consumer researchers have discovered that they can reduce the costs of research quite significantly by using the Internet to recruit and present questionnaires. Sometimes the decrease in cost can be quite significant. As a consequence, a great deal of market research, especially concept testing, is moving to the Internet. With the growing bandwidth, package design testing is also migrating to the Internet, although at a slower pace. Product testing as well is also migrating somewhat, although principally for acquiring the information. The product must still be delivered to the home.

It should come as no surprise that sensory researchers are starting to think seriously about the Internet. Until just recently (2003), many of the sensory researchers could not conceive of Internet research because of the way that they had been working. Sensory researchers are accustomed to tight control, to working with their respondents on a one-to-one basis, or at least in a supervised facility. Market researchers could more easily accept working "remotely" from respondents because they were accustomed to telephone interviews and to mail panels, especially for concept tests and for home use tests. Their counterpart sensory researchers were really not in the remote-testing business.

As Internet research continues to grow, as more types of research are conducted over the Internet, and as software packages are developed to help sensory researchers do remote-based product tests, with consumers at home, we may expect to see dramatic developments by sensory researchers. Consumer researchers are feeling increasingly comfortable with Internet-based research, as they learn how to cope with remote respondents, incentives, dropouts, and so forth (e.g., MacElroy, 2000). Current trends that could blossom into very fertile areas of inspiration and opportunity include the following:

(i) *Better use of home testing*: Creating consumer panels at home, comprising individuals who evaluate in-market products. These consumer panels could act as quality measurement panels, purchasing products in markets, recording the product codes, and rating the product on attributes using questionnaires accessed by the Internet. This first trend is already being used by some sensory and market researchers, and represents simply an extension of other types of research approaches, such as telephone interviews or mail panels.

(ii) *Better use of developing communication technologies to participate in early stage development*: Market researchers are already using focus group technology that allows geographically separated individuals to participate in a conference focus group. The increasing use of cameras mounted atop computers will add video to the Internet focus group. The penetration of wideband technology, allowing fast communication between people and person-to-person communication will allow respondents to evaluate more complex stimuli, such as videos. These technologies will allow the sensory researcher to become a stronger member of the team. The cost of technology continues to drop, and this drop brings into the realm of feasibility hitherto high levels but also unaffordable types of research. The sensory researcher, in turn, has the technical background and can take full advantage of this. In contrast, the market researcher, who lacks the technical background, must often hire out the work, and cannot take full advantage of the technology.

(iii) *Using Internet-based technologies for concept development*: Concept research has traditionally been in the purview of market researchers. However, self-authoring methods for conjoint measurement, and indeed for concept research, fit right in with the self-sufficient world view of the sensory researcher. As more self-authoring technologies emerge, where the researcher can design, execute, and analyze studies on the Internet at low cost, especially concept studies, we may expect to see more activity and contribution from the sensory researcher (see Pawle & Cooper, 2001 for their prescient view of the world as it might be in a decade).

RETHINKING SENSORY EDUCATION—IT'S ALREADY BEING DONE, BUT IS IT ENOUGH?

The sensory analysis division of the IFT (Institute of Food Technologists) has well over 1000 members. A significant proportion of these individuals are from the academic world. Many of them are graduate students, and a few are undergraduates. Go to any IFT Annual Meeting and there will be a large number of excited, enthusiastic graduate students. Many of these students are thrilled to be able to attend. Some of them have submitted their papers and are being given a chance to present the paper at a technical session, perhaps with their professor and fellow graduates looking on as they nervously proceed through the presentation. The Sensory Evaluation Division recognizes its student members, and gives them reinforcement as they are slowly integrated into the community.

What has just been painted is a delightful scenario of a growing field. Indeed, the students in this field are truly excited. Many of them rush into sensory evaluation and are delighted to be members of the community. Unlike their corresponding members in market research, many of today's graduate students want to play a role in sensory analysis when they graduate. In contrast, their colleagues in marketing all too often say that they "fell" into the field, although by this time (2010) there are some departments in business schools that offer a specialization in marketing research. The University of Georgia was the first to offer an MMR or Master's Degree in Marketing Research. That degree, like the sensory analysis specialization first offered by the University of California at Davis, recognizes the importance of consumer research.

The real question is what will happen in the long term. At this writing, all the signs are positive. Unlike the situation 50 years ago, when Rose Marie Pangborn had to fight her way for recognition, the universities are recognizing the importance of sensory analysis as a specialty. This is especially promising for the larger universities, with well-known food science departments. Many of them now have specialties in sensory analysis, and in quite a few of these universities well-known scientists have been given tenured positions. Thus, the future is bright, at least for the formal recognition and acceptance and recognition of sensory analysis in food science departments. That recognition, in turn, will certainly ensure in its wake a stream of students who will keep the field vital and growing.

STATISTICS—ITS ROLE IN SENSORY RESEARCH

One cannot mention sensory analysis without going into an exposition about sensory analysis. For many years one of the key functions of sensory analysis was to test products and to provide an informed opinion about whether or not the product met or failed the acceptance standards. A consequence of this early development is that, for better or worse, sensory analysis still closely identified with statistical tests of difference.

A survey of "zeitgeists" and "fashions" in sensory analysis over the past 50 years reveals the pervasive influence of quantitative methods. For example, in some of the very early statistics, researchers were interested in the probability that a person could discriminate among samples. This question necessarily involves statistical methods. Indeed, Fisher's classic paper on a "Lady tasting tea" (Fisher; reprinted in Newman, 1956) illustrates this type of interpenetration of statistics (or at least quantitative methods) and sensory analysis.

Some years later, the development of the flavor profile involved statistics, or rather, the disavowal of statistical methods. In the mind of A.D. Little's, Stanley Cairncross, Loren (Johnny) Sjostrom, and fellow travelers, it was better to work toward a stated consensus than to use statistics to average data from different people. The professionals developing the profile recognized the pervasiveness of interindividual variability, and were quite well aware of the methods used by statistics to average out the variability. The flavor profile developers were more interested in suppressing variability and walked in the opposite direction from statistics, always, however, recognizing their contrarian view. It would be more than 30 years until researchers at Arthur D. Little would bring themselves to accept statistical analysis, to suppress the inevitable interindividual variability that plagues all sensory research.

By the 1980s, sensory researchers were facing two different paths in the areas of quantitative methods. The first path was *modeling* or creating equations relating variables to each other to predict the level of one variable from values of the other variables. The second path was *multidimensional scaling* or mapping, which shows the relation between variables as locations on a map.

Modeling has a long and venerable history, as discussed earlier in "The search for actionability—beyond purity to practicality." Engineers use modeling to describe relations between variables and to estimate the expected value of the dependent variable as a function of the combination of dependent variables. This is called RSM or *response surface methods*. Modeling is widely used in the physical sciences, less so in the social sciences. The mind-set behind modeling is an engineering mentality—to understand the expected value of a dependent variable, and thus to be able to control the antecedent conditions to achieve that dependent variable. Introduced to the sensory research community about 40 years ago, the modeling way of thinking achieved some degree of popularity but perhaps not as much as one might have be expected. Although sensory analysts took upon themselves the responsibility of providing direction for product development, the use of RSM to generate that direction never took hold, except among those who were trained in mathematical modeling, or perhaps had a psychophysics background. The remainder opted for mapping, as described in the succeeding paragraph.

Mapping is the name given to a collection of methods that, in the main, put stimuli on a geometrical map so that the stimuli that lie close together are similar in a subjective sense. The similarity may result from confusion or discrimination tests so that the stimuli lying close to each other are confused with each or qualitatively similar when directly rated. However the map is created, and by what means it comes to be, mapping has become an exceptionally popular tool among sensory researchers, far more in fact than almost any other advanced analytic tool.

The real question about mapping is: why is it so popular? What does this tell us about the proclivities of the sensory researcher, and most importantly, what does this presage, for the years to come as we look, attempt to discern about the future path of sensory research?

The popularity of mapping can be traced to several sources, including the underlying way that people think about data, the widespread availability and easily used computer programs, and the attractiveness of visual presentation even in the absence of actionability. Mapping began in the 1950s, among psychologists working in language and cognitive processes. The idea was that if the researcher could locate the stimuli on a map so that similar stimuli were located together, then this representation would lead to a deeper understanding of the underlying cognitive processes. Thus, Osgood, Suci, and Tannenbaum (1957) used factor analysis to represent the dimensions of language. In the chemical senses, mapping was very popular in the 1960s when Masaki Yoshida and his colleagues used factor analysis and mapping to locate chemicals and reference standards (Yoshida, 1964, 1968). Mapping further became popular with the advanced programs offered by Shepard, Young, Kruskal, and Torgerson (e.g., Kruskal, Young, & Seery, 1973), and by Carroll and his colleagues subsequently at Bell Laboratories (Carroll & Arabie, 1998).

Ultimately, however, the mapping approach did not lead to the payoff that had been hoped for in experimental psychology. There were no new insights that one obtained from mapping. In the chemical senses, the early researchers using mapping found that the first dimension to emerge was hedonics, followed by other dimensions (Schutz, 1964). These findings, repeated again and again, did little to lay the foundation for scientific growth.

Mapping did, however, find a home among applied researchers in sensory analysis. Beginning with simple mapping, but going to Procrustes mapping, researchers mapped products and simple model stimuli alike, in a frenzy of effort. As the number of different mapping procedures increased and as the number of scientific conferences increased, it became clear that mapping could generate data that one might be able to share with colleagues (e.g., Elmore *et al.*, 1999; Greenhoff & Macfie, 1994). It is no wonder that the use of mapping grew, given the ease of mapping, the availability of mapping programs, the number of different stimuli to study, and the insatiable need of management to show that they were accomplishing something. Everything was mapped, and presented, either in the conference, and/or in papers. There was little to show for it in terms of increased understanding, but the exercise kept going, and continues today, perhaps with just a little less enthusiasm than before (Moskowitz, 2002).

Given the current state of sensory research and the proclivity toward quantification from its history, an interesting question to ask is "where next" in terms of quantitative techniques. There is an interesting evolution going on in sensory analysis that could give us a hint about next steps. About 15 or so years ago, a group of researchers in sensory analysis formed an organization called The Sensometrics Society. The society is similar to the Psychometrics Society in psychology but has a much shorter history. The Sensometrics Society holds regular biennial meetings where they present papers on statistical/quantitative methods applied to sensory problems. Three interesting trends appear to be developing when looking at the development of The Sensometrics Society:

(i) *Sensometrics Society meetings may become specialized, and of less interest to the conventional sensory researcher*. The biennial meetings feature increasingly esoteric papers that may well be beyond the understanding of the average sensory researcher. Most sensory researchers are not quantitatively inclined, or at least their interest in quantitative methods does not go particularly deep. The conventional sensory researcher wants to use quantitative methods but not deeply and profoundly understand them and contribute to them. As a consequence, the Sensometrics Society meetings, once a forum for the quantitative-minded sensory researcher, is beginning to be dominated by the "math types," with the papers hard to comprehend on the part of the standard researcher. This pattern follows the pattern seen for the Association for Chemoreception Sciences (ACHEMS), where psychophysicists and sensory researchers first found a home, until the field and the meeting itself became dominated by molecular biologists. At that point, some years ago, a number of psychophysicists and almost all of the industry sensory researchers dropped out, except for a few diehards.

(ii) *The conventional sensory researcher may opt to create a niche inside the more mainstream meetings to present quantitative-type work*. The aforementioned mapping is making its way through the sensory research field. Many of the researchers do not feel comfortable interacting with high-level, esoterica-minded, sensometric researchers. Instead, these quantitative-minded folk may carve out a place in the more general meetings, where they feel comfortable, and where they do not have to confront (even pleasantly) those individuals who intimidate them intellectually.

(iii) *An appreciation for quantifying the qualitative aspects of behavior is emerging, albeit slowly, in the sensory realm*. There is already a considerable industry in quantitative analysis of qualitative phenomena, and text analysis. This appreciation of the quantitative may be the next big opportunity in sensometrics.

THE STUFF OF SENSORY ANALYSIS

If we were to describe sensory analysis "on one foot" (an old Hebrew statement), what would or could we say? Is sensory analysis simply a collection of methods? Is there a corpus of knowledge underlying sensory analysis? If there is that corpus of knowledge, how much of that knowledge is really attributable to sensory analysis, and how much is attributable to other fields from which sensory analysis has borrowed, or in which sensory analysis functions?

The foregoing question is not limited, of course, to sensory analysis. It has been raised in marketing research, and indeed in all areas of consumer knowledge where the focus is on providing knowledge to answer problems, rather than on providing a corpus of information per se, without considering practical problems to be applied. One could use the same set of questions, applying it to consumer research. In consumer or marketing research are there principles? Or, as is often averred, are there just simply methods, with the field of consumer research similar to the field of sensory analysis—methods in search of problems?

When sensory analysis began there were no issues about the content of the field. In actuality, sensory analysis did not begin as a scientific field at all. It began as a set of test procedures, and began without a unique identity. The researcher was not looking to establish principles of testing or knowledge about the consumer-product interaction. The researcher was simply measuring subjective reactions to test stimuli, much as the organic chemists referenced in Cohn's encyclopedic book *"Die Organischen Geschmacksstoffe"* simply tasted organic compounds and along with their chemical report recorded how they tasted (Cohn, 1914).

What happened then? Why did sensory analysis evolve into a field, and what can we learn from this history? Sensory analysis certainly did not start out with the aim of becoming a science.

The story of sensory analysis as a field, and the lessons for the future, come from the 1960s. Pangborn was already working assiduously as a researcher at the University Of California at Davis. Pangborn's work, always disciplined but not necessarily inspiring for its grand vision, began to build a substantive basis for sensory analysis. Pangborn herself, always the meticulous researcher, would end up training two generations of students. More importantly, however, Pangborn was a collector of information. She read the literature faithfully, made possible by the fact that the density of literature was low, although it was scattered around different journals so tracing it was more the effort than wading through it. Pangborn's unique style in writing, unlike her style in teaching, would also have an effect on the field's development. In person, in her classes, Pangborn would teach students to critique papers. She was a good professor. In her writing, and especially in her major oeuvre, *Principles of Sensory Evaluation of Food*, Pangborn showed her real strength. She presented, almost as separate paragraphs, most of what was known about the different aspects of sensory analysis, up to the 1960s. The different chapters of the book, on psychophysics, test methods, and physiology, brought together a massive amount of information into one place. The field had begun.

Since Pangborn's pioneering volume with Amerine and Roessler (Amerine, Pangborn, & Roessler, 1965) the substantive aspect of the field has increased. Yet, it is still hard to point to what the field actually contains, and what is unique about it. Pangborn's book codified the knowledge up to the 1960s and gave some form to the field. Yet reading the book does not give a sense of where sensory analysis differs from the fields that provide it the basic information. The methods of sensory analysis remain methods from psychophysics. The uniqueness is certainly not to be found in these methods. Knowledge about the way the sensory organs work comes from psychology and physiology. The uniqueness is certainly not to be found in this information.

Perhaps the best we can say is that sensory analysis is an emergent discipline, rather than a fundamental discipline. That is, there may not be a true body of knowledge that belongs to sensory analysis alone, or which can be said to belong more to sensory analysis than to other disciplines. Perhaps it is the applications of methods that define sensory analysis. A piece of information can be said to be a basic part of sensory analysis when it pertains to the application of sensory methods to a sensory problem. It might well be the case that in the end the stuff of sensory analysis is the information obtained by people who call themselves sensory researchers, and who define their information as belonging to sensory research. That is, there may be no fundamental information that represents the essence of sensory analysis per se, but only the intuitive feeling that a variety of different types of information constitutes information *relevant* to sensory analysis. If so, then the articles, books, posters from conferences devoted to sensory analysis, *de facto*, if not *de jure*, constitute the stuff of sensory analysis.

THE ROLE OF JOURNALS AND BOOKS

Journals and books constitute the repository of information for a scientific discipline. Sensory analysis is no different. It is important to note that up to the 1960s, there were really no significant books about sensory analysis, nor any major scientific journals devoted to the topic. Of course there were the ephemera; manuals published by organzations and companies dealing with "how to run" sensory tests (e.g., Ellis, 1966; Larmond, 1977), reports from scientific conferences dealing with issues of sensory testing, and the occasional scientific conference dealing with taste and smell that might be related tangentially to sensory analysis (e.g., ASTM, 1968).

The 1960s were the latent period in sensory analysis when the field was coalescing, when researchers met each other, and began to exchange ideas. There would be the inevitable conferences dealing with chemical senses, at which sensory professionals would discuss their methods. Researchers in taste and smell were often invited to contribute articles to food journals, with the emphasis of these articles primarily on sensory processes, but then having some implications for emerging practice in sensory analysis (e.g., Moulton, 1969). Pangborn and her associates made a major contribution to these fields by providing well-documented bibliographies in her various publications (e.g., Pangborn & Trabue, 1967).

In the early 1970s, author Moskowitz and E.P. Koster recognized that the emerging field of sensory analysis would need a journal. Koster and Moskowitz were approached by the Reidel Publishing Company of Dordrecht, Holland, which some years before had started the *Journal of Texture Studies*. The idea of creating a journal that could incorporate sensory analysis as a major aspect was intriguing to Koster and Moskowitz, but it was clear even in 1972, when they began collaborating, that the topics of sensory analysis were not sufficient. The journal would have to comprise research on chemical senses and on flavor, along with sensory analysis.

Koster and Moskowitz's specific vision of an integrated journal failed, of course, although the journal *Chemical Senses and Flavor*, eventually renamed *Chemical Senses*, succeeded beyond their wildest dreams. *Chemical Senses* became popular because it was embraced by the Association for Chemoreception Research Scientists (ACHEMS), and eventually by both the European Chemoreception Research Organization (ECRO) and the Japanese Chemoreception Research Organization, respectively. Over the succeeding years, the fields of taste and smell research moved far away from psychophysics and became biological, to the point where sensory analysis could no longer be perceived by scientists in chemical senses as having much to contribute. It is interesting, however, that among the first papers to be received for the journal were papers that would today be conventionally classified as belonging to sensory analysis.

In the middle and late 1980s, books were beginning to appear dealing with sensory analysis. Some of these books had a decided practical and "recipe" bent, such as the work on *Sensory Evaluation Practices* by Stone and Sidel (1985) or on *Sensory Evaluation Techniques* (Meilgaard, Civille, & Carr, 1999). The authors had founded a successful research firm in the 1970s to promote sensory profiling using the quantitative descriptive analysis method. Both Stone and Sidel worked in the heritage of food science, although Sidel had a degree in psychology. Their volume came out at the same time as Moskowitz's 1983 book on product testing. Both books clearly revealed the proclivities of their respective authors. The Stone and Sidel book represented a "how to" book, designed for consultation by the practitioner. The first Moskowitz book (1983) and very soon the second (Moskowitz, 1985) both had a decidedly more psychophysical orientation. The Moskowitz book was less of a "how to," and more of "where are we now and what can the field do." All three books departed from the Pangborn tradition of meticulous notes from a historian.

There would be subsequent books, of course. We can classify the books that came after these first three as falling into one of two major categories. The first category comprises those books that attempt to provide knowledge of the perceiver, and only secondarily knowledge of methods. The second category comprises books that focus primarily on methods, and then provide the requisite chapters about the perceiver. This is a subtle difference and comes from the world view of the writer. Those books coming from writers in academia tend to fit into the first category. The academic authors are accustomed to books that present information about a topic. In this case the topic is the perceiver, so the book tries to tell the reader about how we perceive the various aspects of products. Only secondarily does the book deal with test methods, because academics are in the knowledge business, not in the service business. Those books coming from writers in industry tend to fit into the second category. Business and practitioner authors are more accustomed to solving specific problems. They are, of course, familiar with books about fields of knowledge, but their focus is on methods and problem-solving. Books written by these practitioners focus on methods, and on case histories, much as a business book or manual might focus on either descriptions/solutions of problems or "how to" expositions.

THE ROLE OF CONFERENCES

Anyone seriously involved with sensory analysis can be only delighted at the increasing number of conferences devoted in part or in whole to sensory analysis issues. We need only look back 50 years to see a veritable desert, with very few conferences devoted to the issues of product evaluation. Certainly there were the chemical senses conferences, such as the New York Academy of Sciences Conferences in the 1950s, 1960s, 1970s, devoted to odor, some of the topic conferences devoted to the chemical senses and nutrition. Conferences really became popular in the middle 1970s. The annual ACHEMS and biennial ECRO conferences really started the ball rolling, as did the International Symposium on Olfaction and Taste (ISOT) Conferences.

The conferences devoted to topics primarily dealing with sensory analysis issues began later, in the middle and late 1970s. The Pangborn Conference played a major role, because it signified that sensory analysis research had come of age. Indeed, all conferences have that announcement as somewhat of a subtext. The growth of the Pangborn Conference over an 11-year period, starting in 1992, is witness to the interest in the conference.

What do conferences really accomplish? A lot of young scientists sit assiduously in the conference hall, listening to the papers, jotting down notes and, in general, getting ideas. This writer would like to put a different interpretation on conferences, one that may help us understand what may happen down the road, in the next decade or two:

(i) *Conferences provide venues for people to meet each other*. As research becomes increasingly internationalized, people in the same field do not meet each other. Many colleagues do not even know each other. Furthermore, in any research field there are always people who have known each other, people who have heard of each other, people who are just starting and want to meet the "eminences" of the field, etc. Conferences provide a unique personal opportunity for these meetings. Indeed, an informal poll of conference attendees later on will reveal that most people remember who they met far more than what they heard.

(ii) *Conferences allow ideas on the margin to enter the field.* A lot of new ideas in a field come from individuals in other fields. Most individuals in a research area, sensory analysis included, have little enough time to pay to developments in their own field, and precious little time to read the literature of other fields. Even graduate students are becoming increasingly specialized, and do not have enough time to become conversant in areas other than their own specialty, except perhaps at the start of their education when they are unformed. However, and quite fortunately, conferences attract lots of different types of people, who meet up with each other, talk about their work, and in general inspire each other with one or many ideas. Thus, conferences act as venues whereby people can discuss ideas, informally, and take the measure of each other's ideas in a way they could not or would not by simply reading the literature or by staying in written contact, or even by telephone. Often these chance meetings of individuals from different fields, who may know of each other but do not know each other, generate long-term research collaborations on the basis of the emotional bonding that occurs at the conferences. There is something marvelously special about talking at a conference to someone else in an allied field, and having the enthusiasm communicate itself to both parties. This is the substance of scientific discourse.

(iii) *For young researchers, conferences allow safe venues for them to get the necessary reinforcement they need to continue.* Walk around any conference and one quickly sees the pattern of behavior that reveals itself most clearly when people do not listen to the chatter but just act naturally. There are the inevitable young people, filled with anxiety and with drive, talking animatedly to the older, more established professionals. A nod from the professional often means acceptance, as the young researcher attempts to acquire acceptance. It is the purpose of the conference to provide this opportunity for the young and the old, the new and the established to meet. The young gets acceptance from the old. The old gets excitement, vision, even an education into new ideas from the young. That interaction moves the field forward by virtue of the emotional needs met by the chance for young and old to discuss.

(For more on this and related topics, read Moskowitz's *YOU!* book, 2010.)

(iv) *Conferences recognize the legitimacy of, and bestow self-governance and independence on an emerging field.* The first time a conference is held, it is often a specialty conference. Those who plan these specialty conferences often do so because the emerging field needs somewhere to blossom. The newly emerging field cannot blossom easily in the shadow of an existing conference, unless it wishes to occupy a subservient position. Consequently, once the members of the emerging field feel sufficiently emboldened that they can make a conference and that people will come, they look around for sponsors, for topic sessions and for speakers. Ongoing conferences in sensory analysis have this type of history. Since 1973, the Institute of Food Technologists (IFT) has a Sensory Evaluation Division (SED), which now in 2011 numbers over 1200 members. The SED has sponsored yearly symposia since its founding, and now supports several symposia at the Annual IFT meeting. These symposia are well attended, leading one to ask why there are other conferences specializing in sensory analysis. The reason is fairly clear. At IFT the field of sensory analysis is recognized and legitimate but does not occupy a place of honor. The IFT gives the sensory analyst a legitimate representation in the scientific firmament of foods, but it is only a symposium devoted to sensory analysis in its entirety that can give the field its true legitimacy. Sensory analysis becomes self-governing when it has its own conference.

(v) *Conferences provide a venue for nonacademics to mingle with academics.* In every conference there are two groups of people; those who work in the administration of the conference to "make it happen" and those who actually participate in the intellectual give and take. To young scientists, the administration of a conference just "seems to happen," as these young scientists scurry around trying to impress each other, impress older colleagues, and in general create their future. They are blissfully unaware of the exceptional efforts that other people make to put the conference together. Yet, in sensory research as in most other fields, there are many individuals who make their major contribution by helping things to happen. These individuals are the committee members, the volunteers, the individuals who make sure that the invitations go out, the conference is well run, the equipment works, etc. It is important to recognize that the conference serves their need as well and that they are an integral part of the field. Not everyone can contribute new ideas. As Rose Marie Pangborn demonstrated for ACHEMS, the sensory professional can contribute to the field by helping the conferences to function. Pangborn, herself a well-known researcher as has been stated again and again, became the formal secretary treasurer of ACHEMS, and in so doing pointed the way to administrative help as a perfectly valid contribution to a profession.

Given the role of conferences in the development of the field, it is interesting to speculate on the future role of conferences in sensory analysis, and how it will affect the field. The Pangborn Conference is 11-years old and has grown from a triennial to a biennial conference. The Sensometrics Conference is a little younger, and has remained a biennial conference, now held on alternative years so that it does not conflict with the Pangborn Conference. The SED meetings are yearly. It is clear that the demand for conferences is being satisfied by a large number of standard conferences, occurring on virtually a year basis. There are the occasion invitation-only conferences dealing with specialized topics.

The most likely scenario is for the conferences to continue, unabated, with increasing numbers of participants, as the Pangborn Conference has already shown. What may happen, however, is the growth of satellite conferences dealing with specific areas, such as food intake. These areas are not the conventional sensory analysis areas, but rather allied areas. It is likely that these will continue, become stronger, and eventually siphon off some of the sensory researchers, especially those with a stronger bent toward basic research. It does not appear likely for the next 5–10 years that there will be any new, major conferences, however.

THE ROLE OF PROFESSIONALIZATION AND ACCREDITATION—THREAT OR OPPORTUNITY TO SENSORY ANALYSIS

One of the trends that have recently come to the fore in research and consulting is the desire to have professional accreditation as part of one's professional portfolio. Rather than simply being recognized for doing a good job, many professionals in sensory research feel that with professional accreditation the field will become better.

A similar type of movement is occurring in the market research field, and has occurred among those who deal with new products and are members of the PDMA (Product Development and Management Association). The movement is often spearheaded by those individuals in secure and stable positions. Their stated objective is to strengthen the field. Most of those who push for such professional accreditation feel that the field is being overrun by novices and poseurs, or at least act as if the field were undergoing assault by the barbarians at the gate.

It is not clear whether the professionalization of the field will, in fact, accomplish those objectives. A professional accreditation for sensory research means that there is a mechanism that presumably assures that a person so accredited has mastered a body of knowledge. But who in the field is to identify the knowledge, and indeed what are the criteria for the selection of canonical information? It is clear in chemistry, physics, mathematics, that there exists a body of knowledge without which the scientist cannot function. We would not want to entrust our fates to individuals who train students without knowing what they are talking about, or entrust product development production to individuals who do not understand the products with which they work. At an intuitive level having proficiency accreditation for much of the scientific work makes sense.

Let us now turn attention to the field of sensory research on the one hand, and marketing research, its companion, on the other. What is the particular information with which the acolyte is to become familiar? Is this information commonly understood and accepted by the members of the field? Is the information absolutely necessary for the individual to do the job? And, most importantly, is there a possibility that by dictating the body of knowledge to be mastered by the acolyte, the party creating the accreditation is giving preference to one business group over another. In small business everyone searches for advantages. Having one's ideas belong to the knowledge needed for accreditation is certainly an advantage to the small business in sensory research because it means that the small business becomes the "de facto" establishment. Being excluded from the knowledge needed for accreditation means that the small business proffering that knowledge or method has been defined right now, and probably for a long time, as being outside the establishment. Given the power of accreditation to affect one's economic future, the prudent professional in sensory research, and in marketing research, might think twice about the issue of accreditation. It has the power to corrupt the minds of the generation and to bestow unfair, undeserved advantage on those who are fortunate enough be have their knowledge and approaches selected being part of the requisite knowledge base.

REFERENCES

ASTM. 1968. *Manual of Sensory Testing Methods*. ASTM Special Technical Publication 434. Philadelphia: ASTM.

Amerine, M.A., R.M. Pangborn, & E.B. Roessler. 1965. *Principles of Sensory Evaluation of Food*. New York/London: Academic Press.

Carroll, J.D. & P. Arabie. 1998. Multidimensional scaling. In: *Handbook of Perception and Cognition. Volume 3: Measurement, Judgment and Decision Making*, ed. M.H. Birnbaum, pp. 179–250. San Diego, CA: Academic.

Cohn, G. 1914. *Die Organischen Geschmacksstoffe*. Berlin: Siemroth.

Ellis, B.H. 1966. *Guide Book for Sensory Testing*. Chicago, IL: Continental Can Co., Inc.

Elmore, J.R., H. Heymann, J. Johnson, & J.E. Hewett. 1999. Preference mapping: Relating acceptance of "creaminess" to a descriptive sensory map of a semi-solid. *Food Quality and Preference* 10(6):465–476.

Fisher, R.A. 1956. The mathematics of a lady tasting tea In: *The World of Mathematics*, ed. J.R. Newman, pp. 1512–1521. New York: Simon and Shuster.

Greenhoff, K. & H. Macfie. 1994. Preference mapping in practice. In: *Measurement of Food Preferences*, eds MacFie & Thomson. London, UK: Blackie Academic & Professional.

Kruskal, J.B., F.W. Young, & J.B. Seery. 1973. How to use KYST, a very flexible program to do multidimensional scaling and unfolding. Murray Hill, NJ: Bell Laboratories. Unpublished.

Larmond, E. 1977. *Laboratory Methods for Sensory Evaluation of Food*, Pub. 1637, rev. edn. Ottawa, Ontario: Food Research Institute, Canada Department of Agriculture.

MacElroy, B. 2000. Variables influencing dropout rates in Web-based surveys. *Quirks Marketing Research Review*, paper 0605. (Available on www.quirks.com)

Meilgaard, M., G.C. Civille, & B.T. Carr. 1999. *Sensory Evaluation Techniques*, 3rd edn. Boca Raton, FL: CRC Press.

Moskowitz, H.R. 1983. *Product Testing and Sensory Evaluation of Food: Marketing and R&D Approaches.* Westport, CT: Food and Nutrition Press.

Moskowitz, H.R. 1985. *New Directions In Product Testing And Sensory Analysis of Food.* Westport, CT: Food and Nutrition Press.

Moskowitz, H.R. 2002. Mapping in product testing and sensory science: A well lit path or a dark statistical labyrinth? *J. Sensory Studies* 17:207–213.

Moskowitz, H.R. 2010. *YOU! What you MUST know to start your career as a professional.* North Charleston, SC: CreateSpace.

Moulton, D.G. 1969. Sensory perception. *Journal of Dairy Science* 52:811–815.

Osgood, C.E., G.J. Suci, & J. Tannenbaum. 1957. *The Measurement of Meaning.* Urbana, IL: University of Illinois Press.

Pangborn, R.M. & I. Trabue. 1967. Bibliography of the Sense of Taste. In: *The Chemical Senses and Nutrition,* eds M.R. Kare & O. Mailer. Baltimore, MD: The Johns Hopkins Press.

Pawle, J.S. & P. Cooper. 2001. Using Internet research technology to accelerate innovation. *Proceedings of Net Effects*, pp. 11–30. Barcelona, Spain: European Society of Market Research.

Schutz, H.G. 1964. A matching standards method for characterizing odor quality. *Annals of the New York Academy of Sciences*, 116:517–526.

Stone, H. & J.L.H. Sidel. 1985. *Sensory Evaluation Practices.* New York: John Wiley & Sons.

Yoshida, M. 1964. Studies in the psychometric classification of odor. *Japanese Psychological Research* 6(111):124–155.

Yoshida, M. 1968. Dimension of tactual impression (1) (2). *Japanese Psychological Research* 10:123–137, 157–173.

13 Addressable Minds™ and directed innovation: new vistas for the sensory community

INTRODUCTION

An ongoing leitmotif or theme of this book is that the world of sensory research in food and allied fields is expanding into new, hitherto unexplored areas. Sensory professionals have advanced well beyond the stage when they were simply called on to execute "taste" tests. They are now asked to participate in the decision process itself, to draw reasoned conclusions, and to make recommendations (Lawless & Heymann, 1998). Whether they deal with concept or product evaluation, whether with basic knowledge or testing, we continue to witness the increasing growth of new ideas, new methods, and new analyses. Fresh air is blowing through the land of sensory research, providing new visions for many practitioners and new ways to approach old problems (Murray, Delahunty, & Baxter, 2001).

Since the first edition of this book, nearly a half decade ago, and since we began writing nearly a decade ago, a number of new technologies and world views have come to the fore. We alluded to some of these in the first edition; it's time now to more fully flesh them out, and explicate some of their highlights. This chapter does just that explicating. In previous chapters we dealt with the new ways of "knowing what we knew." Now we explicate new ways of measuring what we believed to be important.

Our chapter deals with the use of concepts to understand the mind, and in so doing, to guide innovation. We will deal with development platforms and mashing together new ideas inspired by the world of genomics (Fuller, 1994; Khurana & Rosenthal, 1998). Both topics could each constitute the subject of a full book. Space limitations do not let us expand the size of this book to deal with the topics more fully. Rather, we will confine ourselves to highlights, with some illustrative data from a variety of studies.

WHY THESE NEW DIRECTIONS—TODAY'S IMPETUS FROM BUSINESS

Fifty years ago, around 1960, a lot of the talent in fast moving consumer goods migrated into the food industry. This was the case for sensory analysis as well as for marketing research.

Sensory and Consumer Research in Food Product Design and Development, Second Edition.
Howard R. Moskowitz, Jacqueline H. Beckley, and Anna V.A. Resurreccion.
© 2012 Blackwell Publishing Ltd. Published 2012 by Blackwell Publishing Ltd.

Yet, starting around 1975 and for about 20–25 years, those fast advances grew less frequent. Process supplanted innovation. The old rubric was trotted out again and again—"people always have to eat." And so the advances slowed.

By the start of the new millennium, the year 2000, rumblings could be heard from business. Competition had increased, the world was becoming internationalized and information technology had leveled the playing field, and thus need for innovation was being recognized as a competitive advantage. The pain led to recognition and growth that knowledge, not process, was critical. Thus, the origin of our chapter includes Mind Genomics™, innovation, and innovation platforms. Innovation alone was not the answer. It was also getting the consumer more profoundly involved with innovation, and also better understanding the consumer to turbocharge the innovation activity. In a nutshell, it came down to better products driven by better understanding the consumer.

THE ROLE OF SEGMENTATION IN INNOVATIVE PRODUCTS

Typically, results from the total panel of panelists in a development study do not give "actionable" marching orders for the new product. It is hard to discover a new product that appeals to all the panelists in a study. We're just not all wired alike. What one respondent likes another respondent might well dislike. We saw this interindividual difference in the optimization study for a new product, but also as the foundation for Mind Genomics™, to be presented later.

Markets and the customers who make up those markets are not homogeneous (Claycamp & Massy, 1968). At a time when mass marketing was still in the ascendancy, Wendell Smith (1956) wrote in the *Journal of Marketing* of the need to target homogeneous components of a heterogeneous market rather than the market as a whole. He called this strategy "segmentation," explaining that it was designed for a "heterogeneous market by emphasizing the precision with which a firm's products can satisfy the requirements of one or more distinguishable market segments." Since that time, a rich literature has developed suggesting techniques upon which a single market might be effectively broken into actionable customer segments (Craft, 2004).

When the concept of segmentation is applied to a mass market, the forms it takes are limited only by the imagination of the segmenter and the potential and accessibility of the selected market segment. In the beginning of market segmentation, consumers were segmented by age, sex, race, income, household size, education, and other demographic or socioeconomic characteristics. But from around the mid-1980s, the classic segmentation concept was criticized, and debates in the marketing literature highlighted a concern that consumer markets were becoming increasingly fragmented and consumers were less predictable (Firat & Venkatesh, 1993). Many researchers suggest that classic market segmentation approaches adopting sociodemographic measures are less effective (Firat & Shultz, 1997; Sheth, Banwari, & Newman, 1999).

The industry response to this problem has been an increased popularity of segmentation approaches on the basis of buying behavior, their, personality traits, and emotional needs, and attitudes, interests, opinions, activities, and many other behavioral characteristics (Crawford, Garland, & Ganesh, 1988; Swenson, 1990; Dawar & Parker, 1994; Hassan, Craft, & Kortam, 2003).

Our approach is rather than searching for the single magic bullet, which may or may not exist, and which probably cannot be easily found within the limits of today's corporate environment, it's better to isolate individual "*mind-set segments*" with different preferences. Combining these segments with the ability to "mash up" different products by combining features produces a system to create new product ideas. One doesn't have to search for a needle in a haystack. One simply has to define the products whose features will be combined, and then identify segments of respondents with different patterns of responses to those features.

In this chapter we look at two new paths, sometimes pushing advances separately, sometimes coming together to create a wonderful new opportunity for the food and beverage industry. These two paths or really world views are Mind Genomics™ first, and innovation systems second.

VISTA #1: MIND GENOMICS™ AND THE NOTION OF ADDRESSABLE MINDS™

Consumer research in food products is undergoing an evolution, stimulated in part by the vastly improved capabilities in computation, and stimulated in part by the advances in disparate fields such as biology. In the past decade or so the proliferation of computer technology has changed the once very staid field of consumer research into an exciting hotbed of new developments. Today you cannot be involved in food development without becoming familiar with the Internet, data mining, blogs, social networking, and so forth. Most of these topics lie beyond the purview of these chapters, which deal with food, at the level of product, concept, and package.

In this section, we introduce a newly developing area, which, if not particularly in the world of food right this moment, is nonetheless quite promising, and in fact holds the prospect of entirely changing the world of sensory analysis and consumer research. That area is the notion of 1:1 development and marketing, which we will call by the more catch name of Mind Genomics™. The notion here is that one can begin to understand individuals, what they want, what makes them "tick." Rather than working with the food product with the human being as the bioassay, or working with the concept or package, again with the human being as the assay device, we are going to work with the human being as an individual organism. We take our cue from the research on the human genome and personalized nutrition. The focus there is on the individual and the goal to understand the individual, rather than the individual as a representative of the group.

Let's see how far this notion takes us. As we hope, you will see through this chapter the notion of the individual as the focus of attention rather than the group opens up a lot of new doors and many new opportunities for the food industry.

From individual differences to individual minds

Sensory researchers are well aware of the pervasiveness of individual differences. Instruct consumers to rate the sensory intensity of a set of different products, and more than likely you will get a fair degree of agreement among the different judges. Certainly, there will be variation among the judges in terms of the magnitude of their sensory impressions, but when you look at the data you will feel that the variation is probably in the way that people use the scale, rather than in their basic sensory perception of the product. You can do this experiment quite easily by lining up a variety of pickles of different known sourness, randomizing the order, hiding the product description, and instructing the subject (or panelist) to tell how strong the pickle tastes. People tend to agree with each other.

Now do the same experiment with pickles, but this time instruct the same people to tell you how much each person likes each pickle. You'll get differences this time, as you expect. But, what will seem very different is that there is a lot of disagreement. That is, far more disagreement than you would expect on the basis of differences in how people use numbers and scales. Some of your participants will love the very sour pickles; others will hate them. So you will discover variation in the preferences of people. We see this person-to-person variation all the time when we deal with likes and dislikes. Such variation is profoundly part of our thinking; marketers recognize this built-in, unalterable variation in what people like. Just walk down the aisles of a supermarket to see the recognition of such differences. A significant amount of effort expended by the sensory professional is dedicated to dealing with these interindividual differences, to understand what they really mean in terms of the product, the production, and the market.

With these profound person-to-person differences, let's move beyond products to the communication of product features. We know that there are these interpersonal differences when it comes to the communication of products through concepts. Most of the traditional work on individual differences comes from research on one particular product, study after study. The reason is very simple. Researchers who work with foods and beverages aren't typically scientists who have the funding and the scope to do the large-scale research to understand the nature of person-to-person variability in products. These researchers work with developers in funds-strapped companies, lucky often to have the time and the funding to make a few products to test. So we won't get a profound understanding of the nature of differences in the likes/dislikes of food of individual people across a wide spectrum of different foods. The economics do not permit it.

You can get a sense of what the product developer and the marketer face when creating products for a world of people, many of whom have different preferences. We're not talking of the simple preferences for strong versus weak tasting products, the type of variation in preferences that so simply in the middle 1980s created the burgeoning field of different product flavors (Gladwell, 2004). We're talking rather of a world of true differences in preferences for the many different aspects of the product.

As an example of how the product developer and advertiser today cope with this plethora, consider the "standard" pattern of responses to variation shown in Figure 13.1 (on the next page). The typical company doesn't have time to cover the entire space of different preferences, or even to understand them. It's easiest in such cases simply to divide people by conventional breaks, such as gender, age, market, and even a few convenient general groupings of flavor/texture, create products for those convenient groups, and move on.

From individual differences to mind-set segments

In Chapter 4, we talked about developing concepts for food products, using experimental design. In many of these exercises we work with systematically varied combinations of elements, present the combinations to the respondent, acquire the ratings, and then create a model relating the presence/absence of the elements to the ratings assigned by the respondent on the 9-point scale. Furthermore, since each respondent acts as his or her own control, it is possible

Retailers/Manufacturers ->

...Traditional advertising
    messaging targeted to large
    audiences...based on simple
    classification descriptors and
    what they purchase...it's how
    media is bought

Your customers....advertising
    strategy assumes/hopes for
    group homogeneity...Men
    want "Y," Women want "X"

Result...lots of messages without
    regard to "true/effective"
    messages and motivators "that
    really work" for each customer

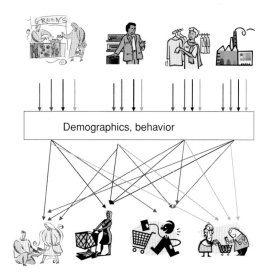

Figure 13.1 The conventional way that developers and marketers cope with individual differences, either for product creation or for product marketing.

to cluster together respondents who show similar patterns of coefficients or utilities. All of the methods are well-established statistical procedures, whether the procedure is estimation of the utilities by ordinary least-squares regression, or clustering using any of many different clustering methods currently available in statistical packages for the personal computer.

Let us now go one step further. We know that we can create these individual-level models and then cluster people. If we were able to cluster individuals, we know that individuals in the same cluster share similar points of view for the particular product on which we did the clustering.

That is, for example, when we do the study on pastry muffins, we can cluster people on the basis of similar mind-sets with respect to pastry muffins. The clustering is typically done using elements or phrases specific to pastry muffins. That is, we don't make general clusters of people with respect to foods, and try to figure out how this general clustering or segmentation may translate itself down to one product, pastry muffins. Instead, we limit our world to pastry muffins, and make all elements granular, specific with respect to pastry muffins.

Once we discover that segmentation, we know a great deal about pastry muffins. We may not know anything about any other food, but it doesn't matter. Our focus is on pastry muffins. We know what to feature in our pastry muffin and how to communicate. Each of the individuals in our study has a known "mind-set" about pastry muffin; hence the name Mind Genomics™. We know exactly how to reach that mind, hence the name Addressable Mind™ (Figure 13.2, next page). We can customize the development and the messaging to the specific segment.

The value of segmentation

We have already seen in Chapter 4, that it's possible to understand person-to-person differences through experimental design. The approach is fairly simple. We'll repeat just the highlights, and then move to the future, to where Mind Genomics™ comes into play.

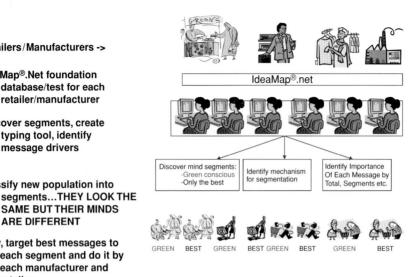

Retailers/Manufacturers ->

IdeaMap®.Net foundation
    database/test for each
    retailer/manufacturer

Discover segments, create
    typing tool, identify
    message drivers

Classify new population into
    segments...THEY LOOK THE
    SAME BUT THEIR MINDS
    ARE DIFFERENT

Now, target best messages to
    each segment and do it by
    each manufacturer and
    retailer

Figure 13.2 The Addressable Minds™ way that developers and marketers cope with individual differences, either for product creation or for product marketing.

As in the standard approach with experimental design, we create different test stimuli, systematically varying in the features that they present. We do this at the up-front, design phase. Figure 13.3 shows an example of a concept created in this way:

(i) *Each panelist generates a vector of numbers, or utilities, showing how the different elements drive the rating.* What is important, here, is that the respondent evaluates different combinations, and that these combinations create an individual-level experimental design. That is, even if we were to work with one or a few respondents, we'd be able to determine what elements are working.

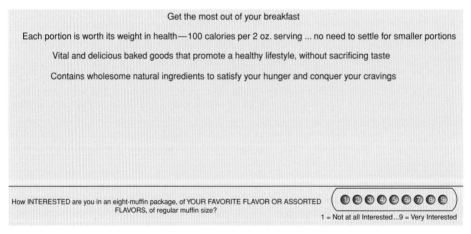

Figure 13.3 A test concept for muffins, created in preparation for a conjoint study. The elements or phrases in the concept are systematically varied. Each respondent evaluates sufficient concepts to create an individual-level model or equation relating the presence/absence of the elements to his or her rating.

Table 13.1 How different elements perform to pastry muffins, for total panel and for two mind-set segments.

	Total	Seg 1	Seg 2
Base size	172	75	97
Additive constant	28	27	29
Winning elements—total			
Contains probiotic bacteria, which treat gastrointestinal ills	7	−5	16
Maximize satisfaction with minimum calories...low fat and high in fiber and nutrients	7	11	3
Winning elements Seg 1			
Get more for your 100 Calories	5	13	0
Maximize satisfaction with minimum calories...low fat and high in fiber and nutrients	7	11	3
Maximum nutrition and pleasure for minimum Calories	5	11	0
Winning elements Seg 2			
Contains probiotic bacteria, which treat gastrointestinal ills	7	−5	16
Contains probiotic bacteria, which may prevent and treat some illnesses, according to *Harvard Medical School Journal*	6	−7	16
Contains probiotic bacteria, which support general wellness	5	−4	12
Contains probiotic bacteria...Northern Europeans consume a lot of foods rich in probiotic bacteria	3	−8	11

(ii) *The approach to finding the segments generates a database, in which people are classified into segments by means of the clustering program.* The resulting data looks very much like Table 13.1. The segments are quite strong polarized. What Segment 1 likes, Segment 2 dislikes, and so forth.

The big opportunity—finding the segments in the population and communicating

Suppose now that the manufacturer wants to create products for each of these two segments. It doesn't pay to appeal to the entire panel; there are no breakthrough ideas. Yet, Segment 1 appears to be interested in taste, Segment 2 appears to be interested in health, and each of them has some elements that perform quite well. Indeed, Segment 2 is very promising; it reacts strongly to new ideas about probiotics. So it is Segment 2 that will comprise the development target, the group to be pleased.

But how does one find them without redoing the entire test? We could look for external factors that predict segment membership. For example, we know that the pastry product comprises two segments, and we know for our 172 respondents specifically which person is in Segment 1, and which person is in Segment 2. We could then try to figure out whether it is males in Segment 1, and females in Segment 2, and so forth. That is, we could create a table of characteristics about the respondent, to look for telltale marks, such as Segment 2 (health-oriented) tends to be older, and so forth.

As reasonable sounding as the approach seems, it generally leads to very little additional prediction power. That is, knowing who the respondents are in each segment does not generate any particular characteristics that we can use to assign a new person to one of the two segments. For many of the characteristics of the people that we can measure, along with their data on reactions to concepts, we find that there is no difference in the "personal profile" of a Segment 1 member versus a Segment 2 member. We'd like there to be that difference, that clear marker, but we're simply not able to find it. There are, of course, situations in which we do find such markers that differentiate segments, but in the food and beverage businesses those situations are the exception rather than the rule. The foregoing, an example of so-called data-mining works in financial services (e.g., mortgage scoring), but not in the world of food and drink, where choices are personal, and not a function of what is able to afford.

Fortunately, it's possible to approach the problem of assigning new members to these "mind-set segments" in a different way. We don't need to hope for markers that we can discover, namely characteristics of our respondents, which somehow link to segment membership. We can create a better testing approach, when we consider the way we have created the segments, and what the segments really mean, in terms of human behavior:

(i) *We know that there are different mind-sets.* We created those mind-sets. For each person, we know the mind-set to which that person belongs.

(ii) *We created those mind-sets by looking at the patterns of utilities across people, and then segregating similar patterns into the same segment.* We used the persuasion model, where the dependent variable is the 9-point rating, and the independent variables are the presence/absence of the 36 elements.

(iii) *Now, however, we could use the persuasion model again.* We can add the constant and the coefficient of a specific element, and through that approach estimate the 9-point rating that the element would have obtained if the element were tested by itself. This way "reconstructs" or actually creates a new, estimated rating on the 9-point scale for each of the 36 elements. The elements were never tested as one-element concepts, but this step creates those 9-point ratings.

(iv) *We now have the necessary data for each respondent.* That data comprises the expected ratings of each of the 36 elements on the 9-point scale, assuming that each element was a concept by itself, and, the segment to which the respondent belongs.

(v) We use the statistical method of discriminant function analysis (DFA) first to identify a limited set of elements that, in common, best assign people from the study to the proper segment. DFA has the 172 respondents in the study on which to practice.

(vi) *There are four elements that suffice to assign the respondent to the proper group.* The four elements appear in Figure 13.3 (page 386). The first element is, *"Get more for your 100 calories,"* and the fourth element is, *"Have the right balance of nutrient, protein, and B vitamins to keep your energy levels throughout the day."*

(vii) DFA creates two equations, called classification functions, linked to these four elements. The classification function comprises a weighting function of the ratings to the four elements. These two equations appear at the right side of Figure 13.4 (page 389).

(viii) *The respondent is presented with these four elements, rates each element one at a time on the 9-point scale.* The classification functions are used in conjunction with the ratings to create two values. These are called the DFA values. Each classification function generates a value.

(ix) *The respondent is assigned to the segment depending upon which classification function shows the highest positive values.*

ADDRESSABLE MINDS™—Classification function to assign to segment	1	2
Additive constant	-2.392	-2.342
Get more for your 100 Calories	0.145	-0.17
Contains probiotic bacteria, which may prevent and treat some illnesses, according to Harvard Medical School Journal	-0.011	0.379
Contains probiotic bacteria ... Northern Europeans consume a lot of foods rich in probiotic bacteria	-0.392	0.223
Has the right balance of nutrients, protein and B vitamins to keep your energy levels high throughout the day	0.81	0.183

	Scratch test—Typing three prospects		
	Per1	Per2	Per3
	2	8	1
	5	4	1
	9	4	6
	7	8	3
Segment 1	►.0	3.6	-2.2
Segment 2	►2.5	0.2	-0.2

Not in a segment ◄— 27

Figure 13.4 The classification functions created by discriminant function analysis.

(x) *Looking at the bottom of Figure 13.4, we see the ratings as assigned by three respondents. Each respondent provided a different profile of ratings for the four elements.* Based on the individual respondent's ratings, the classification functions generate a pair of values, one value for each classification. For a given person, that is, a given pattern of ratings, the person belongs to the segment whose classification function generates the higher positive value, when each classification function is "solved" for the pattern of ratings.

Up to now the approach for segmenting consumers into different mind-sets provides the product developer with a sense of what to create for a consumer, or provides the marketer with a set of messages that will persuade. The classification function further allows the researcher, whether in product development or in marketing, to place individuals in the appropriate segment. At the operational level, a person can complete the typing or mind-set classification in about a minute or less. Figure 13.5 shows an example of a typing screen:

(i) *When an individual types himself, the researcher now knows the segment to which the individual most likely belongs.* During the research and knowledge acquisition phase, the researcher can test the respondent with the stimuli, whether products or concepts or even package designs, obtain the evaluation, and then compare the results from the two segments.

(ii) *More interesting than the knowledge acquisition and research is its application to marketing.* Once a person goes through the typing test, shown in Figure 13.5, the marketer knows the segment to which the individual belongs. That determination of segment membership is very rapid; it is merely plugging the ratings into the two classification

How INTERESTED are you in a pastry product of the following type

1 = Not at interested. 9 = Very interested

Get more for your 100 calories	3
Contains probiotic bacteria, which may prevent and treat some illnesses, according to Harvard Medical School Journal	4
Contains probiotic bacteria...Northern Europeans consume a lot of foods rich in probiotic bacteria	5
Has the right balance of nutrients, protein and B vitamins to keep your energy levels high throughout the day	8

Figure 13.5 Example of a typing screen for the baked good + probiotics study.

functions, and determining the segment to which the person belongs. Afterward, the individual, now a prospective customer who may have come to the Web site and been typed, can be presented with the appropriate web page. That is, the individual does the typing, and then is directed automatically to the appropriate web page, as we see in Figure 13.6, (page 391). Note that the web pages as oriented to appeal to the different mind-sets.

Where next—a vision of Mind Genomics™ powered by experimental design

In traditional work, the notion of the individual's own preferences as key factors to product development and marketing is not generally addressed. Of course, there are movements to have 1:1 marketing (Peppers & Rogers, 1997), and the notion of pharmacogenomics or personalized medicine (Human Genome Project, 2001; Roden *et al.*, 2006; Campos-Outcalt, 2007). On the whole, however, these notions have not taken hold, for various reasons. One of the impediments may be the lack of a standardized method that leads to a "granular" understanding of a person's mind. Or, in simple terms, there's nothing out in the market today, as of this writing (2010) that informs us of the specific responses of a specific individual to a specific set of stimuli. We live in a granular world, but our science tends to be general. We make granular decisions, about specific products and services, but our science is grounded in rules that have to be particularized for a situation, with such particularization more often than not the outcome of reasoned judgment and interpretation.

Addressable Minds™, as described in the paragraphs later, provides a different world view, at least in the domain of communications. It says that there are these segments and that a person can be rapidly identified as being a member of a segment.

The notion of individualizing product design and marketing opens up new frontiers in the world of consumer products. Throughout this book, we have been stressing the notion of general patterns. We begin with understanding the consumer, at the individual level, perhaps, but with a goal of insights that pertain to the entire population. Individual variations in the acceptance of products are treated as either a source of variability, to be partialled out by analysis of variable, or as key subgroups whose individual predilections and preferences are to be understood.

For the world view of Addressable Minds™ we have to move beyond one-product studies and scale-up the approach to an "industrial grade" level. The methods for assigning a person

Here is the personalized landing page for that site
visitor who typed out as Segment 1—Indulgent Dieter

Here is the personalized landing page for that site
visitor who typed out as Segment 2—Health Oriented

Figure 13.6 The two optimized web pages, created to appeal to the two mind-set segments. The pages
are set up to have the same general "look" but the messaging differs.

to a single segment for a single product were described previously. Now, let's see how to
move this approach from one product to a series of products. The approach will be merely
a quantitative extension, not a qualitative extension; nothing new will be added to the mix,
other than a change in the fieldwork.

We follow a series of steps, which in essence will "scale-up" the approach:

(i) *The basic approach to the Addressable Minds™ system is summarized in Figure 13.7.*
    Note that the approach is standard, beginning with an experimental design, going to
    implementation, creating models.

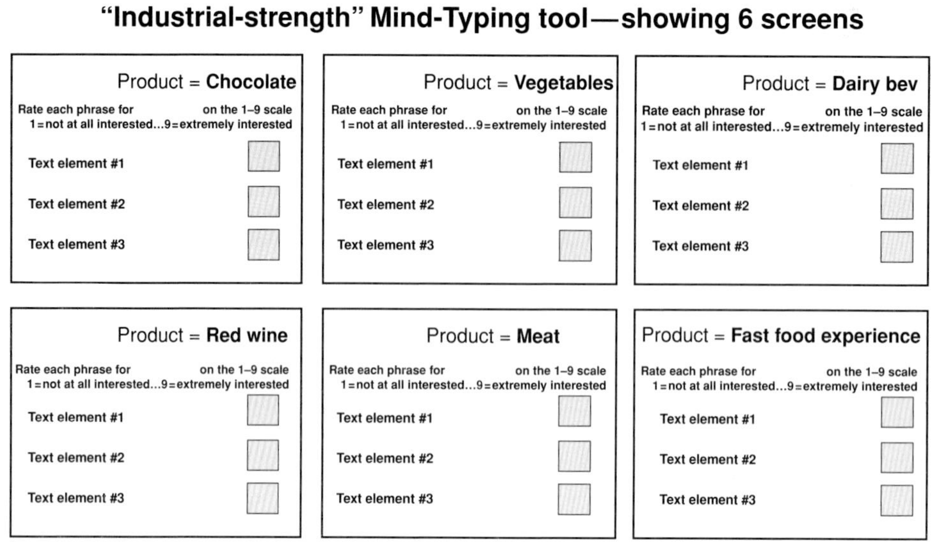

"Industrial-strength" Mind-Typing tool—showing 6 screens

Figure 13.7 Schematic example of six typing "screens," one per product, set up as the basis for a typing system. By mass mind-typing many people on a variety of related product and service areas, one creates the science of Mind Genomics™ for that particular product or service "vertical."

(ii) *Scale-up begins by performing several of these related studies at the same time, to create the granular level science, embedded in each respective study.* The respondents need not be the same and in fact should not be the same. Although the basic study of one topic does use the strategy of within-subjects design, where all respondents evaluate all the elements, albeit in different combinations, different respondents can participate in the discrete studies. Thus, if we were to do as few as two such studies, we could use entirely different and study-appropriate respondents.

(iii) *Moving forward, however, we can imagine a series of some 6 of these studies, or even 30 of these studies, carried out with different groups of respondents.* These studies have already proved feasible in the form of the It!™ studies (Moskowitz, German, & Saguy, 2005). For each of these studies, it is possible to identify the particular segments, and then create a 2–3 question-typing tool for each question. Figure 13.7 shows a schematized version of six of these typing tools, on one sheet. In the actual typing, the respondent would be invited to participate in a typing exercise, be exposed to the six (or even the 30) different screens, with the elements randomized on each. In a matter of 1 minute for one screen, or 30 minutes for 30 screens, it becomes possible to type an individual for one or many different products.

What Addressable Mind™ and Mind Genomics™ could mean for the sensory professional

In the world of "testing," the sensory professional has usually taken the role of guardian of product knowledge. Although there are occasional turf wars between the sensory world and the market research world, the knowledge of the product enjoyed by the sensory professionals gives them a chance to influence the course of product design and development in a way that market researchers could never do.

The world of the product is changing, however. Throughout this book, we have talked about the evaluation of products and, to some extent, the evaluation of concepts about these products. The focus of the material has been on the product. The person who evaluates the product is of secondary interest, but of course still of interest. It is for that reason that good test design considers the nature of the panelist as one of the key factors in the test, whether the test be descriptive analysis, or affective measurement.

The introduction of Mind Genomics™ and Addressable Minds™ brings the panelist into new focus. Responses to the product are no longer simply a combination of signal (what the product "is") and noise (the intractable variability introduced by the panelist). Rather, the responses comprise the stimulus and a more profound understanding of the mind of the panelist.

What is the role of the sensory professional here? It is the sensory professional who can benefit most, by championing the new methods in Mind Genomics™. Rather than focusing on the product alone, it may be quite profitable for the sensory professional to begin a parallel, equally profound investigation into the different ways that people perceive products. Perhaps Mind Genomics™ will reveal different groups of people, those who are most attuned to the visual aspect of products, those who are most attuned to the taste/smell of these products, and still others who are most attuned to the texture/kinesthetic and even acoustical properties of products. This knowledge, of the person-product interaction, may truly present a new world of opportunity to the sensory professional. Time will tell.

VISTA #2: DIRECTED INNOVATION AND AN INNOVATION MACHINE FOR FOOD AND DRINK

Introduction—innovation platforms

Alot is being written today about innovation:

(i) How it goes beyond improvement (Bessant & Caffyn, 1997; Boer & Gertsen, 2003; Chapman & Corso, 2005).
(ii) How it entails seeking and taking risks (Bhatta, 2003; Leavy, 2005).
(iii) How it's all about big ideas and radical departures from convention (O'Connor & Ayers, 2005; Cooper & Edgett, 2009).
(iv) How it means completely scrapping the old system (Cooper & Kleinschmidt, 1990).

Book after book, article after article and, of course, one blog after another is being written about innovation. A lot of the writing deals with the general process of innovation (Rothwell, 1994; Zhu, Kraemer, & Xu, 2006) or the steps one needs to take—three steps (Hauptly, 2007); four steps (Cogliandro, 2007); six steps (Murray, 2009) or ten steps (Brands & Kleinman, 2010). We leave these general issues to the world of business. Our focus here is how the practitioner can use the tools in this book to help innovation.

Now to the world of food and drink

Nothing stays the same. Change is the eternal constant. We hear such words again and again, whether emblazoned on the covers of business magazines, pontificated in *au courant*, must-read blogs, or simply repeated at a mantra as if to ward off some unspeakable evil, the evil of "the same old, same old." With these wonderful words, with the phrase "*innovate or die*" firmly fixed in our mind, what do we do in the food and drink industries? They're not exactly known for innovation. After all, common wisdom tells us that people always have to eat.

But what if we wanted to create an innovation machine for food and drink? Just how would we go about doing it? Is it a matter of serendipity, of hiring an expensive consulting group, or even of systematic competitive intelligence, looking around everywhere for ideas? Of course—it's all of that, and more. Innovation doesn't come so easily in the food industry. We have to wait for and welcome innovations. They don't come along very often. It's a case of chips—not the latest generation of computer chips, which seem to undergo a massive reinvention every few years, but a case of vegetable and potato chips—which seem to take forever to evolve.

That's the downside of innovation. Now, for some better news. What might we do to jumpstart the creative process? And, in doing so, can we stop relying on the truly new, and figure out a system that's sustainable in a modestly evolving industry? The answer is a definite *maybe*, and perhaps a real *yes*. Let's see how.

Reducing the task to a manageable set of operations

At the start of a long journey the prospects are daunting. How, in one's mind, will it be possible to take this journey? There are too many unknowns. What will happen when the weather turns very bad, when we get lost, if/when we run out of money? And so forth. One can conjure up many different horrid situations, ranging from running out of money to one's vehicle breaking down. The rationale, prudent person, looking at such negatives might simply walk away, saying that the "downside" of the journey are so great that it makes more sense to stay put, where one things will be easier, the situation safer.

Perhaps, the foregoing paragraph is a bit overstated. After all, people embark on journeys all the time, and the greatest journey of them all, our individual lives, doesn't come fully furnished with an insurance policy that everything will be fine. Yet, we move on, and we don't move on simply because we have to, wishing we would go on the journey. Rather we move on, often anticipating the next day and the day after. Look only at children, at their excitement about being alive, their hopes about their tomorrows and their futures.

So, back to innovation. How do we make innovation something that we can master? Can we move away from relying on serendipity, on the momentary genius that exits, perhaps, but which eludes us when we try to take its measure and control it? Can we create an innovation machine to develop new-to-the-world ideas? And, of course, can we create the machine in a way that makes its available everywhere, at low cost, at low risk? In other words, can we democratize invention?

Rethinking innovation

A lot of us walk around thinking that innovation is a mystical event, a *deus ex machina*, a rare event to be celebrated. We're taught legends of innovation and invention. We idolize those who create, and tell stories about how it all started. Who hasn't heard the wonderful, by now legendary story of how Hewlett and Packard began in a garage, or how Jobs and Wozniak began in a garage, or how Edison toiled all night, with experiment after experiment.

So let's do this for food. To do so, we first rethink this notion of innovation. Innovation is the recombination of known pieces of knowledge and products into new forms, into new things, new "stuff."

How do we bring the systematic approach of experimental design to innovation? How can we use experimental design to create new products? The answer to this question is fairly simple. Our notion of experimental design of ideas can be likened to a mixing machine. Anyone who has used a Sunbeam Mixmaster™, a food-blending device, realizes that one can blend different products together to create a new combination. Those who are adventurous can create wonderful beverages such as smoothies by blending together yogurt or milk, and different types of fruit. But what happens when we move beyond yogurt and fruit, to other components as well, such as cookies, bread, and vegetables, along with the yogurt and fruit? Of course, the combination may not sound particularly appetizing, but the combination may taste delicious.

Alright, alright you are probably saying, or at least thinking. It makes a lot of sense to focus one's effort on new ideas, because the truth of the matter is that new ideas fuel the future. It's in the future, not in the past, where we're going to live. Furthermore, it makes sense to fund efforts by which to capture, refine, and enhance the creativity of people in the corporation. It is the minds of the people that, in fact, produce these ideas, and a mind is a wonderful thing, especially when that mind manages to come up with winning ideas, again and again. Who wouldn't want to promote such a mind, nourish it, encourage it, and of course where possible clone it, at least in a figurative sense, given today's technology!

CREATING THE INVENTION MACHINE USING RDE (RULE DEVELOPING EXPERIMENTATION)

Let's lay out one possible system, and carry it through to see how it might work. The ideas in the Section "Just what are we going to do—in a nut shell" are not new, by the way. They were developed in the early 1990s, when the senior author worked with a number of companies to accelerate the innovation process. One company was Oral B, a division of Gillette, which used the method to incorporate battery-powered technology with toothbrushes! The result sparked the change in the oral care category, and was written up as a chapter in the author's book on personal products entitled *Consumer Testing & Evaluation of Personal Care Products* (Moskowitz, 1995).

We move from the task of discovering new things in nature to the task of recombining what we know into combinations. That is, the material stuff with which we work may be all we have. Yet, can we create by mixing these raw materials into new, hitherto unexpected combinations? We are transitioning from invention as an act of inspiration to invention as the systematic exploration of mixtures. There is no reason to assume that this systematic exploration of mixtures is any less an act of "invention" than the serendipitous creation of some new ideas, springing forth perhaps as it does out of one's subconscious mind. Naturally, the story of invention springing fully formed from one's subconscious makes interesting reading. Everyone loves reading about the genius of a creator, the leaps into the unknown made by the creator's mind, perhaps acting by itself, and even in defiance of the current norms and modes of thought. Analysis of systematic recombination makes a less interesting story, a boring recitation of what could appear to be a mechanistic, soulless process. Yet, such systematic exploration is what characterizes a great deal of today's science of genomics (sequencing the genome), and drug discovery (exploring molecule after molecule, especially the effect of each molecule on a series of test agents).

Some years back the effort was begun to develop a path for systematic innovation in the food industry. The process was named Innovaid™, for the obvious reason—it combines innovation and aid. The organizing principle behind the effort was that one good way to promote innovation in the food industry was to create a matrix of "elements" (features of food) that one could combine using the web tool, IdeaMap®.Net. The belief behind this effort was that the food industry would eventually accept a systematic approach, but only after the tools and simple processes were in place for several years. Unlike the electronics industry, professionals in the food industry were not yet accustomed to mixing/matching features in a systematic way. The approach had to be laid out so that a generation of younger practitioners would be exposed over several years. Such point of view dictated the strategy; one would not convince by argument, but merely by making the tools available to a new generation of younger professionals. In this chapter, we explore some of the aspects of Innovaid™, which expanded the notion of experimental design to create databases of knowledge and uncover rules (called RDE), moving those databases to generate components that could be mixed and matched to generate new, acceptable mixtures of hitherto unmixed ideas (invention). This is the so-called master idea mixer introduced previously.

JUST WHAT ARE WE GOING TO DO—IN A NUT SHELL

Before we begin with a more or less detailed exposition, it's a good idea to list out the steps that we'll follow. When you understand the steps, and realize that we're going to be mixing ideas from different worlds, the entire approach will suddenly come into focus. Without further ado, here are the six steps. Master them, and you can breeze through the rest of this chapter:

(i) Recognize that we're going to be mixing and matching ideas from different products, into one final idea or product concept. This is our master idea mixer.
(ii) Select the product categories. For this chapter, we're going to select three sweet products; chocolate candy, donut, and cake, respectively.
(iii) For each product category, we run a conjoint analysis study to identify what elements perform well. We'll run separate studies, the master idea mixer, for chocolate candy, another for doughnut, and a third for cake, respectively. This strategy of running three conjoint studies generates three sets of winning ideas or elements, one set of ideas from each product.
(iv) Run the big master idea mixer, combining ideas from each of our three product categories. Respondents will just read these new-to-the world combinations, and rate them.
(v) Identify what new ideas work, and perhaps what segments exist.
(vi) Voila—the master idea mixer, which jumpstarts creative thinking.

A LITTLE BIT OF STRUCTURE TAKES US A LONG WAY ON OUR JOURNEY

We begin by looking at the "wall" of different databases available to the product developer. Figure 13.8 shows us the list of Innovaid™ databases, each of which comes complete with a set of elements that we will see in just a moment. The rationale for the selection of these

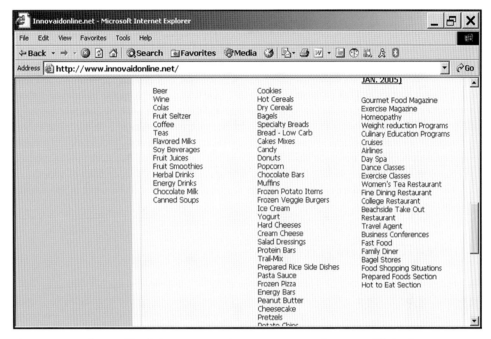

Figure 13.8   The set of foods, beverages, and eating situations in the Innovaid™ database.

databases was that they should include foods that are typically the focus of product developers in corporations, that they should represent a variety of different types of foods and beverages and, finally, in a number of cases, they should move beyond foods into eating situations.

As we'll see, the elements are all meaningful to the product developer. One cannot, of course, capture all of the important factors about a product in a list of 36 elements. However, despite that limitation, the elements provide a good start. Often when one begins the journey of invention in a product, everyone is at first a bit shy and stymied. It's hard to create new ideas, at least at first. Yet give someone a few seed ideas with which to begin, and in just a short time the creative juices begin flowing. The elements in Innovaid™ provide just such particles of ideas, with which one may begin the process of ideation:

(i)   *There are six silos, each with six elements.* The reason for this array is to make the set of elements immediately transferable to IdeaMap®.Net, RDE's web-based tool. The six silos embody specific features, meaningful to product developers:
   (i)   Appearance and texture
  (ii)   Primary ingredients
 (iii)   Special ingredients
 (iv)   Flavors and tastes
  (v)   Packaging
 (vi)   Size and store location

(ii)   *The elements provide more than just a description.* They are written to be interesting, to provide a context, and a human touch. In a lot of research efforts one develops test vignettes that are sterile. These bare-boned vignettes comprise simple, stark phrases, which present to the reader a no-nonsense statement about the package, the ingredients, the taste, and so forth. Whereas one might think that without embellishing words the features of the product "get through to the reader" our experience is just the opposite. Reading test vignettes comprising lists becomes boring, and tedious. We aren't "wired" to be calculating machines, looking at rows of features and at rows of numbers, and then coming up with an answer. We are wired to be engaged with what we read, to look for stories and for patterns. In that spirit, the text of the elements in Innovaid were written to be engaging, to invite and delight, and most of all interest the reader.

(iii) *Our first exploration will give us the materials that we want to use to innovate in our master idea mixer effort.* We are going to extract materials from chocolate candy, cake, and donut. (The ultimate product will be a hybrid of the three.) We need raw materials. Look at Table 13.2, which shows some raw materials (ideas) from each product area that performed well in the specific RDE study, and in turn will be used in the master mixer effort.

PLANNING THE INVENTION

We now move from a description of the "raw materials" of an Innovaid study to the approach. Just how are we to break out of the conventional boundaries, to create new ideas for food? Our approach will be straightforward. We will find winning elements for the three products, by doing a separate study for each. This gives us the raw materials.

Then we will go into master mixer mode. We will mix/match winning elements from three product categories, to create the new idea. Invention and innovation now become combinatorial experiments, which are fairly straightforward to execute. Figure 13.9 shows an example of the logic.

Let's move from theory to practice. The basic idea is to mix together strong performing ideas (i.e., elements) to create newer, hopefully better idea food or beverage products. To the degree that we limit our ideas to "close-in" alternatives to what's currently in the marketplace, we will move our process to one, which produces "extensions," that is, small variations. Being bolder, combining elements from different products, lets us take the same process but make it into an innovation system.

A word about the consumer and extension versus invention

Ask a marketer, a consumer researcher, a sensory specialist about the difference between a line extension and a new product. You are likely to get an earful, some theory, some opinion. The bottom line is that in the minds of these professionals line extensions are "more of the same," whereas invention is new.

Ask a consumer and you won't get such a reasoned, coherent answer. Consumers, our customers who buy products, don't walk around actively saying that a product is a line extension versus a new product. Of course when asked, and presented with yet the latest flavor of ice cream, the consumer will recognize a slight extension (more of the same) versus a different product. For the most part, however, consumers shop, look at offerings, decide whether they want a product, and then either take it or pass it by, going onto the next.

There's an opportunity here. We can create ideas for new products by mixing and matching features from existing products. This new hybrid idea, with aspects from existing products, can be created quickly using RDE, tested with consumers, and then fine-tuned when it appears to be promising.

Table 13.2 Example of strong performing elements for the first Innovaid™ for three products, chocolate, cookies, donuts. These elements will be subsequently mixed and matched in the Mixter® or mash-up rule developing experimentation study, comprising elements from three different products.

SILO 1/appearance/texture

Soft and velvety... melts in your mouth	
Soft and chewy... just like homemade	Cookies
Crisp and crunchy... perfect for dunking	Cookies
A glazed snack that will melt in your mouth	Donuts
A yeast raised snack shaped into a twist... puts a new twist on an old favorite	Donuts
A creamy milk chocolate snack with a soft, chewy center for a satisfying experience	Choc bar
A creamy chocolate snack with a crunchy, nougat center	Choc bar

SILO 2/ingredients (primary)

Real creamery butter for a rich, indulgent taste	Cookies
Made with canola oil which helps lower blood cholesterol levels	Cookies
We only use Kosher ingredients	Donuts
Sweetened with natural fructose for a healthy indulgence	Donuts
Made with the finest Belgian chocolate... for the discerning chocolatier	Choc bar
Made with the finest Swiss chocolate	Choc bar

SILO 3/special ingredients (additives)

Calcium enriched for strong bones	Cookies
Low carb... when you're counting carbs and looking for a great snack	Cookies
Made with no trans fat to fit into your healthy lifestyle	Donuts
A high-fiber snack that boosts your energy level and leaves you feeling full	Donuts
With Soy isoflavones... nutrition at its best	Choc bar
With 4 grams of fiber... a great way to add fiber to your diet	Choc bar

SILO 4/flavor/taste

Comes in spicy flavors... cinnamon, nutmeg and allspice	Cookies
Comes in cool, citrus flavors like orange, lemon and lime... the perfect snack for a ladies afternoon tea	Cookies
Flavored with a variety of palate-teasing spices... cinnamon, nutmeg, allspice,	Donuts
A rich, pure buttery taste... makes your mouth water in anticipation	Donuts
Filled with crunchy cookie chocolate chips	Choc bar
With pieces of dried fruit... for an old world taste	Choc bar

SILO 5/packaging

Each snack is individually wrapped to preserve freshness	Cookies
Comes with a stay fresh inside wrapper... just twist and seal	Cookies
Individually wrapped... take them with you when you're on the run	Donuts
Comes in a crush-proof box	Donuts
Wrapped in aluminum foil... stays fresh longer	Choc bar
Wrapped in see-through plastic wrappers... so you can see what you're getting	Choc bar

SILO 6/size/where found in store

In the bakery section of your supermarket	Cookies
In the frozen foods section of your supermarket... just thaw and serve	Cookies
Find it in the Snack aisle of your local supermarket	Donuts
In the refrigerated section of your supermarket... just thaw and serve	Donuts
Located in the gourmet food section of all fine department stores	Choc bar
Available through the Internet and delivered to your door	Choc bar

Creating our synthesized product

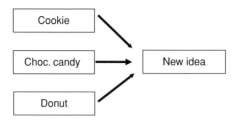

Figure 13.9   Example of the logic to create a new-to-the-world product idea (the Mixmaster® model).

THE MASTER IDEA MIXER

Let us look at the types of elements that we will select from our three IdeaMap™.Net studies, to mix together, that is, to mash-up, in the language of today's technology and genetics. We will select ideas that performed well in the particular product area (first set of three studies mentioned previously), and which seem that will work well together in a mixture. There are a couple of heuristics, or rules of thumb, that we use here:

(i) *We will follow the same approach for this mash-up that we do for the three products.* That is, we will create a set of 36 elements, structured so that we have six silos, each with six elements. See Table 13.2 for these 36 elements.

(ii) *The six silos will be the same ones that we used for the three products.* There's no reason to work with a new structure of vignettes. Our goal remains the same; create a vignette that describes the features of the new, mashed-up product. The six silos that we chose ensure that we end up with "actionable elements." There's no reason to change this structure, which acts as a guide for us.

(iii) *We will choose two elements from each product for each silo.* We could choose the strongest performing six elements across all three products, for each silo. But we don't have to limit ourselves to precisely two elements per product per silo. Doing so, forcing only two elements per product per silo, is a so-called judgment call. Other selection rules could be used with no particular negative consequences.

(iv) *How to introduce our product?* Do we tell the respondent that he is evaluating a new-to-the-world item? Is this the best way to do things? Or, should we simply pass over the issue of new-to-the-world item, and simply present the concepts as just another product to be evaluated? Both approaches work. We'll follow the latter approach, which simply describes the product in minimalist terms, so that the respondent knows that the vignettes refer to a snack product, and we'll stop there. We'll use the term "totally new" to interest the respondent, but go no further than that. We won't tell the respondent that the components of the product come from different realms. Figure 13.10 shows how to orient the respondent in a study, so that the respondent can help to co-create this new food idea.

Respondents have no problem with evaluating different vignettes or test concepts, even when the elements in the vignette may seem to go together a bit "awkwardly." Although marketers and researchers often obsess about the way terms are phrased, about the incompleteness of vignettes, and so forth, the typical respondent is far less perturbed. Whether the elements come from one type of food or many types of food mashed up makes little difference.

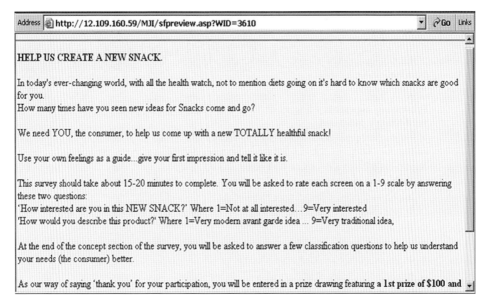

Figure 13.10 The orientation screen in the rule developing experimentation study for the snack product. The screen tells the respondents that the product is new, but does not provide any clues regarding the sources of the component elements.

During the course of the interview the respondent evaluates 40–60 combinations with relatively little difficulty. Respondents simply do not obsess the way professionals do. Furthermore, even when the elements mixed and matched come from different worlds, like foods and equipment in the same combination, the respondent seems to have no difficulty. The only thing that's important is that the elements more or less "go together." That is, they should neither contradict each other in meaning (low calorie versus full calorie in the same vignette), and they should be mixable (the combination should make some sense, even though the interaction of the different parts of the vignette is not well articulated). As long as noncontradiction and some plausibility are satisfied, the consumer simply does what he is told.

We see a small snapshot of the results in Table 13.3. The actual data would comprise a much longer table of results, with columns for the elements, the impact values for the total panel, and the impact values for the different mind-set segments. Of course, the respondents never saw the origin of the different elements. All the respondents saw were the combinations. It is we, the analysts and developers, who know the source. Table 13.3 shows us winning elements for two segments of four that we isolated from the study.

It's pretty straightforward to interpret the numbers. They come from regression:

(i) The additive constant is equal to percent of respondents who would say "Yes, I'm interested in this new product", even without elements. The constant is a measure of readiness.

(ii) Each element is the additive percent of respondents who would change from "not interested" to "interested" in this new product, when the element is introduced into the product description.

(iii) We thus have building blocks. We know what works. It's up to us to figure out our new product. We have the algebra of the mind, or more correctly, the MRI of the mind for this new product.

Table 13.3 Strong performing elements for two of the four segments, the source of these elements, and their performances among the four mind-set segments.

		Seg 1	Seg 2	Seg 3	Seg 4
	Base size	76	72	145	42
Source	Additive constant	36	59	29	39
	Promising elements—Segment 1— traditional, practical				
Candy	A creamy chocolate snack with a crunchy, nougat center	18	−5	6	5
Candy	A creamy milk chocolate snack with a soft, chewy center for a satisfying experience	10	−4	11	−1
Candy	Wrapped in aluminum foil...stays fresh longer	10	4	0	−2
Cookies	Crisp and crunchy...perfect for dunking	10	−1	4	−8
Donuts	Comes in a crush-proof box	9	1	−1	−4
Cookies	Soft and chewy...just like homemade	9	−4	10	−4
Donuts	A glazed snack that will melt in your mouth	8	−7	6	−6
	Promising elements—Segment 3— indulgent				
Candy	Made with the finest Belgian chocolate...for the discerning chocolatier	2	−8	18	19
Candy	Made with the finest Swiss chocolate	1	−17	17	10
Cookies	Real creamery butter for a rich, indulgent taste	−2	−10	11	6
Candy	A creamy milk chocolate snack with a soft, chewy center for a satisfying experience	10	−4	11	−1
Candy	Filled with crunchy cookie chocolate chips	1	−12	10	13
Donuts	Sweetened with natural fructose for a healthy indulgence	−7	−10	10	4
Cookies	Soft and chewy...just like homemade	9	−4	10	−4
Cookies	Made with canola oil, which helps lower blood cholesterol levels	−13	−9	10	−1

So what do we discover? What does Table 13.3 shows us about this new product, which combines features from three existing products?

(i) We really don't have one group of respondents. When we look at the data the respondents fall into four clusters, each with different needs and wants. There are no blockbuster elements for everyone. But there are possible blockbusters for the segments. A blockbuster is an element of +10 or more.
(ii) Segment 1 pays attention to the packaging
(iii) Segment 3 wants a rich, high-quality, natural indulgence
(iv) Just look at the strong elements to get a sense of what's working. This is the raw material from which we have to create the new product idea.

PUTTING IT ALL TOGETHER—THE "OPTIMIZER"

When the concept research tool we present here, IdeaMap^{TM}.Net was first developed in the early 1990s, a great deal of the focus was upon the use of the data to create better concepts. By "better" we mean concepts that score highly in more conventional concept tests, such as concept screens. In those early days, two decades ago, success was measured primarily by performance in other tests and, of course in the toughest test of all, the commercial marketplace. Little attention was given to the science underlying the results, which science now interests us 20 years later.

Let's return then to applying the data, not so much to become smarter, but rather to solve a business problem. Our problem is now to use our data from the mash-up study to create a new, more promising concept. In the simplest of all worlds we could sort our file, looking for the best elements for a product idea. We would then select the elements that "float to the top," that is, the elements that perform well. As long as these elements "go together," or at least do not seem to be discordant, we are fine. The sorting route will give us the answer. There is another issue, however. What happens when we want to optimize for several groups of consumers simultaneously? For example, we may want to appeal to two or more segments, simultaneously. In such a situation we need help from a computer program that can do the sorting for us, in a more directed manner, to maximize two variables simultaneously.

Shortly after the development of IdeaMap, the need arose to develop these winning test concepts, based on the results of the study. In response, the team of programmers at Moskowitz Jacobs, Inc. developed a so-called concept optimizer. The optimizer was a program that could identify winning combinations of elements, when these elements were given impact values from RDE studies. The key to the optimizer was that it could identify the optimum combination of elements when up to five different simultaneous objectives were imposed on the data set. That is, rather than identifying the one combination which maximized a single group, the optimizer would use its internal mathematical routines ("integer optimization") to find a single combination of elements that maximized several variables simultaneously, such as appeal to different groups.

The data used for concept optimization can be augmented by profiling each individual element on a series of attributes, to give that element more information than merely impact values, from the RDE that we just conducted. In this particular mash-up study with three sources of elements, each element was also profiled on a series of bipolar attributes, such as homemade versus store-bought, healthy versus tasty, and so forth. When it is time to optimize a concept, those bipolar scales can also be used to move the optimization beyond just finding the combination, which maximizes acceptance, and toward a combination that also has a specific, desired tonality.

We illustrate the very simplest application of the results in Figure 13.11 (the objective to be met—optimize for Segment 2, the health seeker) and in Figure 13.12 (the output combination). Working the optimizer means simply choosing the number of elements (here four elements), identifying the different groups to be optimized (here one group, the health seeker), and then identifying the combination that emerges.

REVISING ONE'S VIEWPOINT—OPTIMIZATION AT THE EARLY STAGE

Creating new products often requires one to "think out of the box." As companies continue to focus on the opportunities to innovation, they need to create a system that is independent of serendipity, of "Lady Luck." The question then becomes "how?" Some groups of developers prefer to use ideation methods to identify new and novel combinations. Other groups use ethnographic research, observing consumers in their everyday lives, to identify unmet needs for food items. In the two foregoing methods, and others of a similar nature, the implicit,

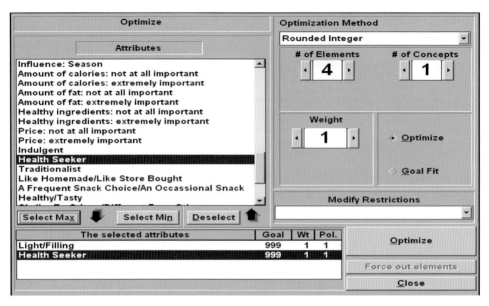

Figure 13.11 Instruction screen for the concept optimizer, to create a single, four-element concept, for one group of respondents (Segment 2, the health seeker), which is also perceived to be "filling" (a tonality, from the tonalities added later on, to each element to give a sense of "shading." The screen shows the selection of two criteria, the specific subgroup and the specific tonality. The resulting concept will attempt to satisfy both objectives.

ingoing assumption is that the insight will produce inspiration, and that inspiration in turn will lead the product.

Traditionally, concept research has been used to evaluate the promise and potential of different products. Most concept research involves close-in stimuli, that is, stimuli representing modest changes away from the current product. Occasionally in the world of concept evaluation, the respondent will be instructed to evaluate concepts pertaining to radically new products. However in that latter case, the concept has already been decided; the job of the respondent is simply to evaluate and to provide feedback as to the different aspects of the concept.

This chapter presents the new use of concept research, and especially conjoint analysis, in order to create novel products. The metaphor that comes to mind is the experimental recombination of ideas. Conjoint analysis as a *master idea mixer* makes the approach increasingly viable, easy to run, quick to analyze.

Using experimental design to combine elements from diverse worlds is not new. It is being used by various researchers in different industries. What is new is a systematic approach, afforded by the computer, guided by the respondent, in a virtuous co-creation process. Rather than mixing and matching elements in a limited set of combinations to create the new products, the approach allows the consumer to evaluate many different combinations. The computer creates the different combinations. While the consumer sorts through them. There is no need to know the right answer up front. With more than 300 respondents, with 48 vignettes, and with experimental design, the respondent is presented with combinations that might never have been seen before. Elements emerging from this "torture test" of ideas are likely winners. Elements from different products mixed together become the basis of winning ideas for the new-to-the-world products, or better said, the new-to-the-world combinations of older ideas.

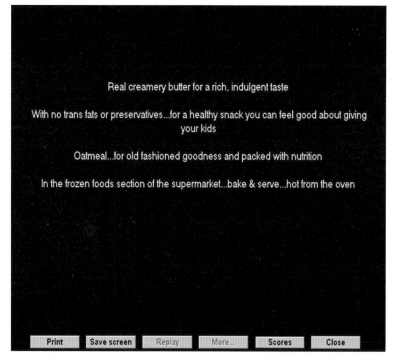

Figure 13.12 The four-element concept that comprises one element from each of four silos, and which satisfies the two criteria of "maximize for the health seeker segment" and be perceived as "filling" on the dimension of light versus filling.

SUMMING UP—THE VISTA OF INNOVATION OPPORTUNITIES IN THE WORLD OF SENSORY

This section of the chapter on new vistas for the sensory professional moves invention and innovation in a different direction, a more mechanistic one, capable of being sustained in a corporation, and scaled-up to industrial strength. The assumption here is that innovation can also be considered as the combination of old elements into next mixtures. Furthermore, to the degree that the elements come from different products, and even different product categories, the combinations should be new to the world.

The task of invention and thus innovation, therefore, is not so much to come up with the new-to-the-world idea, but rather to create the situation where new ideas can be mashed-up, evaluated, successful ideas understand in terms of "what's working," and then new product ideas created from knowledge.

REFERENCES

Bessant, J. & S. Caffyn. 1997. High-involvement innovation through continuous improvement. *International Journal of Technology Management* 14(1):7–28.

Bhatta, G. 2003. Don't just do something, stand there! Revisiting the issue of risks in innovation in the public sector, *The Innovation Journal: A Special Issue on Innovation in Governance* 8(2):1–12.

Boer, H., & F. Gertsen. 2003. From continuous improvement to continuous innovation: a (retro)(per)spective. *International Journal of Technology Management* 26(8):805–827.

Campos-Outcalt, D. 2007. Personalized medicine: The promise, the reality. *The Journal of Family Practice* 56:621.

Chapman, R.L. & M. Corso. 2005. From continuous improvement to collaborative innovation: the next challenge in supply chain management. *Production Planning & Control* 16(4):339–344.

Claycamp, H.J. & W.F. Massey. 1968. A theory of market segmentation. *Journal of Marketing Research* 5:34–45.

Cogliandro, J.A. 2007. *Intelligent Innovation: Four Steps to Achieving a Competitive Edge*. Fort Lauderdale, FL: J. Ross Publishing.

Cooper, R. & E. Kleinschmidt. 1990. *New Products: the Key Factors in Success*. Chicago, IL: American Marketing Association.

Cooper, R.G. & S.J. Edgett. 2009. *Generating Breakthrough New Product Ideas: Feeding the Innovation Funnel*. Ancaster, ON: Product Development Institute.

Craft, S.H. 2004. The international consumer market segmentation managerial decision-making process. *SAM Advanced Management Journal* 69(3):40.

Crawford, J.C., B. Garland, & G. Ganesh. 1988. Identifying the global pro-trade consumer. *International Marketing Review*, Winter:25–33.

Dawar, N. & P. Parker. 1994. Marketing universals: Consumers' use of brand name, price, physical appearance, and retailer reputation as signals of product quality. *Journal of Marketing* 58(2):81–95.

Firat, A.F. & C.J. Shultz. 1997. From segmentation to fragmentation: markets and marketing strategy in the postmodern era. *European Journal of Marketing* 31(4/3):183–207.

Firat, A.F. & A. Venkatesh. 1993. Postmodernity: the age of marketing. *International Journal of Research in Marketing* 10(3):227–249.

Fuller, G.W. 1994. *New Food Product Development: From Concept to Marketplace*. Boca Raton, FL: CRC Press.

Hassan S.S., S.H. Craft, & W. Kortam. 2003. Understand the new bases for global market segmentation. *Journal of Consumer Marketing* 20(5):446–460.

Hauptly, D.J. 2007. *Something Really New: Three Simple Steps to Creating Truly Innovative Products*, New York: AMACOM.

Gladwell, M. 2004. The Ketchup Conundrum. *The New Yorker*, Sept. 6.

Human Genome Project. 2001. *Genomics 101: A primer*. http://www.ornl.gov/sci/techresources/Human_Genome/publicat/primer2001/1.shtml. Accessed 04 May, 2010.

Khurana, A. & S.R. Rosenthal. 1998. Towards holistic 'front end' in new product development. *Journal of Product Innovation Management* 15:57–74.

Lawless, H.T. & H. Heymann. 1998. *Sensory evaluation of food: principles and practices*. New York: Chapman & Hall.

Leavy, B. 2005. A leader's guide to creating an innovation culture. *Strategy & Leadership* 33(4):38–45.

Marketing Association, Chicago.

Moskowitz, H.R. 1995. *Consumer Testing & Evaluation of Personal Care Products*. New York: Marcel Dekker, Inc.

Moskowitz, H.R., J.B. German, & I.S. Saguy. 2005. Unveiling health attitudes and creating good-for-you foods: The genomics metaphor, consumer innovative web based technologies. *CRC Critical Review of Food Science and Nutrition* 45:165.

Murray, D.K. 2009. *Borrowing Brilliance: The Six Steps to Business Innovation by Building on the Ideas of Others*. New York: Gotham.

Murray, J.M., C.M. Delahunty, & I.A. Baxter. 2001. Descriptive sensory analysis: past, present and future. *Food Research International* 34(6):461–471.

O'Connor, G.C. & A.D. Ayers 2005. Building a radical innovation competency. *Research-Technology Management* 48(1):23–31.

Peppers, D. & M. Rogers. 1997. *The One to One Future: Building Relationships One Customer at a Time*. New York: Random House.

Roden, D.M. *et al.* 2006. Pharmacogenomics: Challenges and opportunities. *Annals of Internal Medicine* 145:749.

Rothwell, R.Y. 1994. Towards the Fifth-generation innovation process. *International Marketing Review* 11(1):7–31.

Sheth, J., M. Banwari, & B. Newman. 1999. *Customer Behavior: Consumer Behavior and Beyond.* New York: Dryden.

Smith, W.R. 1956. Product differentiation and market segmentation as alternative marketing strategies. *Journal of Marketing* 21(July):3–8.

Swenson, C.A. 1990. *Selling to a Segmented Market: The Lifestyle Approach.* New York: Quorum Books.

US News. 2010. Personalized Medicine. *US News Health* January 28. Available from http://health.usnews.com/health-conditions/cancer/personalizedmedicine. Accessed 10 December, 2010.

Wikipedia. Definition of Probiotics. en.wikipedia.org/wiki/Probiotics. Accessed 09 December, 2010.

Zhu, K., K.L. Kraemer, & S. Xu. 2006. The process of innovation assimilation by firms in different countries: A technology diffusion perspective on e-business. *Journal of Management Science* 52(10).

Index

Note: Page references in *italics* refer to illustrative material.

Sensory and Consumer Research in Food Product Design and Development, Second Edition.
Howard R. Moskowitz, Jacqueline H. Beckley, and Anna V.A. Resurreccion.
© 2012 Blackwell Publishing Ltd. Published 2012 by Blackwell Publishing Ltd.